The Hierarchy of Fluid Dynamic Equations

Claus Weiland

The Hierarchy of Fluid Dynamic Equations

Foundations of the Second Mathematization Wave of Fluid Dynamics

Springer

Claus Weiland
Bruckmühl, Bayern, Germany

ISBN 978-3-031-71197-8   ISBN 978-3-031-71198-5 (eBook)
https://doi.org/10.1007/978-3-031-71198-5

This Springer imprint is published by the registered company Springer Nature Switzerland AG
The registered company address is: Gewerbestrasse 11, 6330 Cham, Switzerland

Ut conclave sine libris, ita corpus sine anima.
Marcus Tullius Cicero (106–43 B.C.)

Preface

Since the early days of the human race, people were confronted with moving and flowing matter. Rain drops from the clouds, water streams along creeks and rivers, moving air was realized as wind, etc., to mention just a few. Of course, for a long time, around the birth of Christ, water- and wind wheels were built and the experience had shown how they work, but an analytical insight into the flow nature, which forces the wheels, was not existing. Thousands of years long, there was only little progress with respect to the theoretical understanding about the mechanics of such motions. And this is true also in the light of the great occidental thinkers and philosophers Aristotle and Archimedes.

The fifteenth century had to come to see the first subtle attempts of the description of flow phenomena. Observations, empirical studies and simple experimental simulations, for example the study of the outflow of a vessel, started the activities for more coordinated investigations.

The quantities in every physical relation have dimensions, which are necessary beside others for the communication of possible results. But what are the dimensions in the sixteenth, seventeenth or eighteenth century? This illustrates the problems that previous researchers were faced with. Nevertheless, with advancing time for some physical phenomena, mathematical relations were formulated, which consisted in algebraic equations of scalars.

It was not really known at that time that flows cannot be described completely by scalars, but have quantities, which are vectors like the velocity or the heat flux or even tensors like the shear stress. Therefore, there was a need for the development of the necessary mathematical skills with respect to vectors and tensors. Vector algebra was evolved mid of the nineteenth century by H. G. Graßmann (1809–1877) and later vector calculus, which includes the infinitesimal calculus (developed in the seventeenth century by G. W. Leibniz (1646–1716) and I. Newton (1643–1727)), were established by J. W. Gibbs (1839–1903) and others based on the ideas of W. R. Hamilton (1805–1865). This was the origin of the field theory, where one of the first applications was given by the development of the electromagnetic field equations of J. C. Maxwell (1831–1879).

Despite the fact that the three-dimensional momentum equations for "ideal" fluids by L. Euler (1755) and for viscous fluids by C. L. M. H. Navier (1822), S. D. Poisson (1831), J. C. Barré de Saint-Venant (1843) and G. G. Stokes (1845) were already known mid of the nineteenth century, a compact description of these equations, the so-called symbolic form, was assembled not before the first part of the twentieth century.

Then a boost in the understanding and interpretation of flow fields could be observed perhaps driven by the coming up of airplanes, air-based propulsion systems, and other technical devices with flow functions. This has led to the "first wave of mathematization" which has encompassed the period between the first part of the eighteenth and the first part of the twentieth century. An additional impetus of this boost was the installation of experimental facilities like water tunnels, wind tunnels and subsequently propulsion test facilities, etc.

The potential of solving the three-dimensional governing equations, which came up in the second half of the twentieth century, mainly pushed by the enormous progress in the discrete numerical integration methods and the computer capabilities, opened the door for a deep understanding of complex flow fields. This period is called the "second wave of mathematization", which continues till this day.

This book is devoted to the most general governing equations of the fluid mechanics, namely the Navier-Stokes equations and their derivatives. These equations are presented in various manners: for several coordinate systems, for laminar and turbulent flows, for different thermodynamic states of gases, in dimensional and non-dimensional forms, and in an incompressible state. All that is valid also for the variants of the Navier-Stokes equations, where appropriate.

We have already stated that today the only way to solve the fluid dynamic equations for complex three-dimensional problems is given by the numerical integration methods. For this request, it is very helpful to formulate the complete set of governing equations in a vector or vector-matrix form. This is also true for the two equations turbulence models or the equations describing the thermodynamic non-equilibrium effects. These requirements are fully addressed in this book.

Graduate and doctoral students who are concerned with numerical solutions of the fluid dynamic equations for specific problems may find in this book the suggestions regarding the degree of approximation which could be adequate for the task they consider.

Further, persons who are interested in the evolution of the mathematical description of fluid dynamic issues, both from the scientific and also the historical side, may discover suggestions, advices and motivations in this book. Finally, this book can serve as a work of reference.

The preparation of every manuscript will always include shortcomings, deficiencies, misinterpretations and writing errors. It is therefore very important that a manuscript is proofread. This task was performed by my colleague Ernst Heinrich Hirschel, who read several times all the chapters of the book and my daughter Yanina

Weiland, who read part of the book. I am very much indebted to these both for their suggestions and inputs, which have definitely enhanced the readability of the book.

Bruckmühl, Germany Claus Weiland
April 2024

Contents

1 Introduction 1
- 1.1 The Fluid Dynamic Discipline 1
- 1.2 The Role of Mathematics in Fluid Dynamics 3
- 1.3 The Volume of the Governing Equations 5
- 1.4 The Objectives of the Book 6
- References 7

2 The Historical Background of the Theory of Fluid Dynamics 9
- 2.1 Early Activities 9
- 2.2 The Infinitesimal Calculus 10
- 2.3 The Continuum Concept 11
- 2.4 The Establishment of the Conservation Equations 12
- 2.5 The Navier-Stokes Equations and Their Derivatives 15
- 2.6 The Numerical Solutions of the Governing Equations 17
- 2.7 Three Engineering Examples 20
  - 2.7.1 First Example 20
  - 2.7.2 Second Example 21
  - 2.7.3 Third Example 23
  - 2.7.4 Final Remarks 24
- References 26

3 The Fluid Dynamic Equations in Symbolic Form 29
- 3.1 The General Equations as Conservative Laws 30
- 3.2 Integral Forms of the Fluid Dynamic Equations 31
  - 3.2.1 The Conservation of Mass 31
  - 3.2.2 The Conservation of Momentum 32
  - 3.2.3 The Conservation of Energy 33
- 3.3 Differential Forms of the Fluid Dynamic Equations 34
  - 3.3.1 The Continuity Equation 34
  - 3.3.2 The Momentum Equation 34
  - 3.3.3 The Energy Equation 36
- References 40

4 The Fluid Dynamic Equations in Component Form for Three Basic Coordinate Systems 41
4.1 The Equations in Cartesian Components 42
4.2 The Equations in Cylindrical Components 49
4.3 The Equations in Spherical Components 56
References 65

5 The Navier-Stokes Equations in Cartesian Coordinates for Laminar and Turbulent Flow 67
5.1 The Navier-Stokes Equations for Laminar, Perfect Gas Flow 69
5.2 Averaging Process for the Governing Equations 70
5.3 The Reynolds-Averaged-Navier-Stokes Equations (RANS), Perfect Gas Flow 74
5.4 The Reynolds-Averaged Navier-Stokes Equations (RANS), Non-perfect Gas Flow 76
5.4.1 Chemical Non-Equilibrium State 76
5.4.2 Complete Thermodynamic Non-Equilibrium State 76
References 79

6 The Navier-Stokes Equations in Arbitrary Coordinates for Laminar and Turbulent Flow 81
6.1 Arbitrary Coordinates 82
6.2 The Navier-Stokes Equations for Laminar, Perfect Gas Flow 83
6.3 The Navier-Stokes Equations for Turbulent (RANS), Perfect Gas Flow 87
References 89

7 Dimensionless Navier-Stokes Equations 91
7.1 RANS Equations in Cartesian Coordinates for Chemical Non-Equilibrium Flow 93
7.2 RANS Equations in Arbitrary Coordinates for Chemical Non-Equilibrium Flow 96
7.3 The Thermal Non-Equilibrium State in Arbitrary Coordinates for Laminar Flow 100
References 104

8 Turbulence Models 105
8.1 Turbulence Models in Cartesian Coordinates 106
8.2 A Turbulence Model in Arbitrary Coordinates 110
References 112

9 Thermodynamics for Real Gas Flow 115
9.1 Flow with Gases in Chemical Non-Equilibrium 116
9.2 Flow with Gases in Thermal Non-Equilibrium State 119
9.3 Flow with Gases Including Excited Electron Modes 121
9.4 Flow with Gases in Thermodynamic Equilibrium 122
9.5 Species Continuity Equations in Arbitrary Coordinates 124
References 125

10 The Thin Layer Approximation (TLA) of the Navier-Stokes Equations 127
10.1 The TLA Equations for Laminar Perfect Gas Flow in Cartesian Coordinates 128
10.2 The TLA Equations for Turbulent Perfect Gas Flow in Cartesian Coordinates 129
10.3 The TLA Equations for Laminar Perfect Gas Flow in Arbitrary Coordinates 130
10.4 The TLA Equations for Turbulent Perfect Gas Flow in Arbitrary Coordinates 131
10.5 The TLA Equations for Turbulent Chemical Non-equilibrium Flow in Arbitrary Coordinates 132
10.6 Typical Thin Layer Navier-Stokes Solutions 132
References 134

11 The Parabolized Navier-Stokes Equations (PNS) 135
11.1 The PNS Equations for Laminar Perfect Gas Flow in Cartesian and Cylindrical Coordinates 136
11.2 The PNS Equations for Laminar Perfect Gas Flow in Arbitrary Coordinates 137
11.3 The PNS Equations for Turbulent Perfect Gas Flow in Arbitrary Coordinates 142
11.4 A Two-Equation Turbulence Model for the PNS Equations in Arbitrary Coordinates 144
11.5 Typical Parabolized Navier-Stokes Solutions 145
References 146

12 Incompressible Forms of the Fluid Dynamic Equations 149
12.1 The Incompressible Navier-Stokes Equations 149
12.2 A Typical Solution of the Incompressible Navier-Stokes Equations 150
12.3 The Vorticity Equation 151
Reference 153

13 Boundary Layer Equations 155
13.1 The Derivation of the Boundary Layer Equations 155
13.2 A Three-Dimensional Boundary Layer Solution for an Industrial Project 160
References 162

14 Inviscid Forms of the Fluid Dynamic Equations 163
14.1 The Euler Equations 165
14.1.1 General Considerations 165
14.1.2 Conservative Formulation of the Euler Equations 166
14.1.3 Quasi-Conservative Formulation of the Euler Equations 167
14.1.4 Non-conservative Formulation of the Euler Equations 169
14.1.5 Diagonalization of Quasi-Linear Forms of the Euler Equations 170
14.1.6 Euler Equations for Gases in Thermodynamic Equilibrium 180
14.1.7 Euler Equations for Gases in Chemical Non-equilibrium 181
14.2 The Potential Equation 182
14.2.1 Potential Equations in Perturbation Form 186
14.2.2 Linearized Small Perturbation Equation 187
14.2.3 Transonic Small Perturbation Equation 187
References 188

15 Conservation Forms 191
15.1 Formulations of the Euler Equations 192
15.2 Integral Versions of the Various Forms of the Euler Equations 194
15.3 Numerical Approximations of the Differential and Integral Versions of the Euler Equations 196
15.4 Numerical Test Case 199
References 204

16 Flow Discontinuities 205
16.1 General Considerations 206
16.2 Strong Discontinuities 208
16.3 Weak Discontinuities 211
16.4 Contact Discontinuities 213
References 218

17 Boundary Conditions 221
17.1 Physical Boundary Conditions 222
17.2 Characteristic Variable Boundary Conditions for Euler Equations 224
17.3 Boundary Conditions for Problems Solved by Numerical Methods 227
17.3.1 Euler Equations 227
17.3.2 Boundary Conditions for the Navier-Stokes Equations 240
17.3.3 Boundary Conditions for Rarefied Gases 241
References 242

18 Significant Scientists 245
18.1 Galileo Galilei 247
18.2 Johannes Kepler 248
18.3 René Descartes 249
18.4 Pierre de Fermat 250
18.5 Christiaan Huygens 251
18.6 Isaac Newton 252
18.7 Gottfried Wilhelm Leibniz 253
18.8 Daniel Bernoulli 254
18.9 Leonhard Euler 255
18.10 Jean-Baptiste le Rond d'Alembert 256
18.11 Siméon-Denis Poisson 257
18.12 Claude Louis Marie Henri Navier 258
18.13 Jean Claude Barré de Saint-Venant 259
18.14 William Rowan Hamilton 260
18.15 Hermann Günther Grassmann 261
18.16 George Gabriel Stokes 262
18.17 William John Macquorn Rankine 263
18.18 Hermann Ludwig Ferdinand von Helmholtz 264
18.19 James Clerk Maxwell 265
18.20 Ernst Mach 266
18.21 Osborne Reynolds 267
18.22 Pierre Henri Hugoniot 268
18.23 Martin Hans Christian Knudsen 269
18.24 Ludwig Prandtl 270
18.25 Theodore von Kármán 271
References 272

Appendix A: Symbols 275

Appendix B: Acronyms 283

Index 285

Chapter 1
Introduction

Abstract Fluid dynamics has a long tradition. It is a discipline which concerns a lot of occurrences which we daily encounter. Liquid flows as well as gas flows escort us in our domestic home, come along with moving cars, trains, airplanes and ships and their propulsion systems, are the dominating effect of the formation of weather events, determine often the processes of chemical engineering systems and are the main drivers in numerous power plants, just to mention some of their scopes of influence. With the development of the appropriate mathematical skills, which started somewhere along the seventeenth century, equation systems for more or less all possible flow fields were established. Today, with the numerical integration methods we have means available to solve rigorously all these equation systems.

1.1 The Fluid Dynamic Discipline

The physics of fluids is an old discipline. In ancient times humans did observe the flow in rivers, creeks, lakes and oceans and with proceeding time they wanted to receive awareness of the laws and rules about the behavior of liquid matter. Fundamental laws like that fluid always moves from a higher location to a lower one were exploited for the construction of irrigation systems already established in Mesopotamia (Nineveh at the Tigris river) and Egypt (Nile river) thousands of years before Christ. Later, the Romans did construct aqueducts in order to supply cities like Rome with water. Such aqueducts had open and closed channels and even pressurized pipes. Considering the motion of fluids was called hydraulics, which was primarily based on empirical observations.

Some of the ideas and thoughts of Aristotle (384–322 B.C.) and Archimedes (287–212 B.C.) were seminal and are valid until this day, for example the description of a continuum (Aristotle) and the buoyancy law of a body in a fluid (Archimedes). Over a long time the question was discussed, what does a body experience when moved through air or water and what could be a mechanism of resistance. Hundreds of years later Edme Mariotte (1620–1684) and Christiaan Huygens (1629–1695) argued that there exists a proportionality of the resistance to the square of the velocity.

C. Weiland, *The Hierarchy of Fluid Dynamic Equations*,
https://doi.org/10.1007/978-3-031-71198-5_1

Isaac Newton (1643–1727) then gave the first theoretical explanation of the squared velocity law, [1–3].

In the earlier times, most of the physical processes were described by words and not by mathematical formulas. When the scientific aspects came into play, which means mathematics and to some lower extent theoretical physics, the flow phenomena were prescribed by mathematical relations using scalars for the variables, which today we call algebraic equations. Despite the fact that already in previous days some of the mathematical disciplines like geometry were well developed, the mathematical tools for the description of evolutionary processes of vector fields were not available. Since it was more and more recognized that flows behave like vector fields (field theory) an extension of the mathematical discipline was indispensable. Therefore, during the emerging new age (the sixteenth century and beyond) first ideas came up regarding the expansion of mathematics by, for instance, new disciplines like the infinitesimal calculus and the analysis [4]. There is a long list of scientists,[1] who substantially have contributed to the evolution of these new disciplines, see Chap. 2. This was the origin of the hydrodynamic discipline, which in the beginning was far from solving practical fluid dynamic problems.

It was an outstanding moment for the field of hydrodynamics (fluid dynamics) when Leonhard Euler in 1755 published the set of non-linear partial differential equations for the conservation of momentum in inviscid flow, which today are named the Euler equations. This system of equations over decades and centuries had inspired many mathematicians to find rules and laws for the solutions of these non-linear equations. The theory of characteristics is one of such a solution strategy, if the system is hyperbolic, which means that the problems are either time-dependent or the velocity component in the direction of integration is larger than a specific signal speed (speed of sound). Another one is the potential theory developed for elliptical partial differential equation systems. But this theory only is valid for linear partial differential equations and therefore requires a reduction and linearisation of the Euler equations, for example the potential equation for small perturbations of flow variables or the linearised wave propagation equation (acoustic wave approach). The potential equation which describes incompressible flow originally is linear.

As already mentioned it has a long tradition to assume and also to claim that moving bodies in fluids and fluids moving through devices suffer an additional force, which obviously is not included in the momentum equations of Euler for "ideal" fluids. Today we know that this is due to the wall-friction effects.

Therefore, 67 years after Euler had presented his momentum equations Claude Louis Marie Henri Navier (1785–1836) extended these equations by terms representing friction forces. Despite the fact that the physical explanation regarding the additional terms, which Navier gave, was not correct, namely that the friction forces only originate from interactions between the molecules of the fluid, the equations themselves looked like the ones we know today. It took further 21 years until Jean Claude Barré de Saint-Venant (1797–1886) in 1843 stated that the friction force is propor-

[1] The original ideas for the evolution of the infinitesimal calculus date back to Gottfried Wilhelm Leibniz (1646–1716) and Isaak Newton (1643–1727).

tional to the derivative of the velocity components multiplied with the viscosity. This statement characterizes the shear stress, which correctly describes the appearance of friction. Independently George Gabriel Stokes (1819–1903) two years later published the same approach.[2] Since then these equations bear the name Navier-Stokes equations. First solutions of the Navier-Stokes equations, when strongly simplified mostly by linearization and/or by dropping specific terms, were already established in the nineteenth century, for example the Couette flow, the Hagen-Poiseuille problem or the Stokes flow [5].

With the advent of airplanes in the beginning of the last century a new aspect of the fluid dynamics the aerodynamics did originate. The goal of aerodynamics was the determination of the forces lift and drag, and the moments pitch, yaw and roll of airplanes. During the first part of the twentieth century this task could theoretically be treated only using the potential theory, where singularity conditions, for example the Kutta condition, care for a partially modeling of viscous effects. An advanced concept in this regard is given with the "local vorticity content of a shear layer" which is described in [6].

1.2 The Role of Mathematics in Fluid Dynamics

We have already stated how the mathematical discipline had evolved since the days of Leibniz and Newton. There were a lot of mathematical inventions and far-reaching ideas created which at first had allegedly no application in physics, but played later a distinguished role in the description of natural mechanics. A good example is given by the mathematical work of Jean-Baptiste le Rond d'Alembert (1717–1783). d'Alembert established the wave equation and presented a solution of it in 1747 with mathematically pleasing simplifications which however were at odds with observations. This was due to the fact that he believed that mechanics was based on metaphysical principals and not on experimental evidence. In contrast Newton had based his laws of motion on experimental experience. But today the value of the wave equation for physics is beyond any controversy.

Also Euler had an opposite view compared to d'Alembert, when he presented his equations of motion for ideal fluids. The mathematical science and the physical science often went difference routes. But with the works of Euler, Navier, de Saint-Venant, Stokes, Helmholtz, just to mention some of the scientists representing the fluid dynamic discipline, a partially fruitful merger of the two disciplines came into being. Nevertheless, there is no doubt that some branches of mathematics have no significance for application in natural sciences. The progression of mathematics also in the days around the end of the nineteenth and at the beginning of the twentieth century had evolved differently. This reveals for example also the life of David Hilbert (1862–1943). David Hilbert was a prominent German mathematician and probably the greatest at his time. He spent much efforts on pure mathematical questions, like

[2] See also Chap. 2.

logic and axiomatic methods, which were without any perspective of application in natural sciences. Later, he turned to the mathematics relevant for physical applications with some eminent contributions and the publication of a famous textbook [7]. To the end of his life again he returned to the questions of pure mathematics [8].

These examples reflect the complexity of the two disciplines, but also give rise to state that a "first wave of mathematization" of fluid dynamics began in the first part of the eighteenth century and lasted up to the mid of the twentieth century. Major contributions to this wave also were given by the works of Ludwig Prandtl (1875–1953) and his scholars in the first part of the twentieth century, which did revolutionize the understanding of realistic flow problems in particular in the field of aerodynamics. During this time a breakthrough regarding solutions of the non-linear forms of the fluid dynamic equations (Euler and Navier-Stokes equations) came up. With the quality of the mathematical findings solutions for unsteady one-dimensional flows, see Chap. 15, or steady two-dimensional supersonic flows with $M > 1$ in nozzles and ducts were possible, where the latter were found graphically by the application of the method of characteristics.

The time was gone that the mathematicians had utilized the Euler equations, the Navier-Stokes equations or their simplifications like the inviscid or viscous Burgers equations only for providing solution procedures often without a correlation to reality. Since the last four to five decades with the advent of advanced numerical approximation methods and the accompanying expansion of the computer power, practical flow problems with more and more increased complexity are treated by the numerical integration of the fluid dynamic equations. The determination of the aerodynamics of civil and military airplanes and the aerothermodynamics of space vehicles bears witness to this development. Today, the flow field past flight configurations including elevons, deflected wing and body flaps as well as rudders, can successfully be simulated. In addition, the impact of propulsion systems on the flow field, for airplanes the turbojets (or propeller drives) and for space vehicles the rocket motors, can also be considered numerically.

This period in [9] is called the "second wave of mathematization" of fluid dynamics.
Of course, the "first mathematization wave" also had captured other physical disciplines in particular electrodynamics, thermodynamics and later the general theory of relativity and quantum mechanics. These are aspects of the "second wave of mathematization", too.

From the fluid dynamic point of view the "second wave of mathematization" encompasses many disciplines of natural science, where the fluid dynamical equations play a dominant role. Below we list some of them [10].

- Physics of atmosphere with the subset weather prediction, where the unsteady Reynolds-Averaged Navier-Stokes equations (URANS) or scale resolving methods for turbulent flow are employed, see [6].
- Astrophysics, where the formation of structures of the Universe (galaxy formation) is simulated using hydrodynamic equations. In such a case it is assumed that the flow of matter can be constituted as an interstellar or intergalactic gas. For other

problems if plasma effects come into play, for example gamma-ray bursts or active galactic nuclei, the magneto hydrodynamic equations (MHD) are applied. The same is true for the treatment of plasmadynamics of solar systems.

- Geophysical problems like plate tectonics, mantle flow or mantle circulation are successfully simulated by adjusted fluid dynamic equations, when in the latter case the convective terms are neglected (Stokes problem).
- Combustion chamber flow for example in rocket motors is characterized by interactions between gasdynamics, thermodynamics and chemical reactions. Besides the combustion models computational fluid dynamics (CFD) with non-equilibrium thermodynamics play a major role in the simulation of such flow.

1.3 The Volume of the Governing Equations

What we call today Navier-Stokes (N-S) equations in their most general form consist of

- the three components of the momentum equation,
- the continuity equation,
- the energy equation,
- the equations of the thermodynamic state,

and therefore have six unknowns: the three velocity components, the density, the temperature and the pressure.

The equations are formulated for

- time dependent,
- three-dimensional,
- and compressible or incompressible flow.

The mechanical physics regarding

- laminar flow,
- turbulent flow.[3]

and the thermodynamic physics regarding

- perfect thermodynamic state (perfect gas),
- equilibrium thermodynamic state,
- non-equilibrium thermodynamic state.

are addressed.
The equations are formulated in

[3] Of course, we are aware of the laminar–turbulent transition problem, but do not treat it in this book.

- Cartesian coordinates,
- cylindrical coordinates,
- spherical coordinates,
- arbitrary non-orthogonal curvilinear coordinates.

1.4 The Objectives of the Book

A summary of the history of the fluid dynamic discipline is given in Chap. 2. Chapter 3 is devoted to the fluid dynamic equations in symbolic form, while Chap. 4 highlights these equations in several coordinate systems. The full N-S equations including laminar and turbulent flows are presented in Chaps. 5–7 for all thermodynamic states (high temperature flows) in Cartesian as well as in arbitrary non-orthogonal curvilinear coordinates. The equations of various turbulence models,[4] which are applied for the computation of practical flow problems in aeronautics (and also in other application cases), are given in Chap. 8. For gas flow with high temperature, for example in rocket motors or around space planes during re-entry from Earth orbit or the Moon, the thermodynamic state of the gas can be non-perfect, which in detail is covered in Chap. 9.

Reductions of the N-S equations, but still describing viscous flows, are presented in Chap. 10 with the thin-layer approximation (TNS), in Chap. 11 with the parabolisation of the N-S equations (PNS), in Chap. 12 with the constant density approach of the N-S equations (incompressibility) and in Chap. 13 with the derivation of the boundary layer equations.

Inviscid forms of the fluid dynamic equations are treated in Chap. 14, where in particular a focus is laid on the Euler equations, because their diagonalization forms are the basis for the success of today's discrete numerical approximation methods. The specific properties of conservation forms of the governing equations with regard to design appropriate numerical discretisation procedures are addressed in Chap. 15. Figure 1.1 gives an overview over the hierarchy of these fluid dynamic equations.

Flow fields can feature discontinuities with respect to one or more scalar or vectorial flow variables. In particular, flow fields with Mach numbers larger than unity are distinguished by strong discontinuities, the shock waves. Weak discontinuities, for example shear layers, can also appear in flow fields at Mach numbers lower than unity, whereas in shock tubes contact discontinuities of several kinds are found. Chapter 16 is treating all these discontinuities.

Initial and boundary conditions complete the description of flow problems. Chapter 17 is devoted to this topic. Because the historical view on the establishment of the fluid dynamic equations is part of the book, in Chap. 18 the most relevant scientists concerning this matter are portrayed.

---

[4] Not treated are approaches for the determination of laminar–turbulent transition.

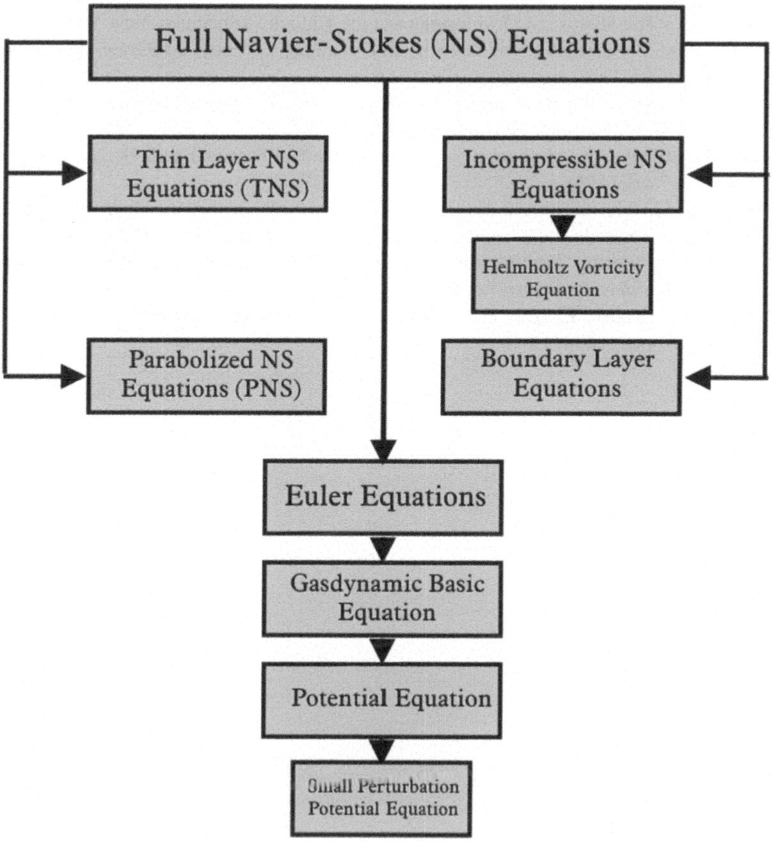

Fig. 1.1 The general hierarchy of the governing equations for fluid dynamic flows

In contrast to other textbooks, we put in front of every chapter, when appropriate, a list of the variables and quantities used. This is done because today some readers, interested in the matter of the book, only look at a single chapter, that is important for them. Hence, they do not have access to the appendix of the book, where the symbols are explained and defined.

References

1. Eckert, M.: The Dawn of Fluid Dynamics. Wiley-VCH, Weinheim (2006)
2. Anderson, J.D.: Brief history of the early development of theoretical and experimental fluid dynamics. In: Encyclopedia of Aerospace Engineering. Wiley, New York (2010)
3. Rizzi, A., Hirschel, E.H.: General developments of numerical fluid mechanics until the middle of the 20th century. In: Hirschel, E.H., Krause, E. (Eds.) 100 Volumes of 'Notes on Numerical Fluid Mechanics'. Springer, Heidelberg (2009)

4. Edwards, C.H.: The Historical Development of the Calculus. Springer, New York (1979)
5. Anderson, J.D.: Fundamentals of Aerothermodynamics, 5th edn. McGraw-Hill, New York (2011)
6. Hirschel, E.H., Rizzi, A., Breitsamter, Ch., Staudacher, W.: Separated and Vortical Flow in Aircraft Wing Aerodynamics. Springer, Heidelberg (2021)
7. Courant, R., Hilbert, D.: Methoden der Mathematischen Physik I und II, 2nd and 3rd edn. Springer, New York (1968)
8. von Wallwitz, G.: Meine Herren, dies ist keine Badeanstalt. Berenberg Verlag, Berlin (2017)
9. Hirschel, E.H.: Towards the virtual product in aircraft design. In: CIMNE Handbooks on Theory and Engineering Applications of Computational Methods, pp. 453–464. Barcelona, Spain (2003). http://www.ehhirschel.com
10. Hirschel, E.H., Krause, E. (Eds.): 100 Volumes of 'Notes on Numerical Fluid Mechanics'. Springer, Heidelberg (2009)

Chapter 2
The Historical Background of the Theory of Fluid Dynamics

Abstract It is part of the history of humans that they were fascinated by the motion of fluids in creeks, rivers and channels and the movement of bodies in fluids like boats and ships. They have thought and speculated about the properties and principles of such motions. This has led to the physical discipline hydrodynamics. The same was true for some matters of classical mechanics like a rock riprap or simply a falling rock. A long tradition also have the disciplines acoustics beginning with the work of Pythagoras (570–500? B.C.) and optics with which Euclid (about 300 B.C.) was firstly concerned. But other physical disciplines like electrodynamics, thermodynamics and the very modern quantum mechanics and solid-state physics etc. were very far away to play a role in the life of the inhabitants of the ancient world. Hydrodynamics has a long history and the first laws were established hundreds of years before Christ. In this chapter, we highlight the growth of the physical discipline hydrodynamics from antiquity until our days. During this race we encounter outstanding persons with decisive contributions to the scientific disciplines of nature in particular the fluid mechanics. In Chap. 18, twenty five of these scientists are portrayed.

2.1 Early Activities

A long time ago humans began to deal with the static and dynamic behavior of fluids. In ancient Greek society, at least two representatives of the various philosophy and science schools contributed to the understanding of the static and dynamic of fluids. The first one was Aristotle (384–322 B.C.), who formulated a concept that we call a continuum today. Further, he argued that a body moving through air experiences a resistance.

The other person was Archimedes (287–212 B.C.) and we all have knowledge of his law regarding the buoyancy of a body in a fluid, which presumably interconnects with his statement that all points of a body in a fluid are pressed by the column of the fluid above it. A second contribution to the behavior of fluids is his statement that compressed parts of fluids move in the direction of parts of fluids which are less compressed. This is the first statement about the role of a pressure gradient.

C. Weiland, *The Hierarchy of Fluid Dynamic Equations*,
https://doi.org/10.1007/978-3-031-71198-5_2

Then, over a large time period including the Middle Ages no corresponding intellectual activities are documented in the known world. But at the beginning of the fifteenth century Leonardo da Vinci (1452–1519) formulated for the first time the continuity of a moving fluid. In the sixteenth and seventeenth centuries, Galileo Galilei (1564–1642), Evangelista Torricelli (1608–1647), René Descartes (1586–1650), Edme Mariotte (1620–1684) and Christiaan Huygens (1629–1695) intensively studied the nature of fluids and some of their conclusions are valid up to these days, for example that the force acting on a moving body is directly proportional to the square of the velocity, which was first stated by Edme Mariotte and Christiaan Huygens [1–3].

Since this book is concerned with the hierarchy of the governing equations of fluid motion, we focus in the following on the scientists who established these equations. But in doing so, these scientists needed the mathematical tool "infinitesimal calculus"[1] and a precise definition of the term "continuum". Both are essential for the set up of the governing equations as we know them today. Therefore, we first give a short historical survey about the infinitesimal calculus and the continuum concept.

2.2 The Infinitesimal Calculus

Infinitesimal calculus has its origin already in the ancient world when for example Archimedes tried to determine the area of curvilinear bounded surfaces or the volume of hollow bodies. He did it with the "exhaustion method" where the surfaces or bodies are divided into surface or volume elements, which are calculable and afterwards summed up to give an approximate solution to the problem. Generally, scientists in earlier times have formulated and discussed

- the determination of arbitrary bounded surfaces and volumes,
- the determination of tangent lines on curves,
- the problem of limiting values of functions (maxima and minima).

The effort of finding mathematical procedures for solving these three topics constituted the main impetus for the development of the infinitesimal calculus. The geometrical approach as addressed above was evolved further and applied by Johannes Kepler (1571–1630), who determined surface areas and volumes of hollow bodies (Kepler'sche Fassregel (rule of barrel)). In this sense, Bonaventura Francesco Cavalieri (1598–1647) introduced the "indivisibles", which are infinitesimal small structures (principle of Cavalieri). With them surfaces and volumes are approximately calculable. But all these geometrical methods were vague procedures and had no real potential for further development. Therefore, the next step on the road to the infinitesimal calculus was to find an algebraic solution of the above topics. An essential move in this direction was performed by Pierre de Fermat (1607–1665),

[1] Today also called "Analysis".

who considered the tangent line problem with much success and he also found a rule for evaluating extreme values of various functions [4, 5].

The essential breakthrough came with the activities of Isaac Newton (1643–1727) and Gottfried Wilhelm Leibniz (1646–1716). Newton developed his "fluxion" calculus (about 1671), which describes the movement of a point on a curve, given for example by $y = f(x)$ (x, y are called the fluents), by the means of the shift of time-dependent infinitesimal elements ξdt and ηdt (ξ, η are called the fluxions) in both coordinate directions with $x(t_0 + dt) = x(t_0) + \xi dt$, $y(t_0 + dt) = y(t_0) + \eta dt$.

The rate of change of the coordinates is then

$$\frac{x(t_0 + dt) - x(t_0)}{dt} = \xi \left(\approx \frac{dx}{dt} \right) \quad \text{and} \quad \frac{y(t_0 + dt) - y(t_0)}{dt} = \eta \left(\approx \frac{dy}{dt} \right), \tag{2.1}$$

leading, if the time parameter is eliminated, to the differential operator $\frac{dy}{dx}$ as we know it today. The components of the velocity, with which the point of the curve is moved, are $\mathbf{v}(t) = (\xi, \eta)^T$. The determination of the fluxions from the fluents corresponds to the differential operation and conversely the determination of the fluents by a relation of fluxions is the integral operation [6].

Gottfried Wilhelm Leibniz developed his infinitesimal calculus (about 1684) when he examined the tangent problem. He found that a tangent can be determined from a "slope triangle" positioned at the point on the curve considered. The sides of the triangle are given by the infinitesimal changes of the values of the functions (the constitution of differences). Leibniz denoted this operation by $\frac{dy}{dx}$, known today as the differential operator. Further, he realized that the approximate calculation of surfaces is an inverse operation of the constitution of differences and denoted the summation of the infinitesimal small surface elements with the symbol $\int$. These notations are used today.

Despite the fact, that there were heated disputes about the credit for the origin of the infinitesimal calculus in the last quarter of the seventeenth century, the general opinion today is that G. W. Leibniz and I. Newton developed the infinitesimal calculus independently of each other more or less at the same time.

2.3 The Continuum Concept

Continuum mechanics as a discipline of the general field of mechanics deals with the motion of bodies of deformable matter. All materials have discrete structures and their behavior with respect to external impacts depends on the interactions of atoms and molecules. The mathematical description of such complex systems is very challenging and so far a solution of related initial and boundary value problems is more or less impossible.

Within the scope of continuum mechanics the material model is transferred into a hypothetical phenomenological model, where the discrete structures of the matter are ignored. This continuum consists of a point set, that fills continuously at every instance of time a space, where all points have the same material properties. In other words, the characteristics of the material are averaged (homogenized). This is valid for all matters like solids, fluids and gases. Applying these conditions, physical properties like density, temperature and velocity are assigned to every point inside a macroscopic body. This makes continuum mechanics a field theory. The first mathematical formulation of continuum mechanics was presented by Augustin Louis Cauchy (1789–1857), who also introduced the fundamental definitions of the strain and stress tensors [7].

The equations of continuum mechanics can be divided into two groups

- relations independent of material,
- relations describing material properties (*c*onstitutive principles).

Kinematic relations and balance equations belong to the first group, where for liquid and gaseous matters conservation equations (particular cases of the balance equations) are derived.

To the second group belongs the theory of matter, where the material properties are specified by *c*onstitutive principles. Prominent principles or laws are the relations for the pressure in gases. Gas can be in a thermodynamically perfect, equilibrium or non-equilibrium state. Newton's law of viscosity and Hooke's law of linear behavior between strain and stress are other important principles [8].

2.4 The Establishment of the Conservation Equations

With the above specified prerequisites, namely the infinitesimal calculus and the characterisation of the continuum, Sects. 2.2 and 2.3, the conservation equations for mass, momentum and energy for liquid and gaseous fluids can be formulated.

In the last quarter of the seventeenth century, Isaac Newton presented in 1687 in his publication "Philosophiae Naturalis Principia Mathematica" a new vision on physical laws in particular for mechanics. Besides his famous three laws of motion, he first gave the theoretical derivation of the squared velocity law of the resistance of moved bodies in fluids. He further formulated the hypothesis that the shear stress is proportional to the velocity derivative normal to the flow direction multiplied by the viscosity. In addition, he presented the sine-squared law,[2] which is a simple equation for calculating the drag of an inclined surface and created a general mathematical basis for physical problems.

[2] The sine-squared law was not very successful regarding flows past vehicles or elements of vehicles at subsonic, transonic and low supersonic Mach numbers. But for hypersonic Mach numbers reasonable results can be achieved due to the nature of the flow, that allows a direct transfer of the momentum of the particles to the surface.

Jacob Bernoulli (1655–1705) and his brother Johann Bernoulli (1667–1748) contributed major developments to the infinitesimal calculus and were also active in the distribution of the new mathematical findings to the scientific world. This was the situation when Daniel Bernoulli (1700–1782), the son of Johann Bernoulli, and Leonhard Euler (1707–1783) came into play. Both were very successful mathematicians and fluid dynamicists.[3]

Daniel Bernoulli in 1724 accepted a position as mathematician in St. Peterburg, Russia, where he worked on the motion of fluids amongst other topics. One of the results of these activities was the formulation of conservation of energy for incompressible pipe flows with the balance between the specific kinetic energy $1/2\rho v^2$, the specific energy of pressure p and the specific potential energy $\rho g h$, from which follows $1/2\rho v^2 + p + \rho g h = \text{const.}$ Today, this equation is known by every engineering student as Bernoulli equation.[4]

In 1738, Bernoulli published the distinguished book "Hydrodynamics", which was a reference for most of his successors then. Daniel Bernoulli and Leonhard Euler had an amicable relationship. On the recommendation of Daniel Bernoulli the Imperial Russian Academy of Science offered Leonhard Euler an employment. Leonhard Euler went to St. Petersburg in 1727 and this was the starting point of a very fruitful six year period of cooperation between the two scientists.[5] Leonhard Euler had an extensive interest in the fields of mathematics, physics and astronomy and his contributions to these disciplines were pivotal [11].

In the beginning of the eighteenth century the theory of the infinitesimal calculus evolved very fast and Leonhard Euler also had a part in this. In this environment, he conceived and established, first published in 1755, the partial differential equations for the conservation of momentum of inviscid fluid flow,[6] [12]. He benefited from the findings of Jean-Baptiste le Rond d'Alembert (1717–1783), who came up in 1744 by connecting a moving fluid element with the balance equation for masses, now called the continuity equation.

Euler's equation was a milestone in the history of fluid mechanics and opened the door for prospective examinations of very complex flow fields, for example laminar and turbulent viscous flows, flows in imperfect thermodynamic state (high temperature non-equilibrium effects), flows with magnetic properties (magnetohydrodynamics) and so on.

For completeness, we mention that in the second half of the eighteenth century and the first half of the nineteenth century significant enhancements of the infinitesimal calculus were made by the work of the mathematicians Joseph Louis Lagrange (1736–

---

[3] And were successful also in other fields of mechanics.

[4] It seems that Leonhard Euler earlier derived the zero-dimensional equation of energy conservation with a successful integration of his set of partial differential equations [2, 3].

[5] Daniel Bernoulli left St. Peterburg in 1733 due to a physical indisposition.

[6] In vector notation this relation, the Euler equation, has the form $\rho \frac{D\mathbf{v}}{Dt} + \text{grad}\, p = \mathbf{F}$ and is completed for incompressible flow by the incompressibility condition $\text{div}\, \mathbf{v} = 0$.

1813), Pierre Simon Laplace (1749–1829), Augustin Louis Cauchy (1789–1857) and Bernhard Riemann (1826–1866).

The experience with moving bodies[7] in fluids or with fluids passing through pipes, tubes, etc. at that time had shown that obviously a physical property exists, which influences the ideal flow situation in the sense that in the momentum equation an additional force has to be included. The source of this force is the friction. Roughly seventy years after Euler had published his system of partial differential equations regarding ideal fluid motion (frictionless fluids), more and more scientists, for example most of the practitioners and to a lesser degree the theorists, came to the conclusion that realistic fluid motion can only be described if the forces of friction are part of the hydrodynamic equations.

Claude Louis Marie Henri Navier (1785–1836) was the first who extended Euler's equations to viscous flow (1822) by adding terms, which are based on second derivatives of the velocity components multiplied by a constant. Navier developed these terms from the assumption that the forces originate from the interaction between the molecules of the fluid. Today, we know that this is not true, instead the concept of Newton's shear stress[8] gives the right answer to the properties of friction. Also Siméon Denis Poisson (1781–1840) presented 1831 an extension of Euler's momentum equations by the friction force derived on the basis of the interaction between molecules as Navier did. Despite the fact that the physical explanation of friction by these scientists was incorrect, the equations itself looked like the ones we know today.

Twelve years later it was Jean Claude Barré de Saint-Venant (1797–1886) who published 1843 the first correct derivation of Euler's equations with friction. He indicated as the origin of friction the derivatives of the velocity components multiplied by the viscosity, called the shear stress. Finally, in 1845 George Gabriel Stokes (1819–1903) presented his derivation of Euler's equations with friction. He utilized, as de Saint-Venant did, that friction comes up by internal shear stresses acting on the faces of the fluid elements. The historians agree, that Georg Gabriel Stokes was not aware of the publications of the three Frenchman referred to above and that he independently performed his work. We conclude with the statement that Euler's equations with friction got henceforth the name Navier-Stokes equations. This naming is somewhat surprising at least with respect to the credit of Barré de Saint-Venant.

Flow can adopt two states: the laminar one and the turbulent one. It was O. Reynolds (1842–1912) who put this issue into perspective. In a paper presented 1883 he reported about experimental observations of flow in a pipe, which have adopted either a laminar state or a turbulent state depending on the dimensionless parameter[9] vD/ν. He further realized that friction in a turbulent flow is higher than in a laminar one, which has the consequence that the Navier-Stokes equations must be adapted to this circumstances. In 1894, Reynolds presented his time-averaged

---

[7] Often ballistic projectiles.

[8] Newton's linear viscosity approach.

[9] With v the velocity, D the pipe diameter and ν the kinematic viscosity. A. Sommerfeld proposed in 1908 to call this parameter Reynolds number Re.

version of the Navier-Stokes equations (RANS), which allow for turbulent flow. But the time-averaging had produced extra terms, which must be modelled in order to close the system. A task occupying the scientists until these days.

For a long time, mathematicians and physicists concerned with the Euler equations believed that the velocity field of inviscid flow always has a potential. Hermann von Helmholtz (1821–1894) conceived flows where this is not true, for example in the flow of a jet in air. His hypothesis was that a discontinuity surface separates the jet flow and the air flow, which he called a vortex sheet or shear layer. The physics of such a discontinuity surface has nothing to do with friction.[10] On this basis von Helmholtz developed his three laws of vortex motion, which are valid until these days, if inviscid flow can approximately be assumed, e.g. in astronomy.

During the turn towards the twentieth century there was a great gap between the results of theoretical hydrodynamics and the practical experiences. One point was that the mechanism how friction influences the resistance of a body moving in a fluid was not really understood. This was the environment when Ludwig Prandtl (1875–1953) presented his paper "Über Flüssigkeitsbewegung bei sehr kleiner Reibung (On fluid motion at very small friction)" in 1904. Prandtl assumed that there exists a no-slip condition of the fluid on the body surface and that the shear stress is only active in a very small region perpendicular and along to the surface: the boundary layer. This was the breakthrough in understanding the role friction plays for the determination of the resistance (aerodynamically: the drag) of moved bodies. With this hypothesis, Prandtl reduced the Navier-Stokes equations to find the boundary layer equations [14, 15].

2.5 The Navier-Stokes Equations and Their Derivatives

Originally, only the viscous momentum equations in three dimensions were named Navier-Stokes equations. But with the upcoming numerical methods for solving these equations the set of equations was extended by the equations for continuity and energy (for compressible flows), which are then called the Navier-Stokes equations.

The Euler equations were established about 265 years ago and the Navier-Stokes equations roughly 175 years ago. Since then a lot of mathematicians, physicists and engineers have spent much effort to find analytical solutions for these equations. In particular, the physicists and engineers are strongly interested in solutions that specify real flow problems as realistic as possible. Until approximately the sixties of the last century only analytical solutions could be achieved.[11] To arrive at those solutions the governing equations had to be reduced rigorously. For example, the

[10] In [13] the concept of the vorticity content of a shear layer is introduced. That concept permits to explain viscous flow phenomena in terms of inviscid flow theory and vice versa. In particular the circulation theory of lifting airfoils and finite-span lifting wings can be connected to viscous flow phenomena having either a kinematically active or inactive content.

[11] A series of analytical solutions, amongst others the ones of Couette, Poiseuille and Hagen, can be found in [9, 10].

reduction of the Euler equations by the streamtube assumption has led to the Bernoulli equation, the Rankine-Hugoniot shock relations, the algebraic energy equation, etc. and also, if the streamtube has a variable cross section, to one-dimensional relations characterizing the flow in a Laval nozzle. An example of an analytical solution of the Navier-Stokes equations was given by G.G. Stokes for creeping flows with very low Reynolds numbers ($Re \ll 1$). The integration could be performed because the convective terms in the Navier-Stokes equations were neglected [11].

Other activities are concerned with irrotational flow where a potential exists. The Laplace equation describes incompressible flow and, if a small perturbation approach can be applied, also compressible flow. For subsonic flow such equation can typically be solved by a singularity method.[12] For two-dimensional supersonic flow the graphical method of characteristics is an option.

Today, novel flow problems arise during the development of advanced and sometimes highly sophisticated industrial projects like civil and combat aircrafts, space planes, all sort of propulsion systems, mechanical power supplies and other devices. These require reliable solutions for the inherent fluid mechanical and aerodynamic problems.

The analytical methods or the approaches on the basis of small perturbation theory in the past gave no sufficiently reliable answers. Therefore, with the beginning somewhere in the 1960s activities started to solve the two- and three-dimensional Euler and Navier-Stokes equations numerically. Of course, at that time computer power and storage capability were modest and the complete three-dimensional equations, and even less the time dependent ones, were not treatable.

This led to the search for reductions of the governing equations, for example of the dimensionality (two-dimensional, axisymmetric), of the time dependency (steady flows and steady integration methods), to the decrease of the shear stress tensor (boundary layer like approximations), to the irrotationality (potential flow) as well as to the combined equation systems (boundary layer solutions together with potential or Euler solutions).

Figure 2.1 exhibits the hierarchy of the fluid dynamic equations. We demonstrate exemplarily for aerospace applications the significance that the various forms of the governing equations in the past had.

The potential equation in small perturbation form (linearized and non-linearized) was widely applied for airfoil flow as well as for flows around wings and wing-body configurations by panel methods. There were also full potential methods, but they did not reach a broader distribution. In the 1980s the first three-dimensional Euler methods were developed and applied to flow past fuselages and fuselage-wing configurations. In parallel, numerical solutions of the boundary layer equations became available either as integral or as finite-difference methods. The next step in the second part of the 1980s was to add some friction terms to the Euler equations in the reduced form of the thin layer approximation (TLA equation). Simultaneously, numerical methods on the basis of the parabolized Navier-Stokes equations (PNS)

[12] Of course, in mathematics the conformal mapping method is often used, but this method is restricted to two dimensions.

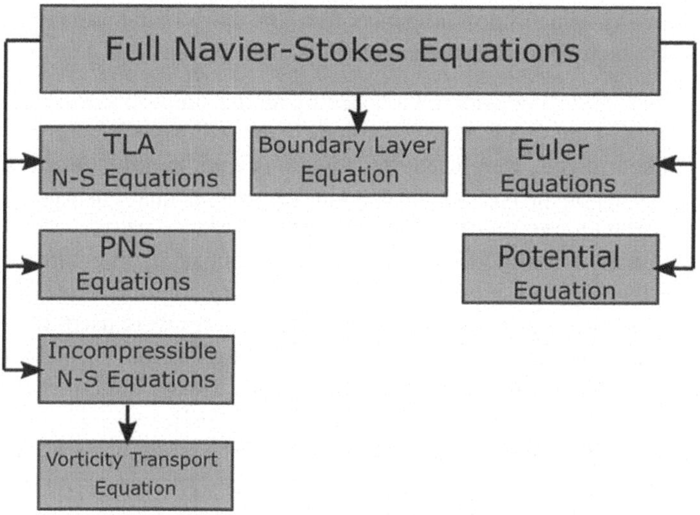

Fig. 2.1 Hierarchical set up of equation systems in fluid dynamics

were developed, which had the ability to march in space. These PNS methods were employed for projects with supersonic flight Mach numbers $M_\infty > 1$ (e.g. for projectiles, missiles, inlets).

With increasing computer power and storage capacity around the 1990s the application of full Navier-Stokes methods became possible.[13] Of course, the possibility to apply the above discussed methods for industrial purposes is strongly dependent on the complexity of the configuration and the capability of the grid generation procedures. All the above mentioned methods are employed to determine steady flow fields. Actually, today also solutions of the unsteady Navier-Stokes equations for instationary flow problems are feasible.

2.6 The Numerical Solutions of the Governing Equations

In the 1960s and the early 1970s only few persons were working on the numerical integration of the governing equations of physically relevant problems. Some of the papers they wrote were seminal. Later the number of people concerned with these issues grew more and more and beyond the second half of the 1980s the activities were extended so much that it was very difficult and almost impossible to completely overlook the published outcome.

[13] In the literature the acronyms RANS and URANS have been naturalized. RANS stands for "Reynolds Averaged Navier-Stokes Equations" and URANS for "Unsteady Reynolds Averaged Navier-Stokes Equations". Both equation systems are able to handle flow fields that include laminar and turbulent flow regimes, whereby statistical turbulence models are employed.

In the following some important papers from the dawn of the numerical methods are discussed, which addressed the numerical solutions of physical problems.[14]

- Babenko et al. (1964) considered a space marching finite-difference method for the integration of the steady Euler equations in three dimensions; curvilinear non-orthogonal coordinates were used; the equations were in non-conservative form; bow shock fitting on the basis of the Rankine-Hugoniot equations was applied; the gas was regarded to be perfect; $M_\infty > 1$ [16].
- Rusanow and Lyubimow (1970) considered a time marching finite-difference method for the integration of the unsteady Euler equations in three dimensions; curvilinear time-dependent non-orthogonal coordinates were used; the equations were in non-conservative form; bow-shock fitting on the basis of the time-dependent Rankine-Hugoniot equations was applied; gases in perfect and equilibrium thermodynamic state were regarded; $M_\infty > 1$ [17].
- Kutler et al. (1972, 1973) considered a space marching finite-difference method for the integration of the steady Euler equations in three dimensions; curvilinear non-orthogonal coordinates were used; the equations were formulated conservatively; shock capturing capability was utilized; gases in thermodynamic equilibrium state were regarded; $M_\infty > 1$ [18, 19].
- Moretti et al. (1966, 1972) considered time marching and space marching finite-difference methods for the integration of the unsteady as well as the steady Euler equations in three dimensions; curvilinear non-orthogonal coordinates were used; the equations were in non-conservative form; shock fitting on the basis of approximated jump conditions was applied; the gas was regarded to be perfect; $M_\infty > 1$ [20, 21].
- Rizzi and Inouye (1973) considered a time marching finite-volume method for the integration of the unsteady Euler equations in three dimensions; curvilinear non-orthogonal coordinates were used; the equations were conservatively formulated; shock capturing capability was utilized; the gas was regarded to be perfect; $M_\infty > 1$ [22].
- Krause and Hirschel (1967, 1971) considered space-marching implicit finite-difference methods for the integration of the steady boundary layer equations in three dimensions; Cartesian coordinates were used; the equations were in non-conservative form; gases in perfect and equilibrium thermodynamic state were regarded; $0 < M_\infty < 25$ [23, 24].
- Pulliam and Steger (1978) considered a time marching finite-difference method for the integration of the unsteady Navier-Stokes equations in three dimensions for laminar flow; curvilinear non-orthogonal coordinates were used; the equations were conservatively formulated; shock capturing capability was utilized; the gas was regarded to be perfect; the shear stresses were exerted similar to boundary-layer theory (thin layer approximation); $M_\infty > 1$ [25].
- Weiland and Pfitzner (1986, 1987, 1991) considered a time marching finite-difference method for the integration of the unsteady Euler equations in three

[14] Of course, this is the author's personal view.

dimensions; curvilinear non-orthogonal coordinates were used; the equations were in quasi-conservative form; bow shock fitting and general shock capturing capability were utilized; the thermodynamic state of the gas was regarded as perfect, in equilibrium and in non-equilibrium; $M_\infty > 1$ [26–28].

There were two new findings in the 1970s and at the beginning of the 1980s that had very successfully shaped the evolution of the numerical methods.

(1) The shock capturing capability requires that the governing equations are formulated conservatively.[15] This is no problem, if Cartesian coordinates are used. But in order to handle complex configurations like airplanes or spaceplanes, it is necessary to introduce curvilinear non-orthogonal coordinates which allow to express the configurational shape to be a coordinate surface.

To transform the independent variables—the coordinates—and the dependent variables—the scalars, vectors and tensors—of a partial differential equation system describing a continuum (field theory), the tensor calculus must be applied. On the one hand this leads to a lot of additional terms representing the metrics of the arbitrary coordinates and on the other hand degrades the conservative behavior of the equations.

Viviand [29], Vinokur [30] and later Warsi [31] have shown that with a specific mix of transformed velocity vectors (contravariant velocities) and original velocity vectors (Cartesian velocities) a formulation of the governing equations in arbitrary coordinates can be achieved, which is strongly conservative and has no additional terms compared to the baseline equations. This was one of the groundbreaking findings in the mid of the 1970s that permitted the successful application of numerical methods.

(**2**) A solution method for hyperbolic partial differential equations is based on the theory of characteristics. The unsteady Euler equations, respectively the Euler part of the Navier-Stokes equations, are hyperbolic systems[16] [32]. In earlier times, the various forms of the discretisation schemes of the differential operators often suffered from stability, accuracy and convergence problems, in particular for flow fields with discontinuities like shocks, shear layers, contact discontinuities, etc. The introduction of the ideas of the theory of characteristics in the governing equations (Flux Vector Splitting, Split-Coefficient Matrix Method) and in the discretized numerical approximation of them (Riemann Solvers), leading to very efficient upwind schemes, has drastically advanced the quality of the numerical solutions [33, 34]. Today, there exist a lot of variants of Riemann Solvers, see for example [35].

Over the years the scientists had a lot of ideas and had pursued many alternatives in order to find the best suited numerical integration methods for the fluid dynamic equations. The early activities used finite-differences and non-conservative formulations (see the list above). The configurations examined were simple and could mostly be described analytically (sphere, hemisphere-cylinder, ogive-cylinder, ellipse, ellipsoid, etc.), which led to structured, unsophisticated numerical grids (mono-block,

[15] ... or quasi-conservatively.

[16] This is also valid for the steady Euler equations if the flow is supersonic and the component of the Mach number in the integration direction x_i is $M_{x_i} > 1$.

grid generation by analytical functions). Since then, the degree of complexity of the configuration was increased permanently, for example in a first step the fuselage of a flight vehicle was extended by wings, vertical fins and tails, in a second step by deflected aerodynamic controls like rudders, elevons and flaps and in a third step by the nacelles of propulsion systems. This has led to a considerable growing of the needs on the discretisation strategy. It was proven that finite-volume schemes could better meet these challenges than finite-difference ones, with the consequence that today presumably all fluid dynamic computer codes are based on the finite-volume approach,[17] which implies the use of conservatively formulated equations.

Further, with mono-block grids no proper resolution of the computational space around complicated configurations could be achieved. Instead, as long as structured grids were employed with much success, multi-block strategies came into play. Finally, codes were developed where the computational space was resolved by unstructured grids. Unstructured grids were typically generated by triangulation procedures (see for example [37]).

In conclusion one can say that today all the fluid dynamic computer codes employed for industrial and practical flow problems are based on

- an upwind differencing for the inviscid part (Euler part),
- a central differencing for the viscous part,
- a conservative formulation,
- a finite-volume method,
- an unstructured grid approach.

2.7 Three Engineering Examples

For the sake of authenticity, we present in this section three examples of numerical solutions achieved by the author and his group during the design and development phase of industrial projects.[18]

2.7.1 First Example

The first example comes from the ARIANE 5 rocket program of the European Space Agency (ESA) and demonstrates how important a high accuracy solution of a fluid problem is for the success of a project. The main engine of the ARIANE 5 rocket was the Vulcain motor (HM60). The thrust of this motor determines the payload to

[17] A comprehensive summary of the fluid dynamic computer codes in Europe that are in use in industry and at research institutes can be found in [36].

[18] The author of this book has headed over roughly three decades scientific departments in research institutes and in aeronautic and space industry (MBB, Dasa, EADS-Astrium, later Airbus). Therefore, the examples are selected from the field of astronautics.

be transported into the orbit. We compare the quantitative results of three approximations:

- analytic relation based on the streamtube approach,
- numerical solution of the Euler equations with a hydrogen-oxygen real gas approach in equilibrium (Fig. 2.2),
- numerical solution of the turbulent Navier-Stokes equations (RANS) with a hydrogen-oxygen real gas approach in equilibrium (Fig. 2.2).

From the streamtube approximation the equation for the thrust F reads

$$F = p_c A_t \gamma \sqrt{\frac{2}{\gamma - 1} \left(\frac{2}{\gamma + 1} \right)^{\frac{\gamma+1}{\gamma-1}}} \sqrt{1 - \left(\frac{p_e}{p_c} \right)^{\frac{\gamma-1}{\gamma}}} + \frac{A_e}{A_t} \frac{(p_e - p_a)}{p_c} \tag{2.2}$$

with

$$\frac{A_e}{A_t} = 45, \quad \gamma = 1.2, \quad p_c = 100\text{bar}, \quad p_e = 0.3\text{bar}, \quad p_a = 1\text{bar},$$

and A_e the nozzle exit area, A_t the throat area, γ the ratio of specific heats for H_2/O_2, p_c the combustion chamber pressure, p_e the exit pressure, p_a the ambient pressure.

The magnitude of the thrust of the main engine determines the success or the failure of a rocket program. The performance of such an engine must be predicted with an accuracy as high as possible. In earlier times, say before the 1980s, the thrust could only be determined on the basis of simple approximations like the Eq. (2.2). The use of such a result requires the application of a safety factor that is extracted usually from experience and/or tests. The magnitude of such a safety factor is often ambiguous. Later, in the 1980s Euler solutions and in the 1990s Navier-Stokes solutions with turbulence modelling (RANS) appropriate for industrial purposes, and with acceptable non-equilibrium real-gas approaches, came into play. The access to results of the quality of a Navier-Stokes solution, as specified above, has dramatically reduced the uncertainties the designers of rockets are faced with.

The upper part of Fig. 2.2 exhibits the Mach number distribution of an Euler solution, whereas in the lower part the lines of constant temperatures for a Navier-Stokes solution (RANS) are plotted [38, 39]. In Table 2.1 the results for the thrust of the three methods, addressed above, are listed. The thrust difference between method 1 and method 3 amounts to 25 kN, from which a payload gap of approximately 750 kg !! would follow for a project like ARIANE 5.

2.7.2 Second Example

The second example deals with the flow in an experimental "Restricted Separation Nozzle (RSS)". This nozzle was designed in order to investigate the different mech-

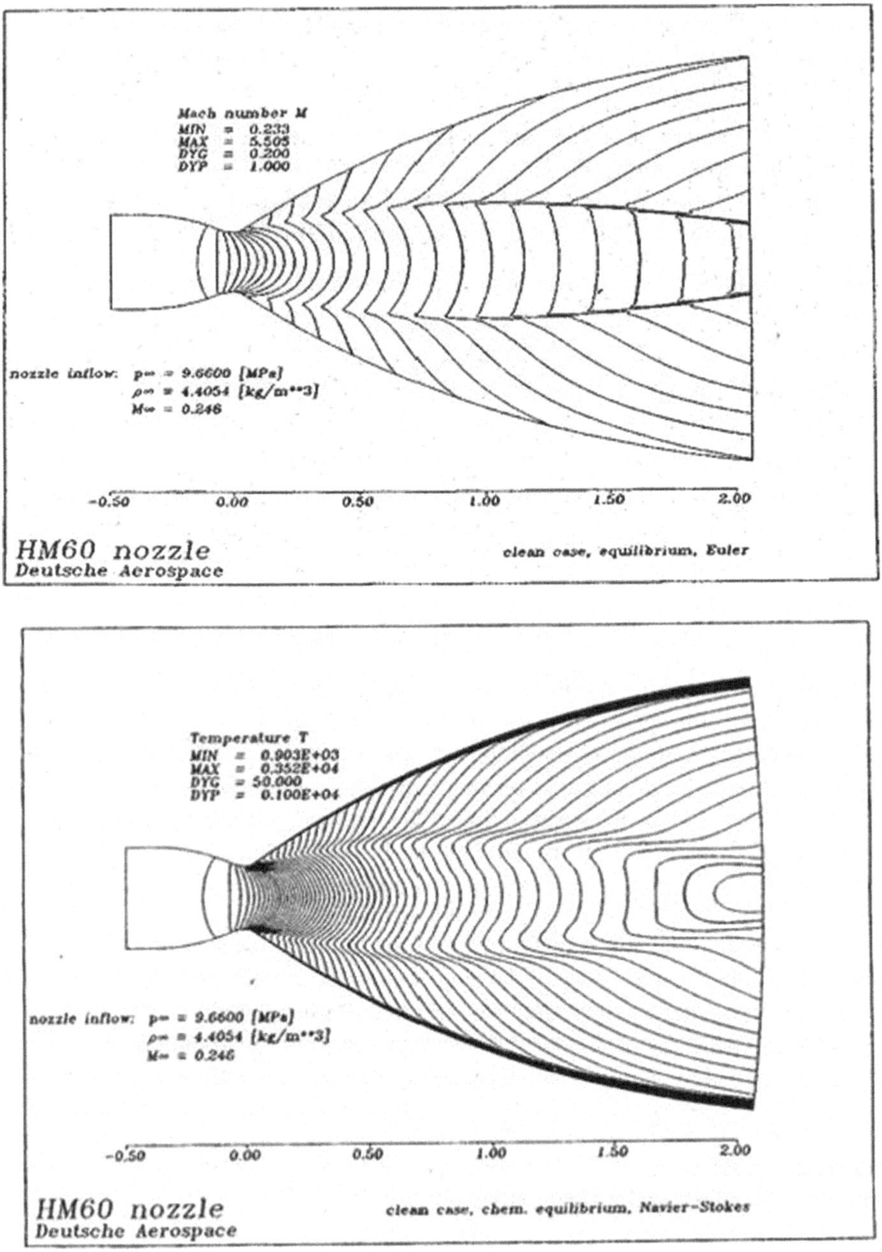

Fig. 2.2 ARIANE 5 Vulcain (HM60) rocket motor nozzle flow. Chamber pressure $p_c = 100$ bar, mixing ratio of the fuel of H_2 to O_2 O/F = 5.89. Equilibrium real gas approach: Euler solution (above), Navier-Stokes solution (RANS) (below) [38, 39]

Table 2.1 Sea-level thrust of Vulcain (HM60) rocket engine

	Method	Sea-level thrust (F) (kN)
1	Eq. (2.2)	790
2	Euler equations	775
3	Navier-Stokes equation	765

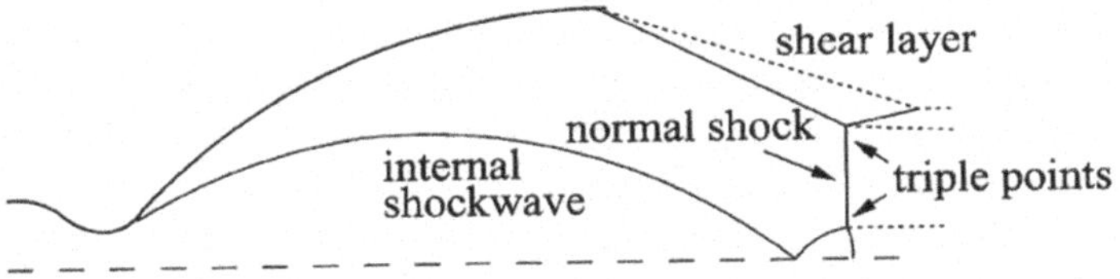

Fig. 2.3 Sketch of restricted separation nozzle (RSS) showing the complicated structure of discontinuities [40]

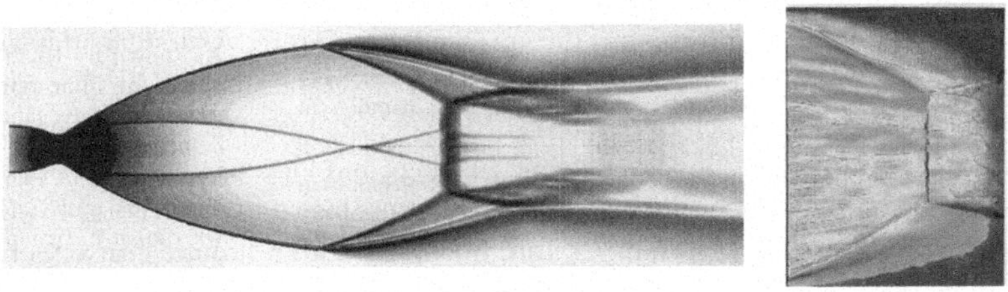

Fig. 2.4 Navier-Stokes solution (RANS) in RSS nozzle including non-equilibrium real gas effects of H_2/O_2 oxidizer/fuel mixture. Pseudo-Schlieren (gradient of density) plots are shown (left), Schlieren image from experiment at DLR (right) [40, 41]

anisms of flow separation in nozzles, which is one of the show-stoppers in nozzle design. The flow downstream of the nozzle exit is extremely complex with interactions between several shocks and shear layers (Fig. 2.3). In Fig. 2.4 (left) a Navier-Stokes solution (RANS) is shown including turbulent flow and non-equilibrium real gas effects for a H_2/O_2 gas mixture [40, 41]. The comparison of the shock and shear layer pattern of an experiment performed at DLR.[19] Figure 2.4 (right) with the computation shows a very convincing agreement.

2.7.3 *Third Example*

In the 1990s the United States' National Aeronautics and Space Administration (NASA) had envisaged to develop a Crew Return Vehicle (CRV) for operational

[19] Deutsches Zentrum für Luft- und Raumfahrt (German Aerospace Center), Lampoldshausen, Germany.

missions in connection with the International Space Station ISS. In the case of illness of crew members or any other emergency, the CRV should be able to bring up to seven astronauts back to Earth. For this mission a configuration on the basis of a lifting body similar to the X-24A was chosen and got the name X-38. A close cooperation between NASA and European industries, agencies and research organisations (Dassault, EADS, ESA, DLR, ONERA) was arranged, particularly for establishing the aerodynamic data base.

X-38 was a single-purpose vehicle. Unfortunately, in 2002 NASA changed its strategy and aimed for a multi-purpose vehicle with the consequence that in June 2002 the X-38 contract was canceled. Nevertheless, the work on the aerothermodynamic data base was brought to a reasonable end.

In Fig. 2.5, some results are shown of a highly sophisticated Navier-Stokes solution (RANS) with a two equation $k - \omega$ turbulence model and a surface radiation wall-boundary condition. Space vehicles experience during the re-entry process, due to the heat transfer from the gas to the wall of the vehicle a strong increase of the surface temperature. But at the same time such hot surfaces radiate heat away [42]. This radiation effect then diminishes the surface temperature. As long as the surface is convex there is no interaction between the elements of the surface. But for concave surfaces the heat radiation is hampered if surface elements are faced by other surface elements, for example the winglets face the fuselage and vice versa and the upperside of the deflected body flap sees the lower side of the fuselage and vice versa [43]. These non-convex effects can lead to temperature hot spots with dramatic consequences.

The freestream conditions of the considered trajectory point were: Flight Mach number $M_\infty = 17.5$, angle of attack $\alpha = 40°$, body flap angle $\eta_{bf} = 20°$, Reynolds number $\text{Re} = 1.02 \cdot 10^6$ and flight altitude $H = 58.8$ km. Figure 2.5(a, b) give views of the X-38 configuration; (c) in the symmetry plane of the rear part with the deflected body flap the temperature distribution and the streamlines are plotted; skin friction lines and pressure distribution in the rear are shown in (d); the influence of the view factor approach in the cavity between deflected body flap and the lower side of the fuselage is demonstrated in (e), where the left cavity part gives the result with the view factor approach leading to higher surface temperatures compared to the right cavity part, where the radiation is not hampered; (f) exhibits an enlargement of the cavity regime [44, 45].

2.7.4 Final Remarks

The three examples discussed above point out how discrete numerical methods have developed. Due to today's powerful discretisation methods, the enormous growth of computer power and storage capacity problems can be treated, that were out of reach only a couple of decades away.

In the future the application of numerical methods is faced with the following three challenges

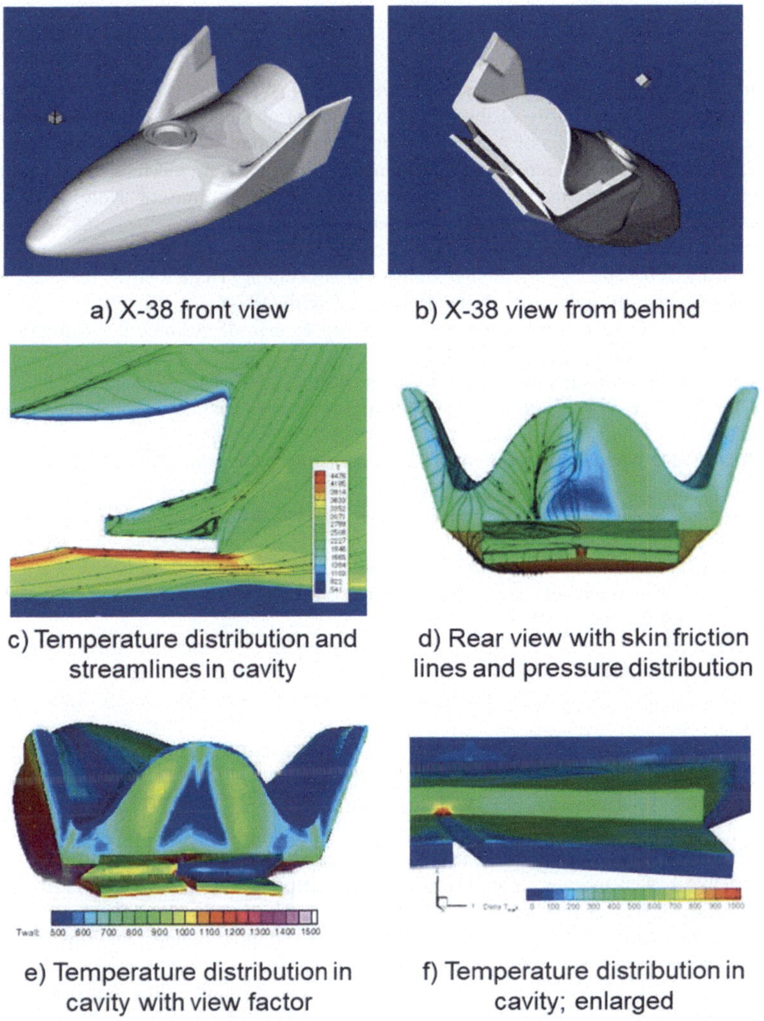

Fig. 2.5 Navier-Stokes solution with $k - \omega$ turbulence model, wall radiation and view factor approach. X-38 space plane. $M_\infty = 17.5, \alpha = 40°, \eta_{bf} = 20°, \ Re = 1.02 \cdot 10^6, \ H = 58.8$ km [43, 44]

- the extension of the fundamental fluid dynamics to problems with advanced thermodynamical requirements including wall catalycity as well as gas and wall radiation,
- the resolution of very complex configurations using appropriate grid generation procedures,
- the modeling and the direct simulation of turbulence including the transition problem.

References

1. Eckert, M.: The Dawn of Fluid Dynamics. Wiley-VCH, Weinheim (2006)
2. Anderson, J.D.: Brief history of the early development of theoretical and experimental fluid dynamics. In: Encyclopedia of Aerospace Engineering. Wiley, New York (2010)
3. Rizzi, A., Hirschel, E.H.: General developments of numerical fluid mechanics until the middle of the 20th century. In: Hirschel, E.H., Krause, E. (Eds.) 100 Volumes of 'Notes on Numerical Fluid Mechanics'. Springer, Heidelberg (2009)
4. Edwards, C.H.: The Historical Development of the Calculus. Springer, New York (1979)
5. Lueroth, J.: Ueber die Geschichte der Infinitesimalrechnung. Heidelberger Texte zur Mathematikgeschichte http://www.ub.uni-heidelberg.de/archiv/12746
6. Newton, I.: The October 1666 Tract on Fluxions, pp.48–63. MS Add.3958.3, Cambridge University Library, Cambridge, UK (2011)
7. Altenbach, H.: Kontinuumsmechanik. Springer, Berlin (2018)
8. Bird, R.B., Stuart, W.E., Lightfoot, E.N.: Transport Phenomena, 2nd edn. Wiley, New York (2002)
9. Anderson, J.D.: Fundamentals of Aerothermodynamics, 5th edn. McGraw-Hill, New York (2011)
10. Schlichting, H., Gersten, K.: Boundary Layer Theorie, 8th revised and enlarged Edition. Springer, Berlin (2000)
11. Weiland, C.: Mechanics of Flow Similarities. Springer, Berlin (2020)
12. Sommerfeld, A.: Mechanik der Deformierbaren Medien. Akademische Verlagsgesellschaft Geest & Portig, Leipzig (1954)
13. Hirschel, E.H., Cousteix, J., Kordulla, W.: Three-Dimensional Attached Viscous Flow. Springer, Berlin (2014)
14. Eckert, M.: Ludwid Prandtl. Strömungsforscher und Wissenschaftsmanager. Springer, Berlin (2017)
15. Rotta, J.C.: Die Aerodynamische Versuchsanstalt in Göttingen, ein Werk Ludwig Prandtls. Vandenhoeck & Ruprecht, Göttingen (1990)
16. Babenko, K.I., Voskresenskii, G.P., Lyubimow, A.N., Rusanow, V.V.: Three-Dimensional Flow of Ideal Gas Around Smooth Bodies (Translated from Russian by "Israel Program for Scientific Translations", Jerusalem (1968)), Isdatel'stoo "Nauka", Moskau (1964)
17. Lyubimow, A.N., Rusanow, V.V.: Gas Flows Past Blunt Bodies. NASA TT F-714 (1973), Isdatel'stoo "Nauka", Moskau (1970)
18. Kutler, P., Reinhardt, W.A., Warming, R.F.: Multishocked three-dimensional supersonic flowfields with real gas effects. AIAA J. **11**, 637–664 (1973)
19. Kutler, P., Lomax, H., Warming, R.F.: Computation of space shuttle flow fields using noncentered finite-difference schemes. AIAA-Paper No. 72-193 (1972)
20. Moretti, G., Abbett, M.: A time-dependent computational method for blunt body flows. AIAA J. **4**(12), 2136–2141 (1966)
21. Moretti, G., Grossmann, B., Marconi, F.: A complete numerical technique for the calculation of three-dimensional inviscid supersonic flows. AIAA Paper No. 72-192 (1972)
22. Rizzi, A., Inouye, M.: A time-split finite-volume technique for three-dimensional blunt-body flow. AIAA J. **11**, 1478–1485 (1973)
23. Krause, E.: Numerical solution of boundary layer equations. AIAA J. **5**, 1231–1236 (1967)
24. Hirschel, E.H.: Ein Implizites Differenzenverfahren zur Berechnung hypersonischer Gleitströmungs-Grenzschichten. DLR FB 71-97 (1971)
25. Pulliam, T.H., Steger, J.L.: On Implicit finite-difference simulation of three dimensional flow. AIAA-Paper 78-10 (1978)
26. Weiland, C., Pfitzner, M.: 3-D and 2-D Solutions of the Quasi-Conservative Euler Equations. Lecture Notes in Physics, vol. 264, pp. 654–659. Springer, Berlin (1986)
27. Pfitzner, M., Weiland, C.: 3-D Euler Solutions for hypersonic freestream Mach numbers. AGARD-CP-428, Paper 22 Bristol, UK (1987)

28. Pfitzner, M.: A 3-D Non-equilibrium shock-fitting algorithm using effective Rankine-Hugoniot relations. AIAA-Paper 91-1467 (1991)
29. Viviand, H.: Formes Conservatives des Equations de la Dynamique des Gaz (Conservative forms of gas dynamics equations). La Recherche Aerospatiale, No. **74–1**, 67–68 (1974)
30. Vinokur, M.J.: Conservation equations of gas dynamics in curvilinear coordinate systems. J. Comp. Phys. **14**, 105–125 (1974)
31. Warsi, Z.U.A.: Conservation form of the Navier-Stokes equations in general nonsteady coordinates. AIAA J. **19**, 240–242 (1981)
32. Courant, R., Hilbert, D.: Methoden der Mathematischen Physik, 2nd edn. Springer, Berlin (1968)
33. Roe, P.L.: Approximate Riemann solvers, parameter vectors, and difference schemes. J. Comp. Phys. **43**, 357–372 (1981)
34. van Leer, B.: Towards the ultimate conservation difference scheme V, a second-order sequel to Godunov's method. J. Comp. Phys. **32**, 101–136 (1979)
35. Yee, H.C.: Upwind and symmetric shock-capturing schemes. NASA Technical Memorandum 89464 (1987)
36. Hirschel, E.H., Krause, E.: 100 Volumes of 'Notes on Numerical Fluid Mechanics'. Notes on Numerical Fluid Mechanics and Multidisciplinary Design, vol. 100. Springer, Heidelberg (2009)
37. Weatherill, N.P.: Delauny triangulation in computational fluid dynamics. Comput. Math. Appl. **24**, 129–150 (1992)
38. Weiland, C., Hartmann, G., Menne, S.: Aerothermodynamics of nozzle flows for advanced hypersonic propulsion systems. AGARD-CP 510, Paper No. 32 (1992)
39. Caporicci, M., Eriksson, L.-E., Pekkari, L.O., Onofri, M., Popp, M., Weiland, C.: Advanced nozzle technologies for the Ariane 5 Vulcain engine. AIAA-Paper 94-3263 (1994)
40. Gross, A., Weiland, C.: Numerical simulation of separated cold gas nozzle flows. J. Propulsion Power **20**(3), 509–519 (2004)
41. Gross, A., Weiland, C.: Numerical simulation of separated hot gas nozzle flows. J. Propulsion Power **20**(5), 879–891 (2004)
42. Hirschel, E.H.: Basics of Aerothermodynamics, 2nd edn. Springer, Berlin (2015)
43. Zeiss, W.: Modelling of hot, radiation cooled surfaces. TET-DASA-21-TN-2402, Dasa, Munich, Germany (1999)
44. Behr, R.: Hot, radiation cooled surfaces. TET-DASA-21 TN-2410, Dasa, Munich, Germany (2002)
45. Hirschel, E.H., Weiland, C.: Selected Aerothermodynamic Design Problems of Hypersonic Flight Vehicles. Springer Verlag, Berlin/Heidelberg and AIAA (Progress in Astronautics and Aeronautics), **229**, Reston, USA (2009)

Chapter 3
The Fluid Dynamic Equations in Symbolic Form

Abstract The fluid dynamic equations as conservation laws can be formulated in symbolic form. The symbolic form is valid for every coordinate system. An integral version and a differential version of these forms are presented in this chapter.

The Variables and Quantities Used in this Chapter

ρ	density
m	mass
V	volume
$\mathbf{v}$	velocity vector
p	pressure
$\mathbf{f}_e$	external forces per unit mass (e.g.: electromagnetic, gravitational, inertial)
Γ	potential of the external forces
$\bar{\bar{\sigma}}$	total stress tensor
$\bar{\bar{\tau}}$	viscous stress tensor. In tensorial notation it reads $\tau_{ij} = \mu(\partial_i v_j + \partial_j v_i) + \lambda(\partial^k v_k \delta_{ij})$ with λ a second viscosity; further the Stokes relation $2\mu + 3\lambda = 0$ holds
μ	dynamic viscosity
κ	thermal conductivity
$\bar{e}$	$= \rho e$ total energy per unit volume
e	$= e_i + e_{kin}$ total energy per unit mass
e_i	internal energy per unit mass
e_{kin}	$= 1/2\lvert\mathbf{v}\rvert^2$ kinetic energy per unit mass
$\mathbf{q}$	Equation (3.10), molecular transport of energy and mass diffusion due to gas mixtures
$\mathbf{j}_i$	Equation (3.10), diffusion mass flux of species i
q_H	other heat sources (radiation, chemical reactions, etc.)
h	$= e_i + p/\rho$ enthalpy per unit mass
h_0	$= e + p/\rho$ total enthalpy per unit mass
h_i	species enthalpy

C. Weiland, *The Hierarchy of Fluid Dynamic Equations*,
https://doi.org/10.1007/978-3-031-71198-5_3

c^2	$= (\partial p/\partial \rho)_s$ speed of sound
T	temperature of the fluid
s	entropy per unit mass

3.1 The General Equations as Conservative Laws

The mathematical description of the motion of a fluid is achieved with the formulation of the conservation of mass, momentum and energy of the flow [1–6]. Mathematics provides us with a general formulation of conservation laws regarding fields of scalar and vector functions. For scalar field functions we have

$$\frac{D\Phi}{Dt} = \frac{\partial}{\partial t}\int_V \phi \, dV + \oint_F \boldsymbol{\psi} \cdot d\mathbf{F} = \int_V S_V \, dV + \oint_F \mathbf{S}_F \cdot d\mathbf{F} \tag{3.1}$$

with the scalars $\phi(t, x_i)$, $\Phi(t)$, S_V, which are defined by[1]

- $\phi(t, x_i)$ scalar field function ,
- $\Phi(t) = \int_V \phi(t, x_i)\, dV$ integration of the scalar field function with respect to the volume V,
- S_V volume sources due to the scalar field function $\phi(t, x_i)$

and the vectors $\boldsymbol{\psi}(t, x_i)$, $\mathbf{S}_F$ defined by

- $\boldsymbol{\psi}(t, x_i) = \phi(t, x_i)\, \mathbf{v}$ flux of the scalar field function,
- $\mathbf{S}_F$ surface sources acting at the surface F of the volume V (see Fig. 3.1).

For vector field functions Eq. (3.1) is valid for each of the components of the vector. But it can also be formulated in vector-tensor notation and one obtains

$$\frac{D\boldsymbol{\Phi}}{Dt} = \frac{\partial}{\partial t}\int_V \boldsymbol{\phi} \, dV + \oint_F \bar{\bar{\psi}} \cdot d\mathbf{F} = \int_V \mathbf{S}_V \, dV + \oint_F \bar{\bar{S}}_F \cdot d\mathbf{F}, \tag{3.2}$$

where $\boldsymbol{\Phi}$, $\boldsymbol{\phi}$, $\mathbf{S}_V$ are vectors and $\bar{\bar{\psi}}$, $\bar{\bar{S}}_F$ are tensors. Further, we need Gauss's theorem and find for the surface integrals of Eq. (3.1)

$$\oint_F \boldsymbol{\psi} \cdot d\mathbf{F} = \int_V \operatorname{div} \boldsymbol{\psi} \, dV \,, \quad \oint_F \mathbf{S}_F \cdot d\mathbf{F} = \int_V \operatorname{div} \mathbf{S}_F \, dV \,, \tag{3.3}$$

[1] Details regarding the derivation and the establishment of Eq. (3.1) can be found within [1, 2].

and for the ones in Eq. (3.2).

$$\oint_F \bar{\bar{\psi}} \cdot d\mathbf{F} = \int_V \operatorname{div} \bar{\bar{\psi}} \, dV \,, \quad \oint_F \bar{\bar{S}}_F \cdot d\mathbf{F} = \int_V \operatorname{div} \bar{\bar{S}}_F \, dV. \tag{3.4}$$

3.2 Integral Forms of the Fluid Dynamic Equations

3.2.1 The Conservation of Mass

The specific mass is the density with $\rho = m/V$. From the physical point of view, it is evident, considering Fig. 3.1, that the temporal change of the mass inside the volume V can only be compensated by the net amount of entering to leaving mass fluxes in the volume V through the surface F (if no mass sources are present). Therefore, we obtain from Eq. (3.1) with $\Phi = m$, $\phi = \rho$, $\boldsymbol{\psi} = \rho \cdot \mathbf{v}$, $S_V = 0$, $\mathbf{S}_F = 0$

$$\frac{Dm}{Dt} = \frac{\partial}{\partial t} \int\limits_V \rho \, dV + \oint\limits_F \rho \mathbf{v} \cdot d\mathbf{F} = 0, \tag{3.5}$$

and with the relations Eq. (3.3) we have

$$\int\limits_V \frac{\partial \rho}{\partial t} \, dV + \int\limits_V \operatorname{div}(\rho \mathbf{v}) \, dV = 0. \tag{3.6}$$

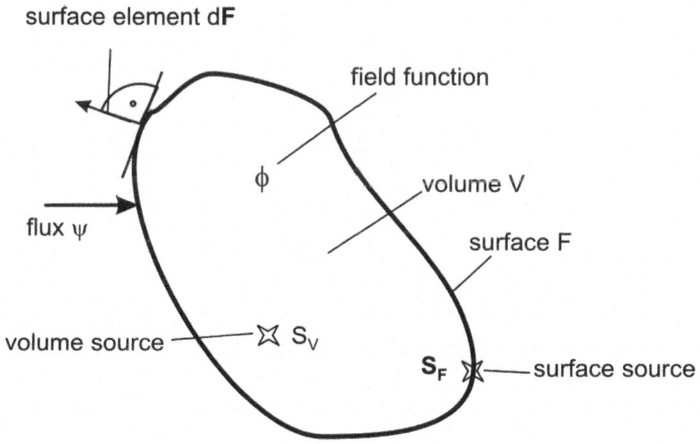

Fig. 3.1 Sketch of the control volume for the definition of conservation laws

3.2.2 The Conservation of Momentum

The vector $\boldsymbol{\phi} = \rho \mathbf{v}$ represents the field function that describes the conservation of momentum. The rate of change of the momentum, due to Newton's law, is equal to the sum of the external forces acting on the volume and the internal forces acting on the surface of this volume. The tensor $\bar{\bar{\psi}}$ is then given by $\bar{\bar{\psi}} = \rho \mathbf{v} \otimes \mathbf{v}$, the external forces by the vector $\mathbf{S}_V = \rho \mathbf{f}_e$, where $\mathbf{f}_e$ are volume forces per unit mass, like the electromagnetic, gravitational and inertial ones, and the internal forces by the total stress tensor $\bar{\bar{S}}_F = \bar{\bar{\sigma}}$ (see Eq. (3.2)). For Newtonian fluids the total stress tensor can be split into the the pressure stress tensor $p\,\bar{\bar{I}}$ with $\bar{\bar{I}}$ the unit tensor and into the viscous shear stress tensor $\bar{\bar{\tau}}$.

$$\bar{\bar{\sigma}} = -p\,\bar{\bar{I}} + \bar{\bar{\tau}}. \tag{3.7}$$

The minus sign of the pressure term is necessary because the unit vector at the surface $d\mathbf{F}$ points outwards while the pressure acts in the opposite direction, Fig. 3.1. With the above definitions, Eq. (3.2) gets the form

$$\begin{aligned}
\frac{D\boldsymbol{\Phi}}{Dt} &= \frac{D}{Dt}\int\limits_V \rho\mathbf{v}dV \\
&= \frac{\partial}{\partial t}\int\limits_V \rho\mathbf{v}\,dV + \oint\limits_F \rho\mathbf{v}(\mathbf{v}\cdot d\mathbf{F}) \\
&= \int\limits_V \rho\mathbf{f}_e\,dV - \oint\limits_F p\,\bar{\bar{I}}\cdot d\mathbf{F} + \oint_F \bar{\bar{\tau}}\cdot d\mathbf{F}\,,
\end{aligned} \tag{3.8}$$

or if we apply Eq. (3.4)

$$\begin{aligned}
&\frac{\partial}{\partial t}\int\limits_V \rho\mathbf{v}\,dV + \int\limits_V \operatorname{div}(\rho\mathbf{v}\otimes\mathbf{v})\,dV \\
&= \int_V \rho\mathbf{f}_e\,dV - \int\limits_V \operatorname{grad} p\,dV + \int\limits_V \operatorname{div}\bar{\bar{\tau}}\,dV\,,
\end{aligned} \tag{3.9}$$

where $\mathbf{v}\otimes\mathbf{v}$ denotes the tensor product of the velocity vector.

3.2.3 The Conservation of Energy

The total energy $\bar{e} = \rho e$ in a volume V is given by the sum of the internal energy of the fluid and the kinetic energy. If we denote the internal energy per unit mass by e_i and the kinetic energy per unit mass by $e_{kin} = 1/2|\mathbf{v}|^2$ we can write

$$\bar{e} = \rho e = \rho \left(e_i + \frac{1}{2}|\mathbf{v}|^2 \right) .$$

The field function is the scalar quantity $\phi = \rho\, e$ and the flux vector is $\boldsymbol{\psi} = \rho\, e\, \mathbf{v}$. The variation of the total energy is balanced by the sum of the volume sources S_V, that is the work of the volume forces $\rho \mathbf{f}_e \cdot \mathbf{v}$ and the heat sources q_H, as well as the sum of the surface sources $\mathbf{S}_F$, which is given by the work done by the pressure stresses $p\mathbf{v}$ and the internal shear stresses $\mathbf{v} \cdot \bar{\bar{\tau}}$. Finally, the molecular transport of energy $\mathbf{q}$ is also part of the energy balance. Heat conduction inside the fluid in the form of Fourier's law is the main part of the molecular transport of energy, but in the case of gas mixtures we have to include also the mass diffusion terms. Therefore, we have

$$\mathbf{q} = -\kappa \text{ grad } T + \sum_i \mathbf{j}_i h_i \, , \tag{3.10}$$

with κ being the thermal conductivity, T the temperature, $\mathbf{j}_i$ the diffusion mass flux (Fick's first law) and h_i the species enthalpy [7].

The energy equation then reads

$$\begin{aligned}
\frac{D\Phi}{Dt} &= \frac{D}{Dt} \int_V \rho e dV \\
&= \frac{\partial}{\partial t} \int_V \rho e \, dV + \oint_F \rho e(\mathbf{v} \cdot d\mathbf{F}) \\
&= \int_V \rho \mathbf{f}_e \cdot \mathbf{v} dV + \int_V q_H \, dV - \\
&- \oint_F p\mathbf{v} \cdot d\mathbf{F} + \oint_F (\bar{\bar{\tau}} \cdot \mathbf{v}) \cdot d\mathbf{F} - \oint_F \mathbf{q} \cdot d\mathbf{F} \, ,
\end{aligned} \tag{3.11}$$

and with Eq. (3.3)

$$\frac{\partial}{\partial t}\int_V \rho e\, dV + \int_V \mathrm{div}(\rho e\mathbf{v}) dV$$
$$= \int_V \rho\mathbf{f}_e \cdot \mathbf{v} dV + \int_V q_H\, dV -$$
$$- \int_V \mathrm{div}(p\,\bar{\bar{I}} \cdot \mathbf{v})\, dV + \int_V \mathrm{div}(\bar{\bar{\tau}} \cdot \mathbf{v})\, dV - \int_V \mathrm{div}\mathbf{q}\, dV\,. \tag{3.12}$$

3.3 Differential Forms of the Fluid Dynamic Equations

In the following we continuously assume that all the integrands in the Eqs. (3.6), (3.9), (3.12) are continuous and continuously differentiable, which means that the related equations in differential form cannot describe singularities like shock waves, shear layers, etc. The governing equations in differential form often are presented using the material or substantial differential operator $\frac{D}{Dt} = \frac{\partial}{\partial t} + \mathbf{v} \cdot \,\mathrm{grad}$.

3.3.1 *The Continuity Equation*

The differential form of the continuity equation (3.6) is given by

$$\frac{\partial \rho}{\partial t} + \mathrm{div}\,(\rho\mathbf{v}) = 0\,, \tag{3.13}$$

or with the material operator

$$\frac{D\rho}{Dt} + \rho\,\mathrm{div}\,\mathbf{v} = 0\,. \tag{3.14}$$

3.3.2 *The Momentum Equation*

From Eq. (3.9) we extract the momentum equation in differential form

$$\frac{\partial \rho\mathbf{v}}{\partial t} + \mathrm{div}\,(\rho\mathbf{v} \otimes \mathbf{v}) = \rho\mathbf{f}_e - \mathrm{grad}\, p + \mathrm{div}\,\bar{\bar{\tau}}\,, \tag{3.15}$$

which can be reformulated by using the material operator

$$\rho \frac{D\mathbf{v}}{Dt} = \rho \mathbf{f}_e - \operatorname{grad} p + \operatorname{div} \bar{\bar{\tau}} \,, \tag{3.16}$$

where we have made use of the relation

$$\begin{aligned} \frac{\partial \rho \mathbf{v}}{\partial t} + \operatorname{div}(\rho \mathbf{v} \otimes \mathbf{v}) &= \rho \frac{\partial \mathbf{v}}{\partial t} + \rho (\mathbf{v} \cdot \operatorname{grad})\mathbf{v} + \mathbf{v} \underbrace{\left(\frac{\partial \rho}{\partial t} + \operatorname{div}(\rho \mathbf{v}) \right)}_{=0} \\ &= \rho \frac{D\mathbf{v}}{Dt} . \end{aligned} \tag{3.17}$$

Crocco's Interpretation of the Momentum Equation

Considering the momentum Eq. (3.15) together with Eq. (3.17) and the formal vector transformation [8]

$$(\mathbf{v} \cdot \operatorname{grad})\mathbf{v} = \operatorname{grad} \frac{\mathbf{v}^2}{2} - (\mathbf{v} \times \operatorname{rot}\mathbf{v}) \,, \tag{3.18}$$

we obtain

$$\frac{\partial \mathbf{v}}{\partial t} + \operatorname{grad} \frac{\mathbf{v}^2}{2} - (\mathbf{v} \times \operatorname{rot}\mathbf{v}) = \mathbf{f}_e - \frac{1}{\rho} \operatorname{grad} p + \frac{1}{\rho} \operatorname{div} \bar{\bar{\tau}} . \tag{3.19}$$

Now, we introduce the well-known relation for the entropy s, defined for a gas in thermodynamic equilibrium and without any mass diffusion.[2]

$$T Ds = Dh - \frac{1}{\rho} Dp = De_i - \frac{p}{\rho^2} D\rho \,, \tag{3.20}$$

with h being the enthalpy and T being the temperature. The first part of Eq. (3.20) is equivalent to a steady change of the thermodynamic state with

$$T \operatorname{grad} s = \operatorname{grad} h - \frac{1}{\rho} \operatorname{grad} p \,. \tag{3.21}$$

Introducing Eq. (3.21) in Eq. (3.19), and if the external forces $\mathbf{f}_e$ have a potential Γ (conservative field of forces), the equation for steady flows can be written as

$$-(\mathbf{v} \times \operatorname{rot}\mathbf{v}) = T \operatorname{grad} s - \operatorname{grad}(h_0 + \Gamma) + \frac{1}{\rho} \operatorname{div} \bar{\bar{\tau}} \,, \tag{3.22}$$

---

[2] The definition and assumptions of this relation can be found in most of the thermodynamic and gasdynamic textbooks, e.g. in [1, 3, 9].

For inviscid and isoenergetic flows, where the term isoenergetic is defined such that $h_0 + \Gamma$ is a constant, we find

$$-\,(\mathbf{v} \times \mathrm{rot}\mathbf{v}) = T \,\mathrm{grad}\, s. \tag{3.23}$$

The energy equation for inviscid flows without heat conduction and $q_H = 0$ reads $Ds/Dt = 0$, Eq. (3.37), which means that the entropy along streamlines (and path lines) is constant (isentropic flow). If, in addition, the flow has no vorticity rot $\mathbf{v} = 0$ Eq. (3.23) says that then grad $s = 0$ which means that the entropy is constant in the whole flow field. On the other hand, if grad $s \neq 0$, the flow must contain vorticity. A special situation is given for a plane flow. There, the vector rot $\mathbf{v}$ is perpendicular to the velocity vector $\mathbf{v}$ and the product ($\mathbf{v} \times \mathrm{rot}\mathbf{v}$) is only zero for rot $\mathbf{v} = 0$. The result is that for plane flow with grad $s = 0$ the flow is irrotational.

3.3.3 The Energy Equation

Turning to Eq. (3.12) we find for the energy equation in differential form

$$\frac{\partial \rho e}{\partial t} + \mathrm{div}\,(\rho e \mathbf{v}) = -\,\mathrm{div}(p\mathbf{v}) + \mathrm{div}(\bar{\bar{\tau}} \cdot \mathbf{v}) - \mathrm{div}\,\mathbf{q} + \rho \mathbf{f}_e \cdot \mathbf{v} + q_H, \tag{3.24}$$

or

$$\frac{\partial \rho e}{\partial t} + \mathrm{div}\,(\rho \mathbf{v} \underbrace{\left(e + \frac{p}{\rho}\right)}_{h_0}) = \mathrm{div}(\bar{\bar{\tau}} \cdot \mathbf{v}) - \mathrm{div}\,\mathbf{q} + \rho \mathbf{f}_e \cdot \mathbf{v} + q_H\,. \tag{3.25}$$

By applying the relation

$$\frac{\partial \rho e}{\partial t} + \mathrm{div}\,(\rho e \mathbf{v}) = \rho \underbrace{\left(\frac{\partial e}{\partial t} + \mathbf{v} \cdot \mathrm{grad}\, e\right)}_{De/Dt} + e \underbrace{\left(\frac{\partial \rho}{\partial t} + \mathrm{div} \rho \mathbf{v}\right)}_{=0}, \tag{3.26}$$

we obtain with Eq. (3.24) for the energy equation described in the material operator form

$$\rho \frac{De}{Dt} = -\,\mathrm{div}(p\mathbf{v}) + \mathrm{div}(\bar{\bar{\tau}} \cdot \mathbf{v}) - \mathrm{div}\,\mathbf{q} + \rho \mathbf{f}_e \cdot \mathbf{v} + q_H. \tag{3.27}$$

Other formulations of the energy equation have a historical background and/or are often helpful for the understanding of reductions of the equation, for example for steady flow, inviscid flow, isoenergetic flow, etc. We therefore discuss below three formulations where the variations of the internal energy e_i, the enthalpy h and the total enthalpy h_0 are used.

The Total Enthalpy Formulation

Introducing the enthalpy per unit mass $h = e_i + p/\rho$ and the total enthalpy per unit mass $h_0 = h + 1/2|\mathbf{v}|^2$ we find with Eqs. (3.24), (3.25)

$$\frac{\partial \rho h_0}{\partial t} + \operatorname{div}(\rho \mathbf{v} h_0) - \frac{\partial p}{\partial t} = \operatorname{div}(\bar{\bar{\tau}} \cdot \mathbf{v}) - \operatorname{div} \mathbf{q} + \rho \mathbf{f}_e \cdot \mathbf{v} + q_H, \tag{3.28}$$

and

$$\rho \frac{D h_0}{D t} = \frac{\partial p}{\partial t} + \operatorname{div}(\bar{\bar{\tau}} \cdot \mathbf{v}) - \operatorname{div} \mathbf{q} + \rho \mathbf{f}_e \cdot \mathbf{v} + q_H. \tag{3.29}$$

The Internal Energy and Enthalpy Formulations

For the reformulation of the energy equation (3.24) to consider the variation of the internal energy e_i we then first use the momentum equation (3.15), which we multiply with the velocity $\mathbf{v}$ and secondly the relation

$$\frac{\partial \rho \mathbf{v}}{\partial t} + \operatorname{div}(\rho \mathbf{v} \otimes \mathbf{v} + p \bar{\bar{I}}) = \rho \frac{\partial \mathbf{v}}{\partial t} + \rho (\mathbf{v} \cdot \operatorname{grad}) \mathbf{v} + \operatorname{grad} p, \tag{3.30}$$

and find

$$\rho \frac{\partial \frac{1}{2}|\mathbf{v}|^2}{\partial t} + \rho (\mathbf{v} \cdot \operatorname{grad} \frac{1}{2}|\mathbf{v}|^2) + \mathbf{v} \cdot \operatorname{grad} p - \mathbf{v} \cdot \operatorname{div} \bar{\bar{\tau}} = \rho \mathbf{f}_e \cdot \mathbf{v}. \tag{3.31}$$

Then the energy equation (3.24) is differentiated, which leads to

$$\begin{aligned} &\rho \left(\frac{\partial e_i}{\partial t} + \mathbf{v} \cdot \operatorname{grad} e_i \right) + \rho \frac{\partial \frac{1}{2}|\mathbf{v}|^2}{\partial t} + \rho (\mathbf{v} \cdot \operatorname{grad} \frac{1}{2}|\mathbf{v}|^2) + \mathbf{v} \cdot \operatorname{grad} p - \\ &- \frac{p}{\rho} \left(\frac{\partial \rho}{\partial t} + \mathbf{v} \cdot \operatorname{grad} \rho \right) + \operatorname{div} \mathbf{q} - \operatorname{div}(\bar{\bar{\tau}} \cdot \mathbf{v}) - q_H = \rho \mathbf{f}_e \cdot \mathbf{v}, \end{aligned} \tag{3.32}$$

where

$$\operatorname{div}(p \mathbf{v}) = -\frac{p}{\rho} \frac{D \rho}{D t} + \mathbf{v} \cdot \operatorname{grad} p$$

was used.

By combining Eqs. (3.31) and (3.32) we obtain for the energy equation based on the variation of the internal energy

$$\rho\frac{De_i}{Dt} - \frac{p}{\rho}\frac{D\rho}{Dt} = -\text{div}\,\mathbf{q} + q_H + (\bar{\bar{\tau}}\cdot\text{grad})\cdot\mathbf{v}, \tag{3.33}$$

with

$$\text{div}(\bar{\bar{\tau}}\cdot\mathbf{v}) - \mathbf{v}\cdot\text{div}\,\bar{\bar{\tau}} = (\bar{\bar{\tau}}\cdot\text{grad})\cdot\mathbf{v} \equiv \tau_{ij}\frac{\partial v_i}{\partial x_j}. \tag{3.34}$$

Equation (3.34) is called the dissipation function and can be construed as an irreversible heat source due to internal friction activated by mechanical energies.

With the enthalpy $h = e_i + p/\rho = c_pT$ (for perfect gas), we reformulate the left-hand term of Eq. (3.33) and find

$$\rho\frac{Dh}{Dt} = \rho c_p\frac{DT}{Dt} = \frac{Dp}{Dt} - \text{div}\,\mathbf{q} + q_H + (\bar{\bar{\tau}}\cdot\text{grad})\cdot\mathbf{v}, \tag{3.35}$$

Equation (3.35) is the basis of similarity considerations when dimensionless numbers are defined with respect to the energy balance [6].
Further, by combining Eq. (3.33) with Eq. (3.20) one obtains

$$\rho T\frac{Ds}{Dt} = -\text{div}\,\mathbf{q} + q_H + (\bar{\bar{\tau}}\cdot\text{grad})\cdot\mathbf{v}. \tag{3.36}$$

For inviscid flows with $\mathbf{q} = 0$ and $q_H = 0$ the energy equation reduces to

$$\frac{Ds}{Dt} = 0. \tag{3.37}$$

Since the definition of the speed of sound is given by $c^2 = \left(\frac{\partial p}{\partial\rho}\right)_{s=const}$, we can formulate an equivalent to Eq. (3.37) by

$$\frac{Dp}{Dt} - c^2\frac{D\rho}{Dt} = 0. \tag{3.38}$$

During the period (approximately the years 1965–1980) when the first numerical solutions of two- and three-dimensional inviscid flow fields for supersonic and hypersonic Mach numbers around blunt bodies appeared, the Eulerian equations were applied often in a non-conservative manner and numerically integrated by finite difference schemes. In these systems of governing equations, the energy equation in the form of Eqs. (3.38) or (3.37) was employed [10–12]. Supersonic and hypersonic flow fields past blunt bodies contain discontinuities. The most prominent of which is the bow shock and eventually several so-called embedded shocks. A very elegant and exact method to deal with these shocks is to use the fitting technique, which treats

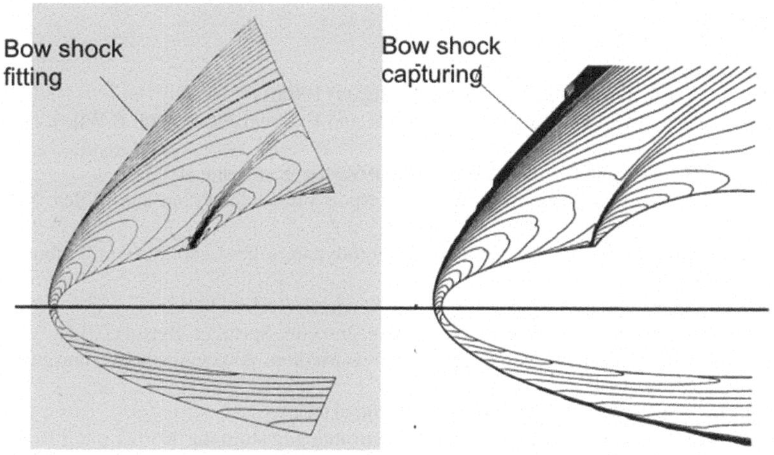

Fig. 3.2 Demonstration of the difference between bow shock fitting and bow shock capturing technique. Two dimensional inviscid flow past a double ellipse. Euler solutions with non-equilibrium real gas approach. Iso Mach lines, $M_\infty = 25,\ \alpha = 30^0,\ p_\infty = 2.52\ Pa,\ T_\infty = 205\ K$. Left figure taken from [17]; right figure taken from [18]

the shock contour by a highly sophisticated geometrical transformation on the basis of the Rankine-Hugoniot equations. With that approach technique high accuracy solutions are possible [13–15].

A drawback of numerical methods of this kind is their reduced robustness, which means that for flow fields past complicated configurations like space planes convergence problems may arise. Generally, today finite volume methods, which identify shock waves by the shock capturing technique have become accepted, even though they have, at times, accuracy problems calculating discontinuities or need intensive mesh refinements in the regimes of them (see also Chap. 16).

Figure 3.2 shows two solutions of a two-dimensional inviscid flow past a double ellipse.[3] The Mach number was $M_\infty = 25$, the angle of attack $\alpha = 30^0$ and the ambient pressure $p_\infty = 2.52\ Pa$. For such conditions, the air behind the bow shock usually is in a non-equilibrium thermodynamic state. The numerical methods applied were formulated for such kind of non-equilibrium flow. The bow shock fitting approach exhibits a sharp shock contour, left part of Fig. 3.2, whereas the shock capturing method shows a more or less blurred bow shock, right part of Fig. 3.2. Typically, such a solution can be improved by a drastic mesh refinement in the regime of the shock contour, but this may boost dramatically the computer costs.

[3] This was a test case in the Antibes Workshop of 1991 [16].

References

1. Becker, E.: Gasdynamik. Teubner Verlag, Stuttgart (1966)
2. Hirsch, C.: Numerical Computation of Internal and External Flows, vol. I. Wiley, New York Reprint (1997)
3. Zierep, J.: Theoretische Gasdynamik. G. Braun Verlag, Karlsruhe (1972)
4. Bird, R.B., Stuart, W.E., Lightfoot, E.N.: Transport Phenomena, 2nd edn. Wiley, New York (2002)
5. Krause, E.: Strömungslehre. Gasdynamik und aerodynamisches Laboratorium. Teubner Verlag, Stuttgart (2003)
6. Weiland, C.: Mechanics of Flow Similarities. Springer, Berlin (2020)
7. Hirschel, E.H.: Basics of Aerothermodynamics, 2nd edn. Springer, Berlin (2015)
8. Sommerfeld, A.: Mechanik der Deformierbaren Medien. Akademische Verlagsgesellschaft Geest & Portig, Leipzig (1954)
9. Becker, R.: Theorie der Wärme. Springer, Berlin (1966)
10. Weiland, C.: Zwei- und dreidimensionale Umströmungen stumpfer Körper unter Berücksichtigung schallnaher Überschallströmungen. Zeitschrift für Flugwissenschaften (ZfW) **24**(5) (1976)
11. Moretti, G., Abbett, M.: A time-dependent computational method for blunt body flows. AIAA J. **4**, 2136–2141 (1966)
12. Lyubimov, A.N., Rusanov, V.V.: Gas Flows Past Blunt Bodies. NASA TT F-714 (1973)
13. Weiland, C.: A contribution to the computation of transonic supersonic flows over blunt bodies. Comput. Fluids **9**, 143–162 (1981)
14. Pfitzner, M., Weiland, C.: 3-D Euler solutions for hypersonic freestream Mach numbers. AGARD-CP-428, Paper 22, Bristol, UK (1987)
15. Weiland, C., Schröder, W., Menne, S.: An extended insight into hypersonic flow phenomena using numerical methods. Comput. Fluids **22**, 407–426 (1993)
16. Abgrall, R., Désidéri, J.-A., Glowinski, R., Mallet, M., Périaux, J. (Eds.): Hypersonic Flows for Reentry Problems. vol. III. Springer Verlag, Berlin/Heidelberg (1992)
17. Menne, S., Weiland, C., Pfitzner, M.: Computation of 3-D hypersonic flows in chemical non-equilibrium including transport phenomena. AIAA Paper 92-2876, 27th Thermophysics Conference, Nashville USA (1992)
18. Fey, M., Jeltsch, R., Karmann, P.: Special aspects of reacting inviscid blunt body flow. In: Abgrall, R., Desideri, J.-A., Glowinski, R., Mallet, M., Periaux, J. (Eds.) Hypersonic Flows for Reentry Problems, vol. III. Springer Verlag, Berlin/Heidelberg (1992)

Chapter 4
The Fluid Dynamic Equations in Component Form for Three Basic Coordinate Systems

Abstract In the previous chapter the fluid dynamic equations are presented in symbolic formation. These formulations are valid for all possible coordinate systems. For practical purposes or, to say, for finding solutions for realistic flow problems, in particular with the aid of modern numerical methods, a frame of reference has to be defined based on a specific coordinate system. This leads to the fluid dynamic equations in component form. In this chapter we present the continuity equation (3.13), the momentum equations (3.15), (3.16), (3.23) and the energy equations (3.24), (3.25), (3.33), (3.35), (3.38) in the three well-known coordinate systems: Cartesian x, y, z, cylindrical r, φ, z, spherical r, θ, φ. The component presentation of the governing equations in the various coordinate systems requires a deep insight into the vector/tensor calculus as it can be found in the literature , see Sect. 4.1. The inspection of the equations treated in Chap. 3 reveals that 11 vector/tensor operations must be provided. For that purpose we use the neutral scalar, vector and tensor quantities c, $\mathbf{a}$, $\bar{\bar{\vartheta}}$. These operations are presented each in the beginning of the Sects. 4.1, 4.2, 4.3.

The Variables and Quantities Used in this Chapter

c	neutral scalar
$\mathbf{a}$	neutral vector
$\bar{\bar{\vartheta}}$	neutral tensor
a_x, a_y, a_z	Cartesian components of neutral vector $\mathbf{a}$
a_r, a_φ, a_z	cylindrical components of neutral vector $\mathbf{a}$
a_θ, a_φ, a_r	spherical components of neutral vector $\mathbf{a}$
$\vartheta_{xx}, \vartheta_{yy}, \vartheta_{zz}$ $\vartheta_{xy}, \vartheta_{xz}, \vartheta_{yz}$	elements of the neutral tensor $\bar{\bar{\vartheta}}$ in Cartesian coordinates
$\vartheta_{rr}, \vartheta_{\varphi\varphi}, \vartheta_{zz}$ $\vartheta_{r\varphi}, \vartheta_{rz}, \vartheta_{\varphi z}$	elements of the neutral tensor $\bar{\bar{\vartheta}}$ in cylindrical coordinates
$\vartheta_{rr}, \vartheta_{\theta\theta}, \vartheta_{\varphi\varphi}$ $\vartheta_{r\theta}, \vartheta_{\theta\varphi}, \vartheta_{\varphi r}$	elements of the neutral tensor $\bar{\bar{\vartheta}}$ in spherical coordinates
v_x, v_y, v_z	Cartesian components of velocity vector $\mathbf{v}$

C. Weiland, *The Hierarchy of Fluid Dynamic Equations*,
https://doi.org/10.1007/978-3-031-71198-5_4

v_r, v_φ, v_z	cylindrical components of velocity vector **v**
v_θ, v_φ, v_r	spherical components of velocity vector **v**
$\tau_{xx}, \tau_{yy}, \tau_{zz}$ $\tau_{xy}, \tau_{xz}, \tau_{yz}$	elements of the viscous stress tensor $\bar{\bar{\tau}}$ in Cartesian coordinates
$\tau_{rr}, \tau_{\varphi\varphi}, \tau_{zz}$ $\tau_{r\varphi}, \tau_{rz}, \tau_{\varphi z}$	elements of the viscous stress tensor $\bar{\bar{\tau}}$ in cylindrical coordinates
$\tau_{rr}, \tau_{\theta\theta}, \tau_{\varphi\varphi}$ $\tau_{r\theta}, \tau_{\theta\varphi}, \tau_{\varphi r}$	elements of the viscous tress tensor $\bar{\bar{\tau}}$ in spherical coordinates
g_x, g_y, g_z	Cartesian components of gravitational acceleration vector **g**
g_r, g_φ, g_z	cylindrical components of gravitational acceleration vector **g**
g_θ, g_φ, g_r	spherical components of gravitational acceleration vector **g**
ρ	density
p	pressure
T	temperature
s	entropy per unit mass
μ	dynamic viscosity
κ	thermal conductivity
e	total energy per unit mass
e_i	internal energy per unit mass
$M_z^{+\varphi}$	matrix: right-handed basic rotation around z-axis
$M_y^{+\theta}$	matrix: right-handed basic rotation around y-axis

4.1 The Equations in Cartesian Components

The Cartesian coordinate system is shown in Fig. 4.1 [1–6]. The vector/tensor operations[1] of the 11 cases are then defined by

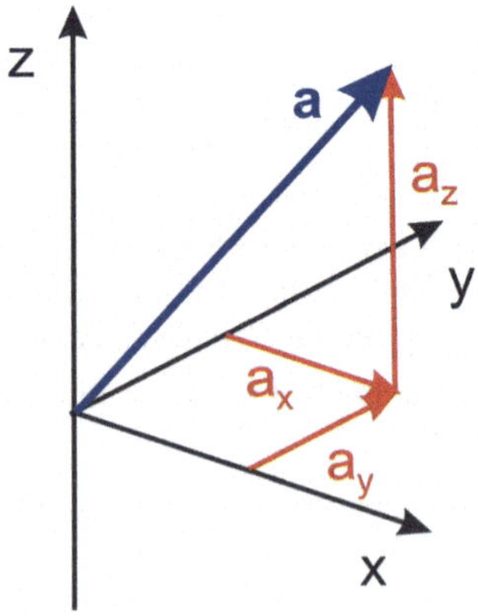

Fig. 4.1 Right-handed Cartesian coordinates x, y, z

[1] The vector/tensor basic operations applied here can be found in the literature [1–6].

1. div **a**: the divergence operation on a vector **a**

$$\operatorname{div}\mathbf{a} = \frac{\partial a_x}{\partial x} + \frac{\partial a_y}{\partial y} + \frac{\partial a_z}{\partial z}, \tag{4.1}$$

2. div (c**a**): the divergence operation on a product of a scalar c with a vector **a**

$$\operatorname{div}(c\mathbf{a}) = \frac{\partial c a_x}{\partial x} + \frac{\partial c a_y}{\partial y} + \frac{\partial c a_z}{\partial z}, \tag{4.2}$$

3. div $\bar{\bar{\vartheta}}$: the divergence operation on a tensor $\bar{\bar{\vartheta}}$

$$\operatorname{div}\bar{\bar{\vartheta}} = \begin{pmatrix} \frac{\partial \vartheta_{xx}}{\partial x} + \frac{\partial \vartheta_{xy}}{\partial y} + \frac{\partial \vartheta_{xz}}{\partial z} \\ \frac{\partial \vartheta_{yx}}{\partial x} + \frac{\partial \vartheta_{yy}}{\partial y} + \frac{\partial \vartheta_{yz}}{\partial z} \\ \frac{\partial \vartheta_{zx}}{\partial x} + \frac{\partial \vartheta_{zy}}{\partial y} + \frac{\partial \vartheta_{zz}}{\partial z} \end{pmatrix}, \tag{4.3}$$

4. div ($c\,\mathbf{a} \otimes \mathbf{a}$): the divergence operation on a product of a scalar c with a tensorial vector product $\mathbf{a} \otimes \mathbf{a}$

$$\operatorname{div}(c\,\mathbf{a} \otimes \mathbf{a}) = \begin{pmatrix} \frac{\partial c\, a_x a_x}{\partial x} + \frac{\partial c\, a_x a_y}{\partial y} + \frac{\partial c\, a_x a_z}{\partial z} \\ \frac{\partial c\, a_y a_x}{\partial x} + \frac{\partial c\, a_y a_y}{\partial y} + \frac{\partial c\, a_y a_z}{\partial z} \\ \frac{\partial c\, a_z a_x}{\partial x} + \frac{\partial c\, a_z a_y}{\partial y} + \frac{\partial c\, a_z a_z}{\partial z} \end{pmatrix}, \tag{4.4}$$

5. div ($\bar{\bar{\vartheta}} \cdot \mathbf{a}$): the divergence operation on a scalar product of a tensor $\bar{\bar{\vartheta}}$ with a vector **a**

$$\begin{aligned} \operatorname{div}(\bar{\bar{\vartheta}} \cdot \mathbf{a}) &= \operatorname{div}\left\{ \begin{pmatrix} \vartheta_{xx} & \vartheta_{xy} & \vartheta_{xz} \\ \vartheta_{yx} & \vartheta_{yy} & \vartheta_{yz} \\ \vartheta_{zx} & \vartheta_{zy} & \vartheta_{zz} \end{pmatrix} \cdot \begin{pmatrix} a_x \\ a_y \\ a_z \end{pmatrix} \right\} \\ &= \frac{\partial}{\partial x}(\vartheta_{xx}a_x + \vartheta_{xy}a_y + \vartheta_{xz}a_z) \\ &\quad + \frac{\partial}{\partial y}(\vartheta_{yx}a_x + \vartheta_{yy}a_y + \vartheta_{yz}a_z) \\ &\quad + \frac{\partial}{\partial z}(\vartheta_{zx}a_x + \vartheta_{zy}a_y + \vartheta_{zz}a_z), \end{aligned} \tag{4.5}$$

6. grad c: the gradient operation on a scalar c

$$\operatorname{grad} c = \begin{pmatrix} \dfrac{\partial c}{\partial x} \\ \dfrac{\partial c}{\partial y} \\ \dfrac{\partial c}{\partial z} \end{pmatrix}, \tag{4.6}$$

7. $(\mathbf{a} \cdot \operatorname{grad})\,\mathbf{a}$: the scalar product of a vector $\mathbf{a}$ with the gradient operator multiplied with the vector $\mathbf{a}$

$$\begin{aligned} (\mathbf{a} \cdot \operatorname{grad})\,\mathbf{a} &= (a_x \frac{\partial}{\partial x} + a_y \frac{\partial}{\partial y} + a_z \frac{\partial}{\partial z}) \begin{pmatrix} a_x \\ a_y \\ a_z \end{pmatrix} \\ &= \begin{pmatrix} a_x \dfrac{\partial a_x}{\partial x} + a_y \dfrac{\partial a_x}{\partial y} + a_z \dfrac{\partial a_x}{\partial z} \\ a_x \dfrac{\partial a_y}{\partial x} + a_y \dfrac{\partial a_y}{\partial y} + a_z \dfrac{\partial a_y}{\partial z} \\ a_x \dfrac{\partial a_z}{\partial x} + a_y \dfrac{\partial a_z}{\partial y} + a_z \dfrac{\partial a_z}{\partial z} \end{pmatrix}, \end{aligned} \tag{4.7}$$

8. $(\bar{\bar{\vartheta}} \cdot \operatorname{grad}) \cdot \mathbf{a}$: the scalar product of a tensor $\bar{\bar{\vartheta}}$ with the gradient operator multiplied scalar with the vector $\mathbf{a}$

$$\begin{aligned} (\bar{\bar{\vartheta}} \cdot \operatorname{grad}) \cdot \mathbf{a} &= \left\{ \begin{pmatrix} \vartheta_{xx} & \vartheta_{xy} & \vartheta_{xz} \\ \vartheta_{yx} & \vartheta_{yy} & \vartheta_{yz} \\ \vartheta_{zx} & \vartheta_{zy} & \vartheta_{zz} \end{pmatrix} \cdot \begin{pmatrix} \dfrac{\partial}{\partial x} \\ \dfrac{\partial}{\partial y} \\ \dfrac{\partial}{\partial z} \end{pmatrix} \right\} \cdot \begin{pmatrix} a_x \\ a_y \\ a_z \end{pmatrix} \\ &= \begin{pmatrix} \vartheta_{xx} \dfrac{\partial}{\partial x} + \vartheta_{xy} \dfrac{\partial}{\partial y} + \vartheta_{xz} \dfrac{\partial}{\partial z} \\ \vartheta_{yx} \dfrac{\partial}{\partial x} + \vartheta_{yy} \dfrac{\partial}{\partial y} + \vartheta_{yz} \dfrac{\partial}{\partial z} \\ \vartheta_{zx} \dfrac{\partial}{\partial x} + \vartheta_{zy} \dfrac{\partial}{\partial y} + \vartheta_{zz} \dfrac{\partial}{\partial z} \end{pmatrix} \cdot \begin{pmatrix} a_x \\ a_y \\ a_z \end{pmatrix} \\ &= \vartheta_{xx} \frac{\partial a_x}{\partial x} + \vartheta_{xy} \frac{\partial a_x}{\partial y} + \vartheta_{xz} \frac{\partial a_x}{\partial z} \\ &\quad + \vartheta_{yx} \frac{\partial a_y}{\partial x} + \vartheta_{yy} \frac{\partial a_y}{\partial y} + \vartheta_{yz} \frac{\partial a_y}{\partial z} \\ &\quad + \vartheta_{zx} \frac{\partial a_z}{\partial x} + \vartheta_{zy} \frac{\partial a_z}{\partial y} + \vartheta_{zz} \frac{\partial a_z}{\partial z}, \end{aligned} \tag{4.8}$$

9. $(\mathbf{a} \times \operatorname{rot} \mathbf{a})$: the cross product of the vector $\mathbf{a}$ with the curl operation on the vector $\mathbf{a}$

$$(\mathbf{a} \times \operatorname{rot} \mathbf{a}) = \begin{pmatrix} a_y \left(\frac{\partial a_y}{\partial x} - \frac{\partial a_x}{\partial y} \right) - a_z \left(\frac{\partial a_x}{\partial z} - \frac{\partial a_z}{\partial x} \right) \\ a_z \left(\frac{\partial a_z}{\partial y} - \frac{\partial a_y}{\partial z} \right) - a_x \left(\frac{\partial a_y}{\partial x} - \frac{\partial a_x}{\partial y} \right) \\ a_x \left(\frac{\partial a_x}{\partial z} - \frac{\partial a_z}{\partial x} \right) - a_y \left(\frac{\partial a_z}{\partial y} - \frac{\partial a_y}{\partial z} \right) \end{pmatrix}, \tag{4.9}$$

10. $\Delta \mathbf{a}$: the Laplace operator of the vector $\mathbf{a}$

$$\begin{aligned} \Delta \mathbf{a} &= \operatorname{grad} \operatorname{div} \mathbf{a} - \operatorname{rot} (\operatorname{rot} \mathbf{a}) \\ &= \begin{pmatrix} \frac{\partial^2 a_x}{\partial x^2} + \frac{\partial^2 a_x}{\partial y^2} + \frac{\partial^2 a_x}{\partial z^2} \\ \frac{\partial^2 a_y}{\partial x^2} + \frac{\partial^2 a_y}{\partial y^2} + \frac{\partial^2 a_y}{\partial z^2} \\ \frac{\partial^2 a_z}{\partial x^2} + \frac{\partial^2 a_z}{\partial y^2} + \frac{\partial^2 a_z}{\partial z^2} \end{pmatrix}, \end{aligned} \tag{4.10}$$

11. $\operatorname{grad} \operatorname{div} \mathbf{a}$: the sequential divergence and gradient operations of the vector $\mathbf{a}$

$$\operatorname{grad} \operatorname{div} \mathbf{a} = \begin{pmatrix} \frac{\partial^2 a_x}{\partial x^2} + \frac{\partial^2 a_y}{\partial x \partial y} + \frac{\partial^2 a_z}{\partial x \partial z} \\ \frac{\partial^2 a_x}{\partial y \partial x} + \frac{\partial^2 a_y}{\partial y^2} + \frac{\partial^2 a_z}{\partial y \partial z} \\ \frac{\partial^2 a_x}{\partial z \partial x} + \frac{\partial^2 a_y}{\partial z \partial y} + \frac{\partial^2 a_z}{\partial z^2} \end{pmatrix}. \tag{4.11}$$

The Continuity Equation

The continuity equation (3.13) in component version by using Eq. (4.2) reads

$$\frac{\partial \rho}{\partial t} + \frac{\partial \rho v_x}{\partial x} + \frac{\partial \rho v_y}{\partial y} + \frac{\partial \rho v_z}{\partial z} = 0. \tag{4.12}$$

The Momentum Equations

We present three versions of the momentum equation in component form. Firstly, Eq. (3.15), which requires the relations Eqs. (4.3), (4.4), and (4.6). As external force we take into account the gravitational force $\rho \mathbf{g}$.

$$
\begin{aligned}
\frac{\partial \rho v_x}{\partial t}+\frac{\partial \rho v_x v_x}{\partial x}+\frac{\partial \rho v_x v_y}{\partial y}+\frac{\partial \rho v_x v_z}{\partial z}&=\rho g_x-\frac{\partial p}{\partial x}+\frac{\partial \tau_{xx}}{\partial x}+\frac{\partial \tau_{xy}}{\partial y}+\frac{\partial \tau_{xz}}{\partial z},\\
\frac{\partial \rho v_y}{\partial t}+\frac{\partial \rho v_y v_x}{\partial x}+\frac{\partial \rho v_y v_y}{\partial y}+\frac{\partial \rho v_y v_z}{\partial z}&=\rho g_y-\frac{\partial p}{\partial y}+\frac{\partial \tau_{yx}}{\partial x}+\frac{\partial \tau_{yy}}{\partial y}+\frac{\partial \tau_{yz}}{\partial z},\\
\frac{\partial \rho v_z}{\partial t}+\frac{\partial \rho v_z v_x}{\partial x}+\frac{\partial \rho v_z v_y}{\partial y}+\frac{\partial \rho v_z v_z}{\partial z}&=\rho g_z-\frac{\partial p}{\partial z}+\frac{\partial \tau_{zx}}{\partial x}+\frac{\partial \tau_{zy}}{\partial y}+\frac{\partial \tau_{zz}}{\partial z}.
\end{aligned}
\tag{4.13}
$$

Secondly, Eq. (3.16) requires the relations Eqs. (4.3), (4.7).

$$
\begin{aligned}
\rho\frac{Dv_x}{Dt}&=\rho\frac{\partial v_x}{\partial t}+\rho v_x\frac{\partial v_x}{\partial x}+\rho v_y\frac{\partial v_x}{\partial y}+\rho v_z\frac{\partial v_x}{\partial z}\\
&=\rho g_x-\frac{\partial p}{\partial x}+\frac{\partial \tau_{xx}}{\partial x}+\frac{\partial \tau_{xy}}{\partial y}+\frac{\partial \tau_{xz}}{\partial z},\\
\rho\frac{Dv_y}{Dt}&=\rho\frac{\partial v_y}{\partial t}+\rho v_x\frac{\partial v_y}{\partial x}+\rho v_y\frac{\partial v_y}{\partial y}+\rho v_z\frac{\partial v_y}{\partial z}\\
&=\rho g_y-\frac{\partial p}{\partial y}+\frac{\partial \tau_{yx}}{\partial x}+\frac{\partial \tau_{yy}}{\partial y}+\frac{\partial \tau_{yz}}{\partial z},\\
\rho\frac{Dv_z}{Dt}&=\rho\frac{\partial v_z}{\partial t}+\rho v_x\frac{\partial v_z}{\partial x}+\rho v_y\frac{\partial v_z}{\partial y}+\rho v_z\frac{\partial v_z}{\partial z}\\
&=\rho g_z-\frac{\partial p}{\partial z}+\frac{\partial \tau_{zx}}{\partial x}+\frac{\partial \tau_{zy}}{\partial y}+\frac{\partial \tau_{zz}}{\partial z}.
\end{aligned}
\tag{4.14}
$$

Thirdly, we write Eq. (3.23) in component form using the relation Eq. (4.9),

$$
\begin{aligned}
v_y\left(\frac{\partial v_y}{\partial x}-\frac{\partial v_x}{\partial y}\right)-v_z\left(\frac{\partial v_x}{\partial z}-\frac{\partial v_z}{\partial x}\right)&=-T\,\frac{\partial s}{\partial x},\\
v_z\left(\frac{\partial v_z}{\partial y}-\frac{\partial v_y}{\partial z}\right)-v_x\left(\frac{\partial v_y}{\partial x}-\frac{\partial v_x}{\partial y}\right)&=-T\,\frac{\partial s}{\partial y},\\
v_x\left(\frac{\partial v_x}{\partial z}-\frac{\partial v_z}{\partial x}\right)-v_y\left(\frac{\partial v_z}{\partial y}-\frac{\partial v_y}{\partial z}\right)&=-T\,\frac{\partial s}{\partial z}.
\end{aligned}
\tag{4.15}
$$

The shear stress tensor in tensorial notation is given by $\tau_{ij}=\mu\left(\partial_i v_j+\partial_j v_i\right)-\frac{2}{3}\mu\,\partial^k v_k\,\delta_{ij}$, if the Stokes' relation $2\mu+3\lambda=0$ with λ being the second viscosity component is valid, [7]. In the case that the dynamic viscosity μ is constant the term div $\bar{\bar{\tau}}$ can be reformulated in a formal vector-tensor notation

$$
\operatorname{div}\bar{\bar{\tau}}=\mu(\Delta\mathbf{v}+\operatorname{grad}\operatorname{div}\mathbf{v}-\frac{2}{3}\operatorname{grad}\operatorname{div}\mathbf{v}).
\tag{4.16}
$$

In components we then have by using the relations Eqs. (4.10), (4.11)

$$
\begin{aligned}
&(\operatorname{div} \bar{\bar{\tau}})_x \\
&= \mu \left(\frac{\partial^2 v_x}{\partial x^2} + \frac{\partial^2 v_x}{\partial y^2} + \frac{\partial^2 v_x}{\partial z^2} + \frac{\partial^2 v_x}{\partial x^2} + \frac{\partial^2 v_y}{\partial x \partial y} + \frac{\partial^2 v_z}{\partial x \partial z} - \frac{2}{3} \left(\frac{\partial^2 v_x}{\partial x^2} + \frac{\partial^2 v_y}{\partial x \partial y} + \frac{\partial^2 v_z}{\partial x \partial z} \right) \right) \\
&= \underbrace{\mu \left(2 \frac{\partial^2 v_x}{\partial x^2} - \frac{2}{3} \left(\frac{\partial^2 v_x}{\partial x^2} + \frac{\partial^2 v_y}{\partial x \partial y} + \frac{\partial^2 v_z}{\partial x \partial z} \right) \right)}_{\partial \tau_{xx}/\partial x} + \underbrace{\mu \left(\frac{\partial^2 v_y}{\partial x \partial y} + \frac{\partial^2 v_x}{\partial y^2} \right)}_{\partial \tau_{xy}/\partial y} \\
&\quad + \underbrace{\mu \left(\frac{\partial^2 v_z}{\partial x \partial z} + \frac{\partial^2 v_x}{\partial z^2} \right)}_{\partial \tau_{xz}/\partial z},
\end{aligned}
\tag{4.17}
$$

and similarily for the y- and z-components. For incompressible flows due to div $\mathbf{v} = 0$ (see Eq. (4.16)) we obtain

$$
(\operatorname{div} \bar{\bar{\tau}})_x = \mu \left(\frac{\partial^2 v_x}{\partial x^2} + \frac{\partial^2 v_x}{\partial y^2} + \frac{\partial^2 v_x}{\partial z^2} \right). \tag{4.18}
$$

The Energy Equation

We consider the two energy equations (3.24) and (3.25), where we have dropped the term for the other heat sources q_H, apply the relations Eqs. (4.2) and (4.5) to them and find the component notations of these equations

$$
\begin{aligned}
&\frac{\partial \rho e}{\partial t} + \frac{\partial \rho e v_x}{\partial x} + \frac{\partial \rho e v_y}{\partial y} + \frac{\partial \rho e v_z}{\partial z} + \frac{\partial p v_x}{\partial x} + \frac{\partial p v_y}{\partial y} + \frac{\partial p v_z}{\partial z} \\
&\equiv \frac{\partial \rho e}{\partial t} + \frac{\partial (\rho e + p) v_x}{\partial x} + \frac{\partial (\rho e + p) v_y}{\partial y} + \frac{\partial (\rho e + p) v_z}{\partial z} \\
&= \frac{\partial}{\partial x} \left(\tau_{xx} v_x + \tau_{xy} v_y + \tau_{xz} v_z \right) + \frac{\partial}{\partial y} \left(\tau_{yx} v_x + \tau_{yy} v_y + \tau_{yz} v_z \right) \\
&\quad + \frac{\partial}{\partial z} \left(\tau_{zx} v_x + \tau_{zy} v_y + \tau_{zz} v_z \right) + \frac{\partial}{\partial x} \left(\kappa \frac{\partial T}{\partial x} \right) + \frac{\partial}{\partial y} \left(\kappa \frac{\partial T}{\partial y} \right) + \frac{\partial}{\partial z} \left(\kappa \frac{\partial T}{\partial z} \right) \\
&\quad + \rho \left(g_x v_x + g_y v_y + g_z v_z \right).
\end{aligned}
\tag{4.19}
$$

The energy equation based on the variation of the internal energy Eq. (3.33) with the relations Eqs. (4.2), (4.8) reads in components

$$
\begin{aligned}
&\rho\left(\frac{\partial e_i}{\partial t}+v_x\frac{\partial e_i}{\partial x}+v_y\frac{\partial e_i}{\partial y}+v_z\frac{\partial e_i}{\partial z}\right)-\frac{p}{\rho}\left(\frac{\partial \rho}{\partial t}+v_x\frac{\partial \rho}{\partial x}+v_y\frac{\partial \rho}{\partial y}+v_z\frac{\partial \rho}{\partial z}\right)\\
&=\frac{\partial}{\partial x}\left(\kappa\frac{\partial T}{\partial x}\right)+\frac{\partial}{\partial y}\left(\kappa\frac{\partial T}{\partial y}\right)+\frac{\partial}{\partial z}\left(\kappa\frac{\partial T}{\partial z}\right)\\
&+\tau_{xx}\frac{\partial v_x}{\partial x}+\tau_{xy}\frac{\partial v_x}{\partial y}+\tau_{xz}\frac{\partial v_x}{\partial z}+\tau_{yx}\frac{\partial v_y}{\partial x}+\tau_{yy}\frac{\partial v_y}{\partial y}+\tau_{yz}\frac{\partial v_y}{\partial z}\\
&+\tau_{zx}\frac{\partial v_z}{\partial x}+\tau_{zy}\frac{\partial v_z}{\partial y}+\tau_{zz}\frac{\partial v_z}{\partial z},
\end{aligned}
\tag{4.20}
$$

and the equation with the variation of the enthalpy Eq. (3.35) then gets the form

$$
\begin{aligned}
&\rho c_p\left(\frac{\partial T}{\partial t}+v_x\frac{\partial T}{\partial x}+v_y\frac{\partial T}{\partial y}+v_z\frac{\partial T}{\partial z}\right)\\
&=\frac{\partial p}{\partial t}+v_x\frac{\partial p}{\partial x}+v_y\frac{\partial p}{\partial y}+v_z\frac{\partial p}{\partial z}\\
&+\frac{\partial}{\partial x}\left(\kappa\frac{\partial T}{\partial x}\right)+\frac{\partial}{\partial y}\left(\kappa\frac{\partial T}{\partial y}\right)+\frac{\partial}{\partial z}\left(\kappa\frac{\partial T}{\partial z}\right)\\
&+\tau_{xx}\frac{\partial v_x}{\partial x}+\tau_{xy}\frac{\partial v_x}{\partial y}+\tau_{xz}\frac{\partial v_x}{\partial z}+\tau_{yx}\frac{\partial v_y}{\partial x}+\tau_{yy}\frac{\partial v_y}{\partial y}+\tau_{yz}\frac{\partial v_y}{\partial z}\\
&+\tau_{zx}\frac{\partial v_z}{\partial x}+\tau_{zy}\frac{\partial v_z}{\partial y}+\tau_{zz}\frac{\partial v_z}{\partial z}.
\end{aligned}
\tag{4.21}
$$

The elements of the shear stress tensor in Cartesian coordinates are

$$
\begin{aligned}
\tau_{xx}&=2\mu\left(\frac{\partial v_x}{\partial x}-\frac{1}{3}\operatorname{div}\mathbf{v}\right),\\
\tau_{yy}&=2\mu\left(\frac{\partial v_y}{\partial y}-\frac{1}{3}\operatorname{div}\mathbf{v}\right),\\
\tau_{zz}&=2\mu\left(\frac{\partial v_z}{\partial z}-\frac{1}{3}\operatorname{div}\mathbf{v}\right),
\end{aligned}
$$

$$
\begin{aligned}
\tau_{xy}&=\tau_{yx}=\mu\left(\frac{\partial v_x}{\partial y}+\frac{\partial v_y}{\partial x}\right),\\
\tau_{xz}&=\tau_{zx}=\mu\left(\frac{\partial v_x}{\partial z}+\frac{\partial v_z}{\partial x}\right),\\
\tau_{yz}&=\tau_{zy}=\mu\left(\frac{\partial v_y}{\partial z}+\frac{\partial v_z}{\partial y}\right).
\end{aligned}
\tag{4.22}
$$

Finally, we present the component version of the Eq. (3.38), which is valid for inviscid perfect gas flow. This equation in the early days was applied in numerical approximations of the Euler equations, when the blunt body problem was investigated and solved [8, 9].

$$\frac{\partial p}{\partial t} + v_x \frac{\partial p}{\partial x} + v_y \frac{\partial p}{\partial y} + v_z \frac{\partial p}{\partial z} - c^2 \left(\frac{\partial \rho}{\partial t} + v_x \frac{\partial \rho}{\partial x} + v_y \frac{\partial \rho}{\partial y} + v_z \frac{\partial \rho}{\partial z} \right) = 0. \quad (4.23)$$

4.2 The Equations in Cylindrical Components

The cylindrical coordinates, Fig. 4.2, are given by r, φ, z, see [1–6]. The relations between the Cartesian coordinates and the cylindrical coordinates are as follows

$$\begin{aligned}
&(r, \varphi, z) \Longrightarrow (x, y, z) && (x, y, z) \Longrightarrow (r, \varphi, z) \\
&x = r \cos \varphi && r = \sqrt{x^2 + y^2} \\
&y = r \sin \varphi && \varphi = \arctan \frac{y}{x} \\
&z = z && z = z
\end{aligned}$$

A vector **a** in Cartesian coordinates $(a_x, a_y, a_z)^T$ and in cylindrical coordinates $(a_r, a_\varphi, a_z)^T$ transforms simply by a basic rotation as defined in Chap. 2 of [10].

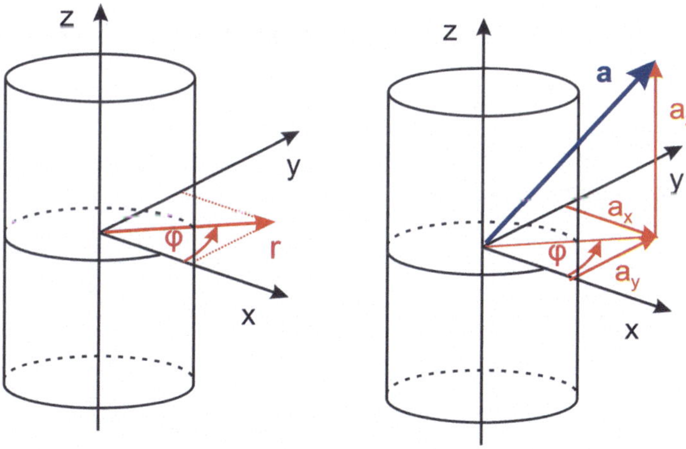

Fig. 4.2 Cylindrical coordinates r, φ, z shown in the left figure. The right figure shows the vector **a** with the components a_x, a_y, a_z given by the transformation Eq. (4.24)

With the matrix $M_z^{+\varphi}$, denoting the right-handed basic rotation around the z-axis, one obtains

$$M_z^{+\varphi} = \begin{pmatrix} \cos\varphi & \sin\varphi & 0 \\ -\sin\varphi & \cos\varphi & 0 \\ 0 & 0 & 1 \end{pmatrix},$$

$$\begin{pmatrix} a_r \\ a_\varphi \\ a_z \end{pmatrix} = M_z^{+\varphi} \begin{pmatrix} a_x \\ a_y \\ a_z \end{pmatrix} = \begin{pmatrix} a_x \cos\varphi + a_y \sin\varphi \\ -a_x \sin\varphi + a_y \cos\varphi \\ a_z \end{pmatrix}.$$

The reverse operation is given by

$$M_z^{+\varphi T} = M_z^{-\varphi} = \begin{pmatrix} \cos\varphi & -\sin\varphi & 0 \\ \sin\varphi & \cos\varphi & 0 \\ 0 & 0 & 1 \end{pmatrix},$$

$$\begin{pmatrix} a_x \\ a_y \\ a_z \end{pmatrix} = M_z^{-\varphi} \begin{pmatrix} a_r \\ a_\varphi \\ a_z \end{pmatrix} = \begin{pmatrix} a_r \cos\varphi - a_\varphi \sin\varphi \\ a_r \sin\varphi + a_\varphi \cos\varphi \\ a_z \end{pmatrix}. \tag{4.24}$$

The vector/tensor operations of the 11 cases are then defined by

1. div **a**: the divergence operation on a vector **a**

$$\operatorname{div} \mathbf{a} = \frac{1}{r}\frac{\partial r a_r}{\partial r} + \frac{1}{r}\frac{\partial a_\varphi}{\partial \varphi} + \frac{\partial a_z}{\partial z}, \tag{4.25}$$

2. div (c**a**): the divergence operation on a product of a scalar c with a vector **a**

$$\operatorname{div}(c\mathbf{a}) = \frac{1}{r}\frac{\partial c r a_r}{\partial r} + \frac{1}{r}\frac{\partial c a_\varphi}{\partial \varphi} + \frac{\partial c a_z}{\partial z}, \tag{4.26}$$

3. div $\bar{\bar{\vartheta}}$: the divergence operation on a tensor $\bar{\bar{\vartheta}}$

$$\operatorname{div} \bar{\bar{\vartheta}} = \begin{pmatrix} \frac{1}{r}\frac{\partial r \vartheta_{rr}}{\partial r} + \frac{1}{r}\frac{\partial \vartheta_{r\varphi}}{\partial \varphi} + \frac{\partial \vartheta_{rz}}{\partial z} - \frac{\vartheta_{\varphi\varphi}}{r} \\ \frac{1}{r^2}\frac{\partial r^2 \vartheta_{\varphi r}}{\partial r} + \frac{1}{r}\frac{\partial \vartheta_{\varphi\varphi}}{\partial \varphi} + \frac{\partial \vartheta_{\varphi z}}{\partial z} \\ \frac{1}{r}\frac{\partial r \vartheta_{zr}}{\partial r} + \frac{1}{r}\frac{\partial \vartheta_{z\varphi}}{\partial \varphi} + \frac{\partial \vartheta_{zz}}{\partial z} \end{pmatrix}, \tag{4.27}$$

4. div ($c\,\mathbf{a} \otimes \mathbf{a}$): the divergence operation on a product of a scalar c with a tensorial vector product $\mathbf{a} \otimes \mathbf{a}$

$$\operatorname{div}(c\,\mathbf{a}\otimes\mathbf{a}) = \begin{pmatrix} \dfrac{\partial c\,a_r a_r}{\partial r} + \dfrac{1}{r}\dfrac{\partial c\,a_r a_\varphi}{\partial \varphi} + \dfrac{\partial c\,a_r a_z}{\partial z} + \dfrac{1}{r} c\,(a_r^2 - a_\varphi^2) \\ \dfrac{\partial c\,a_\varphi a_r}{\partial r} + \dfrac{1}{r}\dfrac{\partial c\,a_\varphi a_\varphi}{\partial \varphi} + \dfrac{\partial c\,a_\varphi a_z}{\partial z} + \dfrac{2}{r} c\,a_r a_\varphi \\ \dfrac{\partial c\,a_z a_r}{\partial r} + \dfrac{1}{r}\dfrac{\partial c\,a_z a_\varphi}{\partial \varphi} + \dfrac{\partial c\,a_z a_z}{\partial z} + \dfrac{1}{r} c\,a_r a_z \end{pmatrix}, \tag{4.28}$$

5. div $(\bar{\bar{\vartheta}} \cdot \mathbf{a})$: the divergence operation on a scalar product of a tensor $\bar{\bar{\vartheta}}$ with a vector $\mathbf{a}$

$$\begin{aligned} \operatorname{div}(\bar{\bar{\vartheta}} \cdot \mathbf{a}) &= \frac{1}{r}\frac{\partial}{\partial r}(r\,(\vartheta_{rr}a_r + \vartheta_{r\varphi}a_\varphi + \vartheta_{rz}a_z)) \\ &+ \frac{1}{r}\frac{\partial}{\partial \varphi}(\vartheta_{\varphi r}a_r + \vartheta_{\varphi\varphi}a_\varphi + \vartheta_{\varphi z}a_z) \\ &+ \frac{\partial}{\partial z}(\vartheta_{zr}a_r + \vartheta_{z\varphi}a_\varphi + \vartheta_{zz}a_z), \end{aligned} \tag{4.29}$$

6. grad c: the gradient operation on a scalar c

$$\operatorname{grad} c = \begin{pmatrix} \dfrac{\partial c}{\partial r} \\ \dfrac{1}{r}\dfrac{\partial c}{\partial \varphi} \\ \dfrac{\partial c}{\partial z} \end{pmatrix}, \tag{4.30}$$

7. $(\mathbf{a} \cdot \text{grad})\mathbf{a}$: the scalar product of a vector $\mathbf{a}$ with the gradient operator multiplied with the vector $\mathbf{a}$

$$\begin{aligned} (\mathbf{a}\cdot\operatorname{grad})\mathbf{a} &= \operatorname{grad}\frac{a^2}{2} - (\mathbf{a} \times \operatorname{rot}\mathbf{a}) \\ &= \begin{pmatrix} a_r\dfrac{\partial a_r}{\partial r} + \dfrac{1}{r}a_\varphi\dfrac{\partial a_r}{\partial \varphi} + a_z\dfrac{\partial a_r}{\partial z} - \dfrac{a_\varphi^2}{r} \\ a_r\dfrac{\partial a_\varphi}{\partial r} + \dfrac{1}{r}a_\varphi\dfrac{\partial a_\varphi}{\partial \varphi} + a_z\dfrac{\partial a_\varphi}{\partial z} + \dfrac{a_r a_\varphi}{r} \\ a_r\dfrac{\partial a_z}{\partial r} + \dfrac{1}{r}a_\varphi\dfrac{\partial a_z}{\partial \varphi} + a_z\dfrac{\partial a_z}{\partial z} \end{pmatrix}, \end{aligned} \tag{4.31}$$

8. $(\bar{\bar{\vartheta}} \cdot \text{grad}) \cdot \mathbf{a}$: the scalar product of a tensor $\bar{\bar{\vartheta}}$ with the gradient operator multiplied scalar with the vector **a**

$$\begin{aligned}(\bar{\bar{\vartheta}} \cdot \text{grad}) \cdot \mathbf{a} &= \vartheta_{rr}\frac{\partial a_r}{\partial r} + \frac{1}{r}\vartheta_{r\varphi}\frac{\partial a_r}{\partial \varphi} + \vartheta_{rz}\frac{\partial a_r}{\partial z} + \frac{1}{r}a_r\vartheta_{\varphi\varphi} \\ &+\vartheta_{\varphi r}\frac{\partial a_\varphi}{\partial r} + \frac{1}{r}\vartheta_{\varphi\varphi}\frac{\partial a_\varphi}{\partial \varphi} + \vartheta_{\varphi z}\frac{\partial a_\varphi}{\partial z} - \frac{1}{r}a_\varphi\vartheta_{\varphi r} \\ &+\vartheta_{zr}\frac{\partial a_z}{\partial r} + \frac{1}{r}\vartheta_{z\varphi}\frac{\partial a_z}{\partial \varphi} + \vartheta_{zz}\frac{\partial a_z}{\partial z},\end{aligned} \tag{4.32}$$

9. $(\mathbf{a} \times \text{rot}\,\mathbf{a})$: the cross product of the vector **a** with the curl operation on the vector **a**

$$(\mathbf{a} \times \text{rot}\,\mathbf{a}) = \begin{pmatrix} a_\varphi\left(\frac{1}{r}\frac{\partial r a_\varphi}{\partial r} - \frac{1}{r}\frac{\partial a_r}{\partial \varphi}\right) - a_z\left(\frac{\partial a_r}{\partial z} - \frac{\partial a_z}{\partial r}\right) \\ a_z\left(\frac{1}{r}\frac{\partial a_z}{\partial \varphi} - \frac{\partial a_\varphi}{\partial z}\right) - a_r\left(\frac{1}{r}\frac{\partial r a_\varphi}{\partial r} - \frac{1}{r}\frac{\partial a_r}{\partial \varphi}\right) \\ a_r\left(\frac{\partial a_r}{\partial z} - \frac{\partial a_z}{\partial r}\right) - a_\varphi\left(\frac{1}{r}\frac{\partial a_z}{\partial \varphi} - \frac{\partial a_\varphi}{\partial z}\right) \end{pmatrix}, \tag{4.33}$$

10. $\Delta\,\mathbf{a}$: the Laplace operator of the vector **a**

$$\begin{aligned}\Delta\,\mathbf{a} &= \text{grad}\,\text{div}\,\mathbf{a} - \text{rot}\,(\text{rot}\,\mathbf{a}) \\ &= \begin{pmatrix} \frac{1}{r}\frac{\partial}{\partial r}\left(r\frac{\partial a_r}{\partial r}\right) + \frac{1}{r^2}\frac{\partial^2 a_r}{\partial \varphi^2} + \frac{\partial^2 a_r}{\partial z^2} - \frac{a_r}{r^2} - \frac{2}{r^2}\frac{\partial a_\varphi}{\partial \varphi} \\ \frac{1}{r}\frac{\partial}{\partial r}\left(r\frac{\partial a_\varphi}{\partial r}\right) + \frac{1}{r^2}\frac{\partial^2 a_\varphi}{\partial \varphi^2} + \frac{\partial^2 a_\varphi}{\partial z^2} - \frac{a_\varphi}{r^2} + \frac{2}{r^2}\frac{\partial a_r}{\partial \varphi} \\ \frac{1}{r}\frac{\partial}{\partial r}\left(r\frac{\partial a_z}{\partial r}\right) + \frac{1}{r^2}\frac{\partial^2 a_z}{\partial \varphi^2} + \frac{\partial^2 a_z}{\partial z^2} \end{pmatrix},\end{aligned} \tag{4.34}$$

11. $\text{grad}\,\text{div}\,\mathbf{a}$: the sequential divergence and gradient operations of the vector **a**

$$\text{grad}\,\text{div}\,\mathbf{a} = \begin{pmatrix} \frac{\partial}{\partial r}\left(\frac{1}{r}\frac{\partial r a_r}{\partial r}\right) + \frac{\partial}{\partial r}\left(\frac{1}{r}\frac{\partial a_\varphi}{\partial \varphi}\right) + \frac{\partial^2 a_z}{\partial r \partial z} \\ \frac{1}{r}\left(\frac{\partial}{\partial \varphi}\left(\frac{1}{r}\frac{\partial r a_r}{\partial r}\right) + \frac{\partial}{\partial \varphi}\left(\frac{1}{r}\frac{\partial a_\varphi}{\partial \varphi}\right) + \frac{\partial^2 a_z}{\partial \varphi \partial z}\right) \\ \frac{\partial}{\partial z}\left(\frac{1}{r}\frac{\partial r a_r}{\partial r}\right) + \frac{\partial}{\partial z}\left(\frac{1}{r}\frac{\partial a_\varphi}{\partial \varphi}\right) + \frac{\partial^2 a_z}{\partial z^2} \end{pmatrix}. \tag{4.35}$$

The Continuity Equation

The continuity equation (3.13) in cylindrical coordinates using Eq. (4.26) reads

$$\frac{\partial \rho}{\partial t} + \frac{1}{r}\frac{\partial \rho r v_r}{\partial r} + \frac{1}{r}\frac{\partial \rho v_\varphi}{\partial \varphi} + \frac{\partial \rho v_z}{\partial z} = 0. \tag{4.36}$$

The Momentum Equations

The three versions of the momentum equation in cylindrical coordinates are presented. Firstly, Eq. (3.15), which necessitates the relations Eqs. (4.27), (4.28), (4.30). As external force we take into account the gravitational force $\rho \mathbf{g}$.

$$\begin{aligned}
\text{r} - \text{component}\quad & \frac{\partial \rho v_r}{\partial t} + \frac{\partial \rho v_r v_r}{\partial r} + \frac{1}{r}\frac{\partial \rho v_r v_\varphi}{\partial \varphi} + \frac{\partial \rho v_r v_z}{\partial z} + \frac{1}{r}\rho(v_r^2 - v_\varphi^2) \\
&= \rho g_r - \frac{\partial p}{\partial r} + \frac{1}{r}\frac{\partial r \tau_{rr}}{\partial r} + \frac{1}{r}\frac{\partial \tau_{r\varphi}}{\partial \varphi} + \frac{\partial \tau_{rz}}{\partial z} - \frac{\tau_{\varphi\varphi}}{r}, \\
\varphi - \text{component}\quad & \frac{\partial \rho v_\varphi}{\partial t} + \frac{\partial \rho v_\varphi v_r}{\partial r} + \frac{1}{r}\frac{\partial \rho v_\varphi v_\varphi}{\partial \varphi} + \frac{\partial \rho v_\varphi v_z}{\partial z} + \frac{2}{r}\rho v_r v_\varphi \\
&= \rho g_\varphi - \frac{1}{r}\frac{\partial p}{\partial \varphi} + \frac{1}{r^2}\frac{\partial r^2 \tau_{\varphi r}}{\partial r} + \frac{1}{r}\frac{\partial \tau_{\varphi\varphi}}{\partial \varphi} + \frac{\partial \tau_{\varphi z}}{\partial z}, \\
\text{z} - \text{component}\quad & \frac{\partial \rho v_z}{\partial t} + \frac{\partial \rho v_z v_r}{\partial r} + \frac{1}{r}\frac{\partial \rho v_z v_\varphi}{\partial \varphi} + \frac{\partial \rho v_z v_z}{\partial z} + \frac{1}{r}\rho v_r v_z \\
&= \rho g_z - \frac{\partial p}{\partial z} + \frac{1}{r}\frac{\partial r \tau_{zr}}{\partial r} + \frac{1}{r}\frac{\partial \tau_{z\varphi}}{\partial \varphi} + \frac{\partial \tau_{zz}}{\partial z}.
\end{aligned} \tag{4.37}$$

Secondly, Eq. (3.16). We work with the relations Eqs. (4.27), (4.31).

$$\begin{aligned}
\rho\frac{Dv_r}{Dt} &= \rho\left(\frac{\partial v_r}{\partial t} + v_r\frac{\partial v_r}{\partial r} + \frac{1}{r}v_\varphi\frac{\partial v_r}{\partial \varphi} + v_z\frac{\partial v_r}{\partial z} - \frac{v_\varphi^2}{r}\right) \\
&= \rho g_r - \frac{\partial p}{\partial r} + \frac{1}{r}\frac{\partial r \tau_{rr}}{\partial r} + \frac{1}{r}\frac{\partial \tau_{r\varphi}}{\partial \varphi} + \frac{\partial \tau_{rz}}{\partial z} - \frac{\tau_{\varphi\varphi}}{r}, \\
\rho\frac{Dv_\varphi}{Dt} &= \rho\left(\frac{\partial v_\varphi}{\partial t} + v_r\frac{\partial v_\varphi}{\partial r} + \frac{1}{r}v_\varphi\frac{\partial v_\varphi}{\partial \varphi} + v_z\frac{\partial v_\varphi}{\partial z} + \frac{v_r v_\varphi}{r}\right) \\
&= \rho g_\varphi - \frac{1}{r}\frac{\partial p}{\partial \varphi} + \frac{1}{r^2}\frac{\partial r^2 \tau_{\varphi r}}{\partial r} + \frac{1}{r}\frac{\partial \tau_{\varphi\varphi}}{\partial \varphi} + \frac{\partial \tau_{\varphi z}}{\partial z}, \\
\rho\frac{Dv_z}{Dt} &= \rho\left(\frac{\partial v_z}{\partial t} + v_r\frac{\partial v_z}{\partial r} + \frac{1}{r}v_\varphi\frac{\partial v_z}{\partial \varphi} + v_z\frac{\partial v_z}{\partial z}\right) \\
&= \rho g_z - \frac{\partial p}{\partial z} + \frac{1}{r}\frac{\partial r \tau_{zr}}{\partial r} + \frac{1}{r}\frac{\partial \tau_{z\varphi}}{\partial \varphi} + \frac{\partial \tau_{zz}}{\partial z}.
\end{aligned} \tag{4.38}$$

Thirdly, we express Eq. (3.23) in cylindrical coordinates using the relation Eq. (4.33).

$$\begin{aligned} v_\varphi\left(\frac{1}{r}\frac{\partial r v_\varphi}{\partial r}-\frac{1}{r}\frac{\partial v_r}{\partial \varphi}\right)-v_z\left(\frac{\partial v_r}{\partial z}-\frac{\partial v_z}{\partial r}\right) &= -T\,\frac{\partial s}{\partial r},\\ v_z\left(\frac{1}{r}\frac{\partial v_z}{\partial \varphi}-\frac{\partial v_\varphi}{\partial z}\right)-v_r\left(\frac{1}{r}\frac{\partial r v_\varphi}{\partial r}-\frac{1}{r}\frac{\partial v_r}{\partial \varphi}\right) &= -T\,\frac{1}{r}\frac{\partial s}{\partial \varphi},\\ v_r\left(\frac{\partial v_r}{\partial z}-\frac{\partial v_z}{\partial r}\right)-v_\varphi\left(\frac{1}{r}\frac{\partial v_z}{\partial \varphi}-\frac{\partial v_\varphi}{\partial z}\right) &= -T\,\frac{\partial s}{\partial z}. \end{aligned} \tag{4.39}$$

The equivalent to Eqs. (4.16) and (4.17) in cylindrical coordinates by employing the relations (4.34) and (4.35) reads

$$\begin{aligned} (\operatorname{div}\bar{\bar{\tau}})_r &= \mu\left\{\frac{1}{r}\frac{\partial}{\partial r}\left(r\frac{\partial v_r}{\partial r}\right)-\frac{v_r}{r^2}+\frac{1}{r^2}\frac{\partial^2 v_r}{\partial \varphi^2}+\frac{\partial^2 v_r}{\partial z^2}-\frac{2}{r^2}\frac{\partial v_\varphi}{\partial \varphi}\right.\\ &\quad+\frac{1}{r}\frac{\partial}{\partial r}\left(r\frac{\partial v_r}{\partial r}\right)-\frac{v_r}{r^2}+\frac{\partial}{\partial r}\left(\frac{1}{r}\frac{\partial v_\varphi}{\partial \varphi}\right)+\frac{\partial^2 v_z}{\partial r\partial z}\\ &\quad\left.-\frac{2}{3}\left(\frac{1}{r}\frac{\partial}{\partial r}\left(r\frac{\partial v_r}{\partial r}\right)-\frac{v_r}{r^2}+\frac{\partial}{\partial r}\left(\frac{1}{r}\frac{\partial v_\varphi}{\partial \varphi}\right)+\frac{\partial^2 v_z}{\partial r\partial z}\right)\right\}\\ &= \underbrace{\mu\left\{2\left(\frac{1}{r}\frac{\partial}{\partial r}\left(r\frac{\partial v_r}{\partial r}\right)-\frac{v_r}{r^2}-\frac{2}{r^2}\frac{\partial v_\varphi}{\partial \varphi}\right)\right.}_{\partial\tau_{rr}/\partial r\ \Longrightarrow}\\ &\quad\underbrace{\left.-\frac{2}{3}\left(\frac{1}{r}\frac{\partial}{\partial r}\left(r\frac{\partial v_r}{\partial r}\right)-\frac{v_r}{r^2}+\frac{\partial}{\partial r}\left(\frac{1}{r}\frac{\partial v_\varphi}{\partial \varphi}\right)+\frac{\partial^2 v_z}{\partial r\partial z}\right)\right\}}_{\Longleftarrow\ \partial\tau_{rr}/\partial r}\\ &\quad+\underbrace{\mu\left\{\frac{\partial}{\partial r}\left(\frac{1}{r}\frac{\partial v_\varphi}{\partial \varphi}\right)+\frac{1}{r^2}\frac{\partial^2 v_r}{\partial \varphi^2}\right\}}_{1/r\,\partial\tau_{r\varphi}/\partial\varphi}+\underbrace{\mu\left\{\frac{\partial^2 v_z}{\partial r\partial z}+\frac{\partial^2 v_r}{\partial z^2}\right\}}_{\partial\tau_{zz}/\partial z}, \end{aligned} \tag{4.40}$$

and similarly for the φ - and z - components (Note the two lines of the term $\partial\tau_{rr}/\partial r$). For incompressible flow due to $\operatorname{div}\mathbf{v}=0$ (see Eq. (4.16)) we obtain

$$(\operatorname{div}\bar{\bar{\tau}})_r = \mu\left(\frac{1}{r}\frac{\partial}{\partial r}\left(r\frac{\partial v_r}{\partial r}\right)-\frac{v_r}{r^2}+\frac{1}{r^2}\frac{\partial^2 v_r}{\partial \varphi^2}+\frac{\partial^2 v_r}{\partial z^2}-\frac{2}{r^2}\frac{\partial v_\varphi}{\partial \varphi}\right). \tag{4.41}$$

The Energy Equation

The two energy equations (3.24), (3.25) in cylindrical coordinates are given below by applying the relations Eqs. (4.25), (4.29)

$$
\begin{aligned}
&\frac{\partial \rho e}{\partial t}+\frac{1}{r}\frac{\partial r\rho e v_r}{\partial r}+\frac{1}{r}\frac{\partial \rho e v_\varphi}{\partial \varphi}+\frac{\partial \rho e v_z}{\partial z}+\frac{1}{r}\frac{\partial r p v_r}{\partial r}+\frac{1}{r}\frac{\partial p v_\varphi}{\partial \varphi}+\frac{\partial p v_z}{\partial z}\\
&\equiv \frac{\partial \rho e}{\partial t}+\frac{1}{r}\frac{\partial r(\rho e+p)v_r}{\partial r}+\frac{1}{r}\frac{\partial(\rho e+p)v_\varphi}{\partial \varphi}+\frac{\partial(\rho e+p)v_z}{\partial z}\\
&=\frac{1}{r}\frac{\partial}{\partial r}\left(r(\tau_{rr}v_r+\tau_{r\varphi}v_\varphi+\tau_{rz}v_z)\right)+\frac{1}{r}\frac{\partial}{\partial \varphi}\left(\tau_{\varphi r}v_r+\tau_{\varphi\varphi}v_\varphi+\tau_{\varphi z}v_z\right)\\
&\quad+\frac{\partial}{\partial z}\left(\tau_{zr}v_r+\tau_{z\varphi}v_\varphi+\tau_{zz}v_z\right)\\
&\quad+\frac{1}{r}\frac{\partial}{\partial r}\left(r\kappa\frac{\partial T}{\partial r}\right)+\frac{1}{r}\frac{\partial}{\partial \varphi}\left(\kappa\frac{1}{r}\frac{\partial T}{\partial \varphi}\right)+\frac{\partial}{\partial z}\left(\kappa\frac{\partial T}{\partial z}\right)\\
&\quad+\rho\left(g_r v_r+g_\varphi v_\varphi+g_z v_z\right).
\end{aligned}
\tag{4.42}
$$

The energy equation based on the variation of the internal energy Eq. (3.33) with the relations Eqs. (4.26), (4.32) reads in cylindrical coordinates

$$
\begin{aligned}
&\rho\left(\frac{\partial e_i}{\partial t}+v_r\frac{\partial e_i}{\partial r}+v_\varphi\frac{1}{r}\frac{\partial e_i}{\partial \varphi}+v_z\frac{\partial e_i}{\partial z}\right)-\frac{p}{\rho}\left(\frac{\partial \rho}{\partial t}+v_r\frac{\partial \rho}{\partial r}+v_\varphi\frac{1}{r}\frac{\partial \rho}{\partial \varphi}+v_z\frac{\partial \rho}{\partial z}\right)\\
&=\frac{1}{r}\frac{\partial}{\partial r}\left(r\kappa\frac{\partial T}{\partial r}\right)+\frac{1}{r}\frac{\partial}{\partial \varphi}\left(\kappa\frac{1}{r}\frac{\partial T}{\partial \varphi}\right)+\frac{\partial}{\partial z}\left(\kappa\frac{\partial T}{\partial z}\right)\\
&+\tau_{rr}\frac{\partial v_r}{\partial r}+\frac{1}{r}\tau_{r\varphi}\frac{\partial v_r}{\partial \varphi}+\tau_{rz}\frac{\partial v_r}{\partial z}+\frac{1}{r}v_r\tau_{\varphi\varphi}+\tau_{\varphi r}\frac{\partial v_\varphi}{\partial r}+\frac{1}{r}\tau_{\varphi\varphi}\frac{\partial v_\varphi}{\partial \varphi}+\tau_{\varphi z}\frac{\partial v_\varphi}{\partial z}\\
&-\frac{1}{r}v_\varphi\tau_{\varphi r}+\tau_{zr}\frac{\partial v_z}{\partial r}+\frac{1}{r}\tau_{z\varphi}\frac{\partial v_z}{\partial \varphi}+\tau_{zz}\frac{\partial v_z}{\partial z},
\end{aligned}
\tag{4.43}
$$

and the equation with the variation of the enthalpy Eq. (3.35) looks in cylindrical coordinates

$$
\begin{aligned}
&\rho c_p\left(\frac{\partial T}{\partial t}+v_r\frac{\partial T}{\partial r}+\frac{1}{r}v_\varphi\frac{\partial T}{\partial \varphi}+v_z\frac{\partial T}{\partial z}\right)\\
&=\frac{\partial p}{\partial t}+v_r\frac{\partial p}{\partial r}+\frac{1}{r}v_\varphi\frac{\partial p}{\partial \varphi}+v_z\frac{\partial p}{\partial z}\\
&+\frac{1}{r}\frac{\partial}{\partial r}\left(r\kappa\frac{\partial T}{\partial r}\right)+\frac{1}{r}\frac{\partial}{\partial \varphi}\left(\kappa\frac{1}{r}\frac{\partial T}{\partial \varphi}\right)+\frac{\partial}{\partial z}\left(\kappa\frac{\partial T}{\partial z}\right)\\
&+\tau_{rr}\frac{\partial v_r}{\partial r}+\frac{1}{r}\tau_{r\varphi}\frac{\partial v_r}{\partial \varphi}+\tau_{rz}\frac{\partial v_r}{\partial z}+\frac{1}{r}v_r\tau_{\varphi\varphi}+\tau_{\varphi r}\frac{\partial v_\varphi}{\partial r}+\frac{1}{r}\tau_{\varphi\varphi}\frac{\partial v_\varphi}{\partial \varphi}+\tau_{\varphi z}\frac{\partial v_\varphi}{\partial z}\\
&-\frac{1}{r}v_\varphi\tau_{\varphi r}+\tau_{zr}\frac{\partial v_z}{\partial r}+\frac{1}{r}\tau_{z\varphi}\frac{\partial v_z}{\partial \varphi}+\tau_{zz}\frac{\partial v_z}{\partial z}.
\end{aligned}
\tag{4.44}
$$

The elements of the shear stress tensor in cylindrical coordinates are determined from the tensorial notation $\tau_{ij} = \mu\left(\partial_i v_j + \partial_j v_i\right) - \frac{2}{3}\mu\, \partial^k v_k\, \delta_{ij}$ if the derivatives in this relation are interpreted as covariant derivatives [4]

$$\tau_{rr} = 2\mu\left(\frac{\partial v_r}{\partial r} - \frac{1}{3}\operatorname{div}\mathbf{v}\right),$$

$$\tau_{\varphi\varphi} = 2\mu\left(\frac{1}{r}\frac{\partial v_\varphi}{\partial \varphi} + \frac{v_r}{r} - \frac{1}{3}\operatorname{div}\mathbf{v}\right),$$

$$\begin{aligned}
\tau_{zz} &= 2\mu\left(\frac{\partial v_z}{\partial z} - \frac{1}{3}\operatorname{div}\mathbf{v}\right),\\
\tau_{r\varphi} &= \tau_{\varphi r} = \mu\left(r\frac{\partial}{\partial r}\left(\frac{v_\varphi}{r}\right) + \frac{1}{r}\frac{\partial v_r}{\partial \varphi}\right),\\
\tau_{rz} &= \tau_{zr} = \mu\left(\frac{\partial v_z}{\partial r} + \frac{\partial v_r}{\partial z}\right),\\
\tau_{\varphi z} &= \tau_{z\varphi} = \mu\left(\frac{\partial v_\varphi}{\partial z} + \frac{1}{r}\frac{\partial v_z}{\partial \varphi}\right).
\end{aligned} \tag{4.45}$$

To conclude this paragraph, we are providing the equivalent to Eq. (4.23) in cylindrical coordinates

$$\frac{\partial p}{\partial t} + v_r\frac{\partial p}{\partial r} + \frac{1}{r}v_\varphi\frac{\partial p}{\partial \varphi} + v_z\frac{\partial p}{\partial z} - c^2\left(\frac{\partial \rho}{\partial t} + v_r\frac{\partial \rho}{\partial r} + \frac{1}{r}v_\varphi\frac{\partial \rho}{\partial \varphi} + v_z\frac{\partial \rho}{\partial z}\right) = 0. \tag{4.46}$$

4.3 The Equations in Spherical Components

The spherical coordinates, Fig. 4.3, are given by θ, φ, r, see [1–6]. The relations between the Cartesian coordinates and the spherical coordinates are as follows

$(\theta, \varphi, r) \Longrightarrow (x, y, z)$	$(x, y, z) \Longrightarrow (\theta, \varphi, r)$
$x = r\sin\theta\cos\varphi$	$\theta = \arctan\dfrac{\sqrt{x^2+y^2}}{z}$
$y = r\sin\theta\sin\varphi$	$\varphi = \arctan\dfrac{y}{x}$
$z = r\cos\theta$	$r = \sqrt{x^2+y^2+z^2}$

A vector **a** in Cartesian coordinates $(a_x, a_y, a_z)^T$ and in spherical coordinates $(a_\theta, a_\varphi, a_r)^T$ simply transforms by two basic rotations as defined in Chap. 2 of [10]. With the matrices $M_z^{+\varphi}$ denoting the right-handed basic rotation around the z-axis

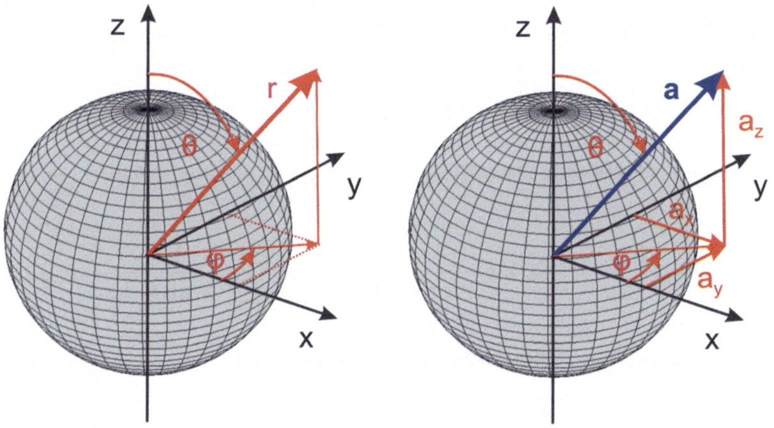

Fig. 4.3 Spherical coordinates r, θ, φ shown in the left figure. The right figure shows the vector **a** with the components a_x, a_y, a_z given by the transformation Eq. (4.47)

and $M_y^{+\theta}$ denoting the right-handed rotation around the y-axis one obtains

$$M_z^{+\varphi} M_y^{+\theta} = M_{\theta\varphi} = \begin{pmatrix} \cos\theta & 0 & -\sin\theta \\ 0 & 1 & 0 \\ \sin\theta & 0 & \cos\theta \end{pmatrix} \begin{pmatrix} \cos\varphi & \sin\varphi & 0 \\ -\sin\varphi & \cos\varphi & 0 \\ 0 & 0 & 1 \end{pmatrix}$$

$$= \begin{pmatrix} \cos\theta\cos\varphi & \cos\theta\sin\varphi & -\sin\theta \\ -\sin\varphi & \cos\varphi & 0 \\ \sin\theta\cos\varphi & \sin\theta\sin\varphi & \cos\theta \end{pmatrix},$$

$$\begin{pmatrix} a_\theta \\ a_\varphi \\ a_r \end{pmatrix} = M_{\theta\varphi} \begin{pmatrix} a_x \\ a_y \\ a_z \end{pmatrix} = \begin{pmatrix} a_x \cos\theta\cos\varphi + a_y \cos\theta\sin\varphi - a_z \sin\theta \\ -a_x \sin\varphi + a_y \cos\varphi \\ a_x \sin\theta\cos\varphi + a_y \sin\theta\sin\varphi + a_z\cos\theta \end{pmatrix}.$$

The reverse operation is given by

$$M_{\theta\varphi}^T = \begin{pmatrix} \cos\theta\cos\varphi & -\sin\varphi & \sin\theta\cos\varphi \\ \cos\theta\sin\varphi & \cos\varphi & \sin\theta\sin\varphi \\ -\sin\theta & 0 & \cos\theta \end{pmatrix},$$

$$\begin{pmatrix} a_x \\ a_y \\ a_z \end{pmatrix} = M_{\theta\varphi}^T \begin{pmatrix} a_\theta \\ a_\varphi \\ a_r \end{pmatrix} = \begin{pmatrix} a_\theta \cos\theta\cos\varphi - a_\varphi \sin\varphi + a_r \sin\theta\cos\varphi \\ a_\theta \cos\theta\sin\varphi + a_\varphi \cos\varphi + a_r \sin\theta\sin\varphi \\ -a_\theta \sin\theta + a_r \cos\theta \end{pmatrix}. \quad (4.47)$$

The vector/tensor operations of the 11 cases are then determined by

1. div **a**: the divergence operation on a vector **a**

$$\operatorname{div}\mathbf{a} = \frac{1}{r^2}\frac{\partial}{\partial r}\left(r^2 a_r\right) + \frac{1}{r\sin\theta}\frac{\partial}{\partial\theta}\left(a_\theta\sin\theta\right) + \frac{1}{r\sin\theta}\frac{\partial}{\partial\varphi}\left(a_\varphi\right), \tag{4.48}$$

2. div(c**a**): the divergence operation on a product of a scalar c with a vector **a**

$$\operatorname{div}(c\mathbf{a}) = \frac{1}{r^2}\frac{\partial}{\partial r}\left(r^2 c a_r\right) + \frac{1}{r\sin\theta}\frac{\partial}{\partial\theta}\left(c a_\theta\sin\theta\right) + \frac{1}{r\sin\theta}\frac{\partial}{\partial\varphi}\left(c a_\varphi\right), \tag{4.49}$$

3. div $\bar{\bar{\vartheta}}$: the divergence operation on a tensor function $\bar{\bar{\vartheta}}$

$$\operatorname{div}\bar{\bar{\vartheta}} = \begin{pmatrix} \frac{1}{r^2}\frac{\partial}{\partial r}\left(r^2\vartheta_{rr}\right) + \frac{1}{r\sin\theta}\frac{\partial}{\partial\theta}\left(\vartheta_{r\theta}\sin\theta\right) + \frac{1}{r\sin\theta}\frac{\partial}{\partial\varphi}\left(\vartheta_{r\varphi}\right) \\ -\frac{\vartheta_{\theta\theta}+\vartheta_{\varphi\varphi}}{r}, \\ \frac{1}{r^2}\frac{\partial}{\partial r}\left(r^2\vartheta_{\theta r}\right) + \frac{1}{r\sin\theta}\frac{\partial}{\partial\theta}\left(\vartheta_{\theta\theta}\sin\theta\right) + \frac{1}{r\sin\theta}\frac{\partial}{\partial\varphi}\left(\vartheta_{\theta\varphi}\right) \\ +\frac{\vartheta_{\theta r} - \cot\theta\,\vartheta_{\varphi\varphi}}{r}, \\ \frac{1}{r^2}\frac{\partial}{\partial r}\left(r^2\vartheta_{\varphi r}\right) + \frac{1}{r}\frac{\partial}{\partial\theta}\left(\vartheta_{\varphi\theta}\right) + \frac{1}{r\sin\theta}\frac{\partial}{\partial\varphi}\left(\vartheta_{\varphi\varphi}\right) \\ +\frac{\vartheta_{\varphi r} + 2\cot\theta\,\vartheta_{\varphi\theta}}{r}. \end{pmatrix}, \tag{4.50}$$

4. div($c\,\mathbf{a}\otimes\mathbf{a}$): the divergence operation on a product of a scalar c with a tensorial vector product $\mathbf{a}\otimes\mathbf{a}$

$$\operatorname{div}(c\,\mathbf{a}\otimes\mathbf{a}) = \begin{pmatrix} \frac{\partial}{\partial r}\left(c a_r a_r\right) + \frac{1}{r\sin\theta}\frac{\partial}{\partial\theta}\left(c a_r a_\theta\sin\theta\right) \\ +\frac{1}{r\sin\theta}\frac{\partial}{\partial\varphi}\left(c a_r a_\varphi\right) + \frac{1}{r}c(2a_r a_r - a_\theta a_\theta - a_\varphi a_\varphi), \\ \frac{\partial}{\partial r}\left(c a_r a_\theta\right) + \frac{1}{r\sin\theta}\frac{\partial}{\partial\theta}\left(c a_\theta a_\theta\sin\theta\right) \\ +\frac{1}{r\sin\theta}\frac{\partial}{\partial\varphi}\left(c a_\theta a_\varphi\right) + \frac{1}{r}c(3a_r a_\theta - a_\varphi a_\varphi\cot\theta), \\ \frac{\partial}{\partial r}\left(c a_r a_\varphi\right) + \frac{1}{r}\frac{\partial}{\partial\theta}\left(c a_\theta a_\varphi\right) \\ +\frac{1}{r\sin\theta}\frac{\partial}{\partial\varphi}\left(c a_\varphi a_\varphi\right) + \frac{1}{r}c(3a_r a_\varphi + 2a_\theta a_\varphi\cot\theta). \end{pmatrix}, \tag{4.51}$$

5. div($\bar{\bar{\vartheta}} \cdot \mathbf{a}$): the divergence operation on a scalar product of a tensor $\bar{\bar{\vartheta}}$ with a vector **a**

$$
\begin{aligned}
\operatorname{div}(\bar{\bar{\vartheta}} \cdot \mathbf{a}) &= \frac{1}{r^2}\frac{\partial}{\partial r}(r^2\,(\vartheta_{rr}a_r + \vartheta_{r\theta}a_\theta + \vartheta_{r\varphi}a_\varphi)) \\
&+ \frac{1}{r\sin\theta}\frac{\partial}{\partial\theta}(\sin\theta(\vartheta_{\theta r}a_r + \vartheta_{\theta\theta}a_\theta + \vartheta_{\theta\varphi}a_\varphi)) \\
&+ \frac{1}{r\sin\theta}\frac{\partial}{\partial\varphi}(\vartheta_{\varphi r}a_r + \vartheta_{\varphi\theta}a_\theta + \vartheta_{\varphi\varphi}a_\varphi),
\end{aligned}
\tag{4.52}
$$

6. grad c: the gradient operation on a scalar c

$$
\operatorname{grad} c = \begin{pmatrix} \dfrac{\partial c}{\partial r} \\ \dfrac{1}{r}\dfrac{\partial c}{\partial\theta} \\ \dfrac{1}{r\sin\theta}\dfrac{\partial c}{\partial\varphi} \end{pmatrix},
\tag{4.53}
$$

7. ($\mathbf{a} \cdot$ grad)**a**: the scalar product of a vector **a** with the gradient operator multiplied with the vector **a**

$$
\begin{aligned}
(\mathbf{a} \cdot \operatorname{grad})\mathbf{a} &= \operatorname{grad}\frac{a^2}{2} - (\mathbf{a} \times \operatorname{rot}\mathbf{a}) \\
&= \begin{pmatrix}
a_r\dfrac{\partial a_r}{\partial r} + \dfrac{a_\theta}{r}\dfrac{\partial a_r}{\partial\theta} + \dfrac{a_\varphi}{r\sin\theta}\dfrac{\partial a_r}{\partial\varphi} - \dfrac{a_\theta^2 + a_\varphi^2}{r} \\
a_r\dfrac{\partial u_\theta}{\partial r} + \dfrac{u_\theta}{r}\dfrac{\partial a_\theta}{\partial\theta} + \dfrac{a_\varphi}{r\sin\theta}\dfrac{\partial a_\theta}{\partial\varphi} + \dfrac{a_r a_\theta - a_\varphi^2\cot\theta}{r} \\
a_r\dfrac{\partial a_\varphi}{\partial r} + \dfrac{a_\theta}{r}\dfrac{\partial a_\varphi}{\partial\theta} + \dfrac{a_\varphi}{r\sin\theta}\dfrac{\partial a_\varphi}{\partial\varphi} + \dfrac{a_r a_\varphi + a_\theta a_\varphi\cot\theta}{r}
\end{pmatrix},
\end{aligned}
\tag{4.54}
$$

8. ($\bar{\vartheta} \cdot$ grad) $\cdot$ **a**: the scalar product of a tensor $\bar{\bar{\vartheta}}$ with the gradient operator multiplied scalar with the vector **a**

$$
\begin{aligned}
&(\bar{\bar{\vartheta}} \cdot \operatorname{grad}) \cdot \mathbf{a} \\
&= \vartheta_{rr}\frac{\partial a_r}{\partial r} + \vartheta_{\theta\theta}\left(\frac{1}{r}\frac{\partial a_\theta}{\partial\theta} + \frac{a_r}{r}\right) + \vartheta_{\varphi\varphi}\left(\frac{1}{r\sin\theta}\frac{\partial a_\varphi}{\partial\varphi} + \frac{a_r + a_\theta\cot\theta}{r}\right) \\
&+ \vartheta_{r\theta}\left(\frac{\partial a_\theta}{\partial r} + \frac{1}{r}\frac{\partial a_r}{\partial\theta} - \frac{a_\theta}{r}\right) + \vartheta_{r\varphi}\left(\frac{\partial a_\varphi}{\partial r} + \frac{1}{r\sin\theta}\frac{\partial a_r}{\partial\varphi} - \frac{a_\varphi}{r}\right) \\
&+ \vartheta_{\theta\varphi}\left(\frac{1}{r}\frac{\partial a_\varphi}{\partial\theta} + \frac{1}{r\sin\theta}\frac{\partial a_\theta}{\partial\varphi} - \frac{a_\varphi\cot\theta}{r}\right),
\end{aligned}
\tag{4.55}
$$

9. $(\mathbf{a} \times \operatorname{rot} \mathbf{a})$: the cross product of the vector $\mathbf{a}$ with the curl operation on the vector $\mathbf{a}$

$$(\mathbf{a} \times \operatorname{rot} \mathbf{a}) = \begin{pmatrix} a_\theta \left(\frac{1}{r}\frac{\partial r a_\theta}{\partial r} - \frac{1}{r}\frac{\partial a_r}{\partial \theta}\right) - a_\varphi \left(\frac{1}{r\sin\theta}\frac{\partial a_r}{\partial \varphi} - \frac{1}{r}\frac{\partial r a_\varphi}{\partial r}\right) \\ a_\varphi \left(\frac{1}{r\sin\theta}\frac{\partial a_\varphi \sin\theta}{\partial \theta} - \frac{1}{r\sin\theta}\frac{\partial a_\theta}{\partial \varphi}\right) - a_r \left(\frac{1}{r}\frac{\partial r a_\theta}{\partial r} - \frac{1}{r}\frac{\partial a_r}{\partial \theta}\right) \\ a_r \left(\frac{1}{r\sin\theta}\frac{\partial a_r}{\partial \varphi} - \frac{1}{r}\frac{\partial r a_\varphi}{\partial r}\right) - a_\theta \left(\frac{1}{r\sin\theta}\frac{\partial a_\varphi \sin\theta}{\partial \theta} - \frac{1}{r\sin\theta}\frac{\partial a_\theta}{\partial \varphi}\right) \end{pmatrix}, \tag{4.56}$$

10. $\Delta\, \mathbf{a}$: the Laplace operator of the vector $\mathbf{a}$

$$\Delta\, \mathbf{a} = \operatorname{grad} \operatorname{div} \mathbf{a} - \operatorname{rot} (\operatorname{rot} \mathbf{a}) = \begin{pmatrix} \frac{1}{r^2}\frac{\partial}{\partial r}\left(r^2 \frac{\partial a_r}{\partial r}\right) + \frac{1}{r^2 \sin\theta}\frac{\partial}{\partial \theta}\left(\sin\theta \frac{\partial a_r}{\partial \theta}\right) + \frac{1}{r^2 \sin^2\theta}\frac{\partial^2 a_r}{\partial \varphi^2} \\ -\frac{2}{r^2}\left(a_r + \frac{\partial a_\theta}{\partial \theta} + a_\theta \cot\theta + \frac{1}{r^2 \sin\theta}\frac{\partial a_\varphi}{\partial \varphi}\right), \\ \frac{1}{r^2}\frac{\partial}{\partial r}\left(r^2 \frac{\partial a_\theta}{\partial r}\right) + \frac{1}{r^2 \sin\theta}\frac{\partial}{\partial \theta}\left(\sin\theta \frac{\partial a_\theta}{\partial \theta}\right) + \frac{1}{r^2 \sin^2\theta}\frac{\partial^2 a_\theta}{\partial \varphi^2} \\ +\frac{1}{r^2}\left(2\frac{\partial a_r}{\partial \theta} - \frac{a_\theta}{\sin^2\theta} - \frac{2\cot\theta}{\sin\theta}\frac{\partial a_\varphi}{\partial \varphi}\right), \\ \frac{1}{r^2}\frac{\partial}{\partial r}\left(r^2 \frac{\partial a_\varphi}{\partial r}\right) + \frac{1}{r^2 \sin\theta}\frac{\partial}{\partial \theta}\left(\sin\theta \frac{\partial a_\varphi}{\partial \theta}\right) + \frac{1}{r^2 \sin^2\theta}\frac{\partial^2 a_\varphi}{\partial \varphi^2} \\ +\frac{1}{r^2 \sin\theta}\left(2\frac{\partial a_r}{\partial \varphi} - \frac{a_\theta}{\sin\theta} + 2\cot\theta \frac{\partial a_\varphi}{\partial \varphi}\right). \end{pmatrix}, \tag{4.57}$$

11. $\operatorname{grad} \operatorname{div} \mathbf{a}$: the sequential divergence and gradient operations of the vector $\mathbf{a}$

$$\operatorname{grad} \operatorname{div} \mathbf{a} = \begin{pmatrix} \frac{\partial}{\partial r}\left[\left(\frac{1}{r^2}\frac{\partial r^2 a_r}{\partial r}\right) + \left(\frac{1}{r\sin\theta}\frac{\partial a_\theta \sin\theta}{\partial \theta}\right) + \left(\frac{1}{r\sin\theta}\frac{\partial a_\varphi}{\partial \varphi}\right)\right] \\ \frac{1}{r}\frac{\partial}{\partial \theta}\left[\left(\frac{1}{r^2}\frac{\partial r^2 a_r}{\partial r}\right) + \left(\frac{1}{r\sin\theta}\frac{\partial a_\theta \sin\theta}{\partial \theta}\right) + \left(\frac{1}{r\sin\theta}\frac{\partial a_\varphi}{\partial \varphi}\right)\right] \\ \frac{1}{r\sin\theta}\frac{\partial}{\partial \varphi}\left[\left(\frac{1}{r^2}\frac{\partial r^2 a_r}{\partial r}\right) + \left(\frac{1}{r\sin\theta}\frac{\partial a_\theta \sin\theta}{\partial \theta}\right) + \left(\frac{1}{r\sin\theta}\frac{\partial a_\varphi}{\partial \varphi}\right)\right] \end{pmatrix}. \tag{4.58}$$

The Continuity Equation

The continuity equation (3.13) in spherical coordinates using Eq. (4.49) reads

$$\frac{\partial \rho}{\partial t} + \frac{1}{r^2}\frac{\partial \rho r^2 v_r}{\partial r} + \frac{1}{r \sin\theta}\frac{\partial \rho v_\theta \sin\theta}{\partial \varphi} + \frac{1}{r \sin\theta}\frac{\partial \rho v_\varphi}{\partial \varphi} = 0. \qquad (4.59)$$

The Momentum Equations

Again we consider the three versions of the momentum equations (3.15), (3.16), (3.23). The presentation of Eq. (3.15) in spherical coordinates requires Eqs. (4.50), (4.51), (4.53)

$$\begin{aligned}
\text{r} - \text{component} \quad & \frac{\partial \rho v_r}{\partial t} + \frac{\partial \rho v_r v_r}{\partial r} + \frac{1}{r\sin\theta}\frac{\partial \rho v_r v_\theta \sin\theta}{\partial \theta} + \frac{1}{r\sin\theta}\frac{\partial \rho v_r v_\varphi}{\partial \varphi} \\
& + \frac{1}{r}\rho(2v_r^2 - v_\theta^2 - v_\varphi^2) = \rho g_r - \frac{\partial p}{\partial r} \\
& + \frac{1}{r^2}\frac{\partial r^2 \tau_{rr}}{\partial r} + \frac{1}{r\sin\theta}\frac{\partial \tau_{r\theta}\sin\theta}{\partial \theta} + \frac{1}{r\sin\theta}\frac{\partial \tau_{r\varphi}}{\partial \varphi} - \frac{\tau_{\theta\theta} + \tau_{\varphi\varphi}}{r}, \\
\theta - \text{component} \quad & \frac{\partial \rho v_\theta}{\partial t} + \frac{\partial \rho v_\theta v_r}{\partial r} + \frac{1}{r\sin\theta}\frac{\partial \rho v_\theta v_\theta \sin\theta}{\partial \theta} + \frac{1}{r\sin\theta}\frac{\partial \rho v_\theta v_\varphi}{\partial \varphi} \\
& + \frac{1}{r}\rho(3v_r v_\theta - v_\varphi^2 \cot\theta) = \rho g_\theta - \frac{1}{r}\frac{\partial p}{\partial \theta} \\
& + \frac{1}{r^2}\frac{\partial r^2 \tau_{\theta r}}{\partial r} + \frac{1}{r\sin\theta}\frac{\partial \tau_{\theta\theta}\sin\theta}{\partial \theta} + \frac{1}{r\sin\theta}\frac{\partial \tau_{\theta\varphi}}{\partial \varphi} + \frac{\tau_{r\theta} - \tau_{\varphi\varphi}\cot\theta}{r}, \\
\varphi - \text{component} \quad & \frac{\partial \rho v_\varphi}{\partial t} + \frac{\partial \rho v_\varphi v_r}{\partial r} + \frac{1}{r}\frac{\partial \rho v_\varphi v_\theta}{\partial \theta} + \frac{1}{r\sin\theta}\frac{\partial \rho v_\varphi v_\varphi}{\partial \varphi} \\
& + \frac{1}{r}\rho(3v_r v_\varphi + 2v_\theta v_\varphi \cot\theta) = \rho g_\varphi - \frac{1}{r\sin\theta}\frac{\partial p}{\partial \varphi} \\
& + \frac{1}{r^2}\frac{\partial r^2 \tau_{\varphi r}}{\partial r} + \frac{1}{r}\frac{\partial \tau_{\varphi\theta}}{\partial \theta} + \frac{1}{r\sin\theta}\frac{\partial \tau_{\varphi\varphi}}{\partial \varphi} + \frac{\tau_{r\varphi} + 2\tau_{\theta\varphi}\cot\theta}{r}. \qquad (4.60)
\end{aligned}$$

The momentum Eq. (3.16) in spherical coordinates are determined by using the relations Eqs. (4.50) and (4.54) and one obtains

$$\begin{aligned}
\rho\frac{Dv_r}{Dt} &= \rho\left(\frac{\partial v_r}{\partial t} + v_r\frac{\partial v_r}{\partial r} + \frac{v_\theta}{r}\frac{\partial v_r}{\partial \theta} + \frac{v_\varphi}{r\sin\theta}\frac{\partial v_r}{\partial \varphi} - \frac{v_\theta^2 + v_\varphi^2}{r}\right) \\
&= \rho g_r - \frac{\partial p}{\partial r} + \frac{1}{r^2}\frac{\partial r^2 \tau_{rr}}{\partial r} + \frac{1}{r\sin\theta}\frac{\partial \tau_{r\theta}\sin\theta}{\partial \theta} + \frac{1}{r\sin\theta}\frac{\partial \tau_{r\varphi}}{\partial \varphi} \\
& \quad - \frac{\tau_{\theta\theta} + \tau_{\varphi\varphi}}{r},
\end{aligned}$$

$$\rho\frac{Dv_\theta}{Dt} = \rho\left(\frac{\partial v_\theta}{\partial t} + v_r\frac{\partial v_\theta}{\partial r} + \frac{v_\theta}{r}\frac{\partial v_\theta}{\partial \theta} + \frac{v_\varphi}{r\sin\theta}\frac{\partial v_\theta}{\partial \varphi} + \frac{v_r v_\theta - v_\varphi^2\cot\theta}{r}\right)$$
$$= \rho g_\theta - \frac{1}{r}\frac{\partial p}{\partial \theta} + \frac{1}{r^2}\frac{\partial r^2\tau_{\theta r}}{\partial r} + \frac{1}{r\sin\theta}\frac{\partial \tau_{\theta\theta}\sin\theta}{\partial \theta} + \frac{1}{r\sin\theta}\frac{\partial \tau_{\theta\varphi}}{\partial \varphi}$$
$$+\frac{\tau_{r\theta} - \tau_{\varphi\varphi}\cot\theta}{r},$$
$$\rho\frac{Dv_\varphi}{Dt} = \rho\left(\frac{\partial v_\varphi}{\partial t} + v_r\frac{\partial v_\varphi}{\partial r} + \frac{v_\theta}{r}\frac{\partial v_\varphi}{\partial \theta} + \frac{v_\varphi}{r\sin\theta}\frac{\partial v_\varphi}{\partial \varphi} + \frac{v_r v_\varphi + v_\theta v_\varphi\cot\theta}{r}\right)$$
$$= \rho g_\varphi - \frac{1}{r\sin\theta}\frac{\partial p}{\partial \varphi} + \frac{1}{r^2}\frac{\partial r^2\tau_{\varphi r}}{\partial r} + \frac{1}{r}\frac{\partial \tau_{\varphi\theta}}{\partial \theta} + \frac{1}{r\sin\theta}\frac{\partial \tau_{\varphi\varphi}}{\partial \varphi}$$
$$+\frac{\tau_{r\varphi} + 2\tau_{\theta\varphi}\cot\theta}{r}. \tag{4.61}$$

The momentum equations on the basis of Crocco's derivation for inviscid flow, given by Eq. (3.23), read in spherical coordinates, where the relation (4.56) is used

$$v_\theta\left(\frac{1}{r}\frac{\partial r v_\theta}{\partial r} - \frac{1}{r}\frac{\partial v_r}{\partial \theta}\right) - v_\varphi\left(\frac{1}{r\sin\theta}\frac{\partial v_r}{\partial \varphi} - \frac{1}{r}\frac{\partial r v_\varphi}{\partial r}\right) = -T\frac{\partial s}{\partial r},$$
$$v_\varphi\frac{1}{r\sin\theta}\left(\frac{\partial v_\varphi\sin\theta}{\partial \theta} - \frac{\partial v_\theta}{\partial \varphi}\right) - v_r\left(\frac{1}{r}\frac{\partial r v_\theta}{\partial r} - \frac{1}{r}\frac{\partial v_r}{\partial \theta}\right) = -T\frac{1}{r}\frac{\partial s}{\partial \theta},$$
$$v_r\left(\frac{1}{r\sin\theta}\frac{\partial v_r}{\partial \varphi} - \frac{1}{r}\frac{\partial r v_\varphi}{\partial r}\right) - v_\theta\frac{1}{r\sin\theta}\left(\frac{\partial v_\varphi\sin\theta}{\partial \theta} - \frac{\partial v_\theta}{\partial \varphi}\right) = -T\frac{1}{r\sin\theta}\frac{\partial s}{\partial \varphi}. \tag{4.62}$$

The divergence of the shear stress tensor in the case that the dynamic viscosity μ is constant has the form (Eq. (4.16))

$$\mathrm{div}\,\bar{\bar{\tau}} = \mu(\Delta\mathbf{v} + \mathrm{grad}\,\mathrm{div}\,\mathbf{v} - \frac{2}{3}\,\mathrm{grad}\,\mathrm{div}\,\mathbf{v}).$$

Utilizing Eqs. (4.57) and (4.58), the aforementioned relation in spherical coordinates reads

$$(\mathrm{div}\,\bar{\bar{\tau}})_r = \mu\left\{\frac{1}{r^2}\frac{\partial}{\partial r}\left(r^2\frac{\partial v_r}{\partial r}\right) + \frac{1}{r^2\sin\theta}\frac{\partial}{\partial \theta}\left(\sin\theta\frac{\partial v_r}{\partial \theta}\right) + \frac{1}{r^2\sin^2\theta}\frac{\partial^2 v_r}{\partial \varphi^2}\right.$$
$$-\frac{2}{r^2}\left(v_r + \frac{\partial v_\theta}{\partial \theta} + v_\theta\cot\theta + \frac{1}{r^2\sin\theta}\frac{\partial v_\varphi}{\partial \varphi}\right)$$
$$+\frac{\partial}{\partial r}\left[\left(\frac{1}{r^2}\frac{\partial r^2 v_r}{\partial r}\right) + \left(\frac{1}{r\sin\theta}\frac{\partial v_\theta\sin\theta}{\partial \theta}\right) + \left(\frac{1}{r\sin\theta}\frac{\partial v_\varphi}{\partial \varphi}\right)\right]$$
$$\left.-\frac{2}{3}\frac{\partial}{\partial r}\left[\left(\frac{1}{r^2}\frac{\partial r^2 v_r}{\partial r}\right) + \left(\frac{1}{r\sin\theta}\frac{\partial v_\theta\sin\theta}{\partial \theta}\right) + \left(\frac{1}{r\sin\theta}\frac{\partial v_\varphi}{\partial \varphi}\right)\right]\right\}, \tag{4.63}$$

where again only the r-component is listed. The other components (θ and φ) can be obtained by similar operations. For incompressible flow due to $\operatorname{div} \mathbf{v} = 0$ one obtains

$$(\operatorname{div} \bar{\bar{\tau}})_r = \mu \left\{ \frac{1}{r^2} \frac{\partial}{\partial r} \left(r^2 \frac{\partial v_r}{\partial r} \right) + \frac{1}{r^2 \sin\theta} \frac{\partial}{\partial \theta} \left(\sin\theta \frac{\partial v_r}{\partial \theta} \right) + \frac{1}{r^2 \sin^2\theta} \frac{\partial^2 v_r}{\partial \varphi^2} \right.$$
$$\left. - \frac{2}{r^2} \left(v_r + \frac{\partial v_\theta}{\partial \theta} + v_\theta \cot\theta + \frac{1}{r^2 \sin\theta} \frac{\partial v_\varphi}{\partial \varphi} \right) \right\}. \quad (4.64)$$

The Energy Equation

The energy equations (3.24) or (3.25) in spherical coordinates by applying the relations Eqs. (4.48),(4.49) and (4.52) are given by

$$\frac{\partial \rho e}{\partial t} + \frac{1}{r^2} \frac{\partial}{\partial r} \left(r^2 \rho e v_r \right) + \frac{1}{r \sin\theta} \left(\frac{\partial}{\partial \theta} \left(\rho e v_\theta \sin\theta \right) + \frac{\partial}{\partial \varphi} \left(\rho e v_\varphi \right) \right)$$
$$+ \frac{1}{r^2} \frac{\partial}{\partial r} \left(r^2 p v_r \right) + \frac{1}{r \sin\theta} \left(\frac{\partial}{\partial \theta} (p v_\theta \sin\theta) + \frac{\partial}{\partial \varphi} (p v_\varphi) \right)$$
$$\equiv \frac{\partial \rho e}{\partial t} + \frac{1}{r^2} \frac{\partial}{\partial r} \left(r^2 (\rho e + p) v_r \right) + \frac{1}{r \sin\theta} \left(\frac{\partial}{\partial \theta} \left((\rho e + p) v_\theta \sin\theta \right) \right.$$
$$\left. + \frac{\partial}{\partial \varphi} \left((\rho e + p) v_\varphi \right) \right) = \frac{1}{r^2} \frac{\partial}{\partial r} \left(r^2 (\tau_{rr} v_r + \tau_{r\theta} v_\theta + \tau_{r\varphi} v_\varphi) \right)$$
$$+ \frac{1}{r \sin\theta} \frac{\partial}{\partial \theta} \left(\sin\theta (\tau_{\theta r} v_r + \tau_{\theta\theta} v_\theta + \tau_{\theta\varphi} v_\varphi) \right)$$
$$+ \frac{1}{r \sin\theta} \frac{\partial}{\partial \varphi} \left(\tau_{\varphi r} v_r + \tau_{\varphi\theta} v_\theta + \tau_{\varphi\varphi} v_\varphi \right)$$
$$+ \frac{1}{r^2} \frac{\partial}{\partial r} \left(r^2 \kappa \frac{\partial T}{\partial r} \right) + \frac{1}{r^2 \sin\theta} \frac{\partial}{\partial \theta} \left(\sin\theta \kappa \frac{\partial T}{\partial \theta} \right) + \frac{1}{r^2 \sin^2\theta} \frac{\partial}{\partial \varphi} \left(\kappa \frac{\partial T}{\partial \varphi} \right)$$
$$+ \rho \left(g_r v_r + g_\theta v_\theta + g_\varphi v_\varphi \right). \quad (4.65)$$

The energy equation based on the internal energy (3.33) can be transformed to spherical coordinates with the help of the relations Eqs. (4.49) and (4.55)

$$
\begin{aligned}
&\rho\left(\frac{\partial e_i}{\partial t}+v_r\frac{\partial e_i}{\partial r}+\frac{v_\theta}{r}\frac{\partial e_i}{\partial\theta}+\frac{v_\varphi}{r\sin\theta}\frac{\partial e_i}{\partial\varphi}\right)\\
&-\frac{p}{\rho}\left(\frac{\partial\rho}{\partial t}+v_r\frac{\partial\rho}{\partial r}+\frac{v_\theta}{r}\frac{\partial\rho}{\partial\theta}+\frac{v_\varphi}{r\sin\theta}\frac{\partial\rho}{\partial\varphi}\right)\\
&=\frac{1}{r^2}\frac{\partial}{\partial r}\left(r^2\kappa\frac{\partial T}{\partial r}\right)+\frac{1}{r^2\sin\theta}\frac{\partial}{\partial\theta}\left(\sin\theta\kappa\frac{\partial T}{\partial\theta}\right)+\frac{1}{r^2\sin^2\theta}\frac{\partial}{\partial\varphi}\left(\kappa\frac{\partial T}{\partial\varphi}\right)\\
&+\tau_{rr}\frac{\partial v_r}{\partial r}+\tau_{\theta\theta}\left(\frac{1}{r}\frac{\partial v_\theta}{\partial\theta}+\frac{v_r}{r}\right)+\tau_{\varphi\varphi}\left(\frac{1}{r\sin\theta}\frac{\partial v_\varphi}{\partial\varphi}+\frac{v_r+v_\theta\cot\theta}{r}\right)\\
&+\tau_{r\theta}\left(\frac{\partial v_\theta}{\partial r}+\frac{1}{r}\frac{\partial v_r}{\partial\theta}-\frac{v_\theta}{r}\right)+\tau_{r\varphi}\left(\frac{\partial v_\varphi}{\partial r}+\frac{1}{r\sin\theta}\frac{\partial v_r}{\partial\varphi}-\frac{v_\varphi}{r}\right)\\
&+\tau_{\theta\varphi}\left(\frac{1}{r}\frac{\partial v_\varphi}{\partial\theta}+\frac{1}{r\sin\theta}\frac{\partial v_\theta}{\partial\varphi}-\frac{v_\varphi\cot\theta}{r}\right),
\end{aligned}
\tag{4.66}
$$

and the equation with the enthalpy variable ($h = c_pT$, perfect gas) Eq. (3.35) reads in spherical coordinates

$$
\begin{aligned}
&\rho c_p\left(\frac{\partial T}{\partial t}+v_r\frac{\partial T}{\partial r}+\frac{v_\theta}{r}\frac{\partial T}{\partial\theta}+\frac{v_\varphi}{r\sin\theta}\frac{\partial T}{\partial\varphi}\right)\\
&=\left(\frac{\partial p}{\partial t}+v_r\frac{\partial p}{\partial r}+\frac{v_\theta}{r}\frac{\partial p}{\partial\theta}+\frac{v_\varphi}{r\sin\theta}\frac{\partial p}{\partial\varphi}\right)\\
&+\frac{1}{r^2}\frac{\partial}{\partial r}\left(r^2\kappa\frac{\partial T}{\partial r}\right)+\frac{1}{r^2\sin\theta}\frac{\partial}{\partial\theta}\left(\sin\theta\kappa\frac{\partial T}{\partial\theta}\right)+\frac{1}{r^2\sin^2\theta}\frac{\partial}{\partial\varphi}\left(\kappa\frac{\partial T}{\partial\varphi}\right)\\
&+\tau_{rr}\frac{\partial v_r}{\partial r}+\tau_{\theta\theta}\left(\frac{1}{r}\frac{\partial v_\theta}{\partial\theta}+\frac{v_r}{r}\right)+\tau_{\varphi\varphi}\left(\frac{1}{r\sin\theta}\frac{\partial v_\varphi}{\partial\varphi}+\frac{v_r+v_\theta\cot\theta}{r}\right)\\
&+\tau_{r\theta}\left(\frac{\partial v_\theta}{\partial r}+\frac{1}{r}\frac{\partial v_r}{\partial\theta}-\frac{v_\theta}{r}\right)+\tau_{r\varphi}\left(\frac{\partial v_\varphi}{\partial r}+\frac{1}{r\sin\theta}\frac{\partial v_r}{\partial\varphi}-\frac{v_\varphi}{r}\right)\\
&+\tau_{\theta\varphi}\left(\frac{1}{r}\frac{\partial v_\varphi}{\partial\theta}+\frac{1}{r\sin\theta}\frac{\partial v_\theta}{\partial\varphi}-\frac{v_\varphi\cot\theta}{r}\right).
\end{aligned}
\tag{4.67}
$$

The elements of the shear stress tensor in spherical coordinates are determined from the tensorial notation $\tau_{ij} = \mu\left(\partial_i v_j + \partial_j v_i\right) - \frac{2}{3}\mu\,\partial^k v_k\,\delta_{ij}$ if the derivatives in this relation are interpreted as covariant derivatives [4]

$$
\begin{aligned}
\tau_{rr} &= 2\mu\left(\frac{\partial v_r}{\partial r} - \frac{1}{3}\operatorname{div}\mathbf{v}\right), \\
\tau_{\theta\theta} &= 2\mu\left(\frac{1}{r}\frac{\partial v_\theta}{\partial \theta} + \frac{v_r}{r} - \frac{1}{3}\operatorname{div}\mathbf{v}\right), \\
\tau_{\varphi\varphi} &= 2\mu\left(\frac{1}{r\sin\theta}\frac{\partial v_\varphi}{\partial \varphi} + \frac{v_r}{r} + \frac{v_\theta\cot\theta}{r} - \frac{1}{3}\operatorname{div}\mathbf{v}\right), \\
\tau_{r\theta} &= \tau_{\theta r} = \mu\left(r\frac{\partial}{\partial r}\left(\frac{v_\theta}{r}\right) + \frac{1}{r}\frac{\partial v_r}{\partial \theta}\right), \\
\tau_{\theta\varphi} &= \tau_{\varphi\theta} = \mu\left(\frac{\sin\theta}{r}\frac{\partial}{\partial \theta}\left(\frac{v_\varphi}{\sin\theta}\right) + \frac{1}{r\sin\theta}\frac{\partial v_\theta}{\partial \varphi}\right), \\
\tau_{\varphi r} &= \tau_{r\varphi} = \mu\left(\frac{1}{r\sin\theta}\frac{\partial v_r}{\partial \varphi} + r\frac{\partial}{\partial r}\left(\frac{v_\varphi}{r}\right)\right),
\end{aligned}
\tag{4.68}
$$

where the term div **v** is given by Eq. (4.48). The equivalent to Eq. (4.23) in spherical coordinates has the form

$$
\frac{\partial p}{\partial t} + v_r\frac{\partial p}{\partial r} + \frac{v_\theta}{r}\frac{\partial p}{\partial \theta} + \frac{v_\varphi}{r\sin\theta}\frac{\partial p}{\partial \varphi} - c^2\left(\frac{\partial \rho}{\partial t} + v_r\frac{\partial \rho}{\partial r} + \frac{v_\theta}{r}\frac{\partial \rho}{\partial \theta} + \frac{v_\varphi}{r\sin\theta}\frac{\partial \rho}{\partial \varphi}\right) = 0.
\tag{4.69}
$$

References

1. Bird, R.B., Stuart, W.E., Lightfoot, E.N.: Transport Phenomena, 2nd edn. Wiley, New York (2002)
2. Aris, R.: Vectors. Tensors and the Basic Equations of Fluid Mechanics. Prentice Hall, Englewood Cliffs, UK (1962)
3. Sommerfeld, A.: Mechanik der Deformierbaren Medien. Akademische Verlagsgesellschaft Geest & Portig, Leipzig (1954)
4. Klingbeil, E.: Tensorrechnung fuer Ingenieure. B.I. Wissenschaftsverlag, Mannheim (1966)
5. Teichmann, H.: Physikalische Anwendungen der Vektor- und Tensorrechnung. B.I. Wissenschaftsverlag, Mannheim (1973)
6. Bronstein, I.N., Semendjajew, K.A.: Taschenbuch der Mathematik. Harri Deutsch Verlag, Frankfurt (1973)
7. Hirschel, E.H.: Basics of Aerothermodynamics, 2nd edn. Springer, Berlin (2015)
8. Weiland, C.: Zwei- und dreidimensionale Umstrómungen stumpfer Körper unter Berücksichtigung schallnaher Überschallströmungen. Zeitschrift für Flugwissenschaften (ZfW), **24(5)** (1976)
9. Lyubimov, A.N., Rusanov, V.V.: Gas Flows Past Blunt Bodies. NASA TT F-714 (1973)
10. Weiland, C.: Computational Space Flight Mechanics. Springer Verlag, Berlin/Heidelberg (2010)

Chapter 5
The Navier-Stokes Equations in Cartesian Coordinates for Laminar and Turbulent Flow

Abstract It has been demonstrated in the past that it is advantageous to write the set of fluid dynamic equations in vector or vector-matrix form. Any transformation of the coordinates (Chap. 6) or diagonalization of the corresponding Jacobi matrices (Chap. 14) can be executed in a homogeneous way. This is also true for statistical two equation turbulence models (Chap. 8) and for the conservation equation of species and species energy (Chap. 9) if the gas is in a non-equilibrium state (vibrational excitation, dissociation, ionisation). Therefore, in this and the following chapters the vector or vector-matrix representation is presented.

The Variables and Quantities Used in this Chapter

$\mathbf{E}, \mathbf{F}, \mathbf{G}$	vector: inviscid part
$\mathbf{E}_{visc}, \mathbf{F}_{visc}, \mathbf{G}_{visc}$	vector: viscous part
$\mathbf{Q}$	vector: conservative variables
$\mathbf{S}_{NS}$	vector: external sources
$\hat{\mathbf{E}}, \hat{\mathbf{F}}, \hat{\mathbf{G}}$	vector: inviscid part of RANS equations
$\hat{\mathbf{E}}_{visc}, \hat{\mathbf{F}}_{visc}, \hat{\mathbf{G}}_{visc}$	vector: viscous part of RANS equations
$\hat{\mathbf{Q}}$	vector: conservative variables of RANS equations
$\hat{\mathbf{E}}_C, \hat{\mathbf{F}}_C, \hat{\mathbf{G}}_C$	vector: inviscid part of RANS equations for a complete non-equilibrium thermodynamic state
$\hat{\mathbf{E}}_{C,visc}, \hat{\mathbf{F}}_{C,visc}, \hat{\mathbf{G}}_{C,visc}$	vector: viscous part of RANS equations for a complete non-equilibrium thermodynamics state
$\hat{\mathbf{Q}}_C$	vector: conservative variables of RANS equations for a complete non-equilibrium thermodynamic state
$\mathbf{S}_C$	vector: source term of chemical and thermal species conservation
e	total energy per unit mass
e_i	internal energy per unit mass
ρ	density
p	pressure
T	temperature

C. Weiland, *The Hierarchy of Fluid Dynamic Equations*,
https://doi.org/10.1007/978-3-031-71198-5_5

q_H	heat sources
v_x, v_y, v_z	components of $\mathbf{v}$ in Cartesian coordinates
v_i, v_j	components of $\mathbf{v}$ in tensor notation
x_i, x_j, x_k	coordinates in tensor notation
g_x, g_y, g_z	Cartesian components of gravitational acceleration $\mathbf{g}$
$\tau_{xx}, \tau_{yy}, \tau_{zz}$ $\tau_{xy}, \tau_{xz}, \tau_{yz}$	elements of viscous stress tensor $\bar{\bar{\tau}}$ in Cartesian coordinates
τ_{ij}	components of $\bar{\bar{\tau}}$ in tensor notation
δ_{ij}	Kronecker symbol
γ	ratio of specific heats
c_p	(mass) specific heat at constant pressure
μ	dynamic viscosity
μ_t	turbulent viscosity
μ^*	$= \mu + \mu_t/\sigma_k$ abbreviation
$\check{\mu}$	$= \mu + \mu_t$ total viscosity
σ_k	coefficient
κ	thermal conductivity
$\check{\kappa}$	$= \kappa(1 + Pr\mu_t/Pr_t\mu)$ total thermal conductivity
q_j	components of $\mathbf{q}$ in tensor notation
φ	dependent variable
$\bar{\varphi}$	mean part of Reynolds averaged dependent variable φ
φ'	fluctuating part of Reynolds averaged dependent variable φ
$\hat{\varphi}$	mean part of Favre averaged dependent variable φ
φ''	fluctuating part of Favre averaged dependent variable φ
k	turbulent kinetic energy
Pr	Prandtl number
Pr_t	turbulent Prandtl number
$\mathbf{j}_i$	vector: diffusion mass flux of species i
$\check{D}_i$	$= D_i + \frac{1}{Sc_t}\frac{\mu_t}{\rho}$ mass diffusion coefficient for turbulent flow
D_i	multicomponent mass diffusion coefficient of species i
Sc_t	turbulent Schmidt number
h_i	enthalpy of species i
ω_i	mass fraction of species i
α_i	source term of conservation equation of species i
ρ_i	density of chemical species i
ρ_j	density of molecular species j
$e_{v,j}$	vibrational energy of molecular species j
p_e	pressure of electrons
e_e	electron energy per unit mass
$\kappa_{v,j}$	frozen thermal conductivity of vibrational energy of molecular species j
κ_e	frozen thermal conductivity of electron energy
$T_{v,j}$	vibrational temperature of molecular species j

T_e	temperature of electrons
$Q_{v,j}$	vibrational coupling term of molecular species j
Q_e	electron coupling term

5.1 The Navier-Stokes Equations for Laminar, Perfect Gas Flow

The five Eqs. (4.12), (4.13) and (4.19) can be compiled in a formal vector notation, which often is the basis of the approximations by numerical methods. Initially, we consider these equations for laminar flow when the gas is in a perfect thermodynamic state

$$\frac{\partial \mathbf{Q}}{\partial t} + \frac{\partial \mathbf{E}}{\partial x} + \frac{\partial \mathbf{F}}{\partial y} + \frac{\partial \mathbf{G}}{\partial z} = \frac{\partial \mathbf{E}_{visc}}{\partial x} + \frac{\partial \mathbf{F}_{visc}}{\partial y} + \frac{\partial \mathbf{G}_{visc}}{\partial z} + \mathbf{S}_{NS}\,, \tag{5.1}$$

with

$$\mathbf{Q} = \begin{pmatrix} \rho \\ \rho v_x \\ \rho v_y \\ \rho v_z \\ \rho e \end{pmatrix}, \quad \mathbf{S}_{NS} = \begin{pmatrix} 0 \\ \rho g_x \\ \rho g_y \\ \rho g_z \\ \rho(g_x v_x + g_y v_y + g_z v_z) + q_H \end{pmatrix},$$

$$\mathbf{E} = \begin{pmatrix} \rho v_x \\ \rho v_x^2 + p \\ \rho v_x v_y \\ \rho v_x v_z \\ (\rho e + p) v_x \end{pmatrix}, \mathbf{F} = \begin{pmatrix} \rho v_y \\ \rho v_y v_x \\ \rho v_y^2 + p \\ \rho v_y v_z \\ (\rho e + p) v_y \end{pmatrix}, \mathbf{G} = \begin{pmatrix} \rho v_z \\ \rho v_z v_x \\ \rho v_z v_y \\ \rho v_z^2 + p \\ (\rho e + p) v_z \end{pmatrix},$$

$$\mathbf{E}_{visc} = \begin{pmatrix} 0 \\ \tau_{xx} \\ \tau_{xy} \\ \tau_{xz} \\ v_x \tau_{xx} + v_y \tau_{xy} + v_z \tau_{xz} - q_x \end{pmatrix},$$

$$\mathbf{F}_{visc} = \begin{pmatrix} 0 \\ \tau_{yx} \\ \tau_{yy} \\ \tau_{yz} \\ v_x \tau_{yx} + v_y \tau_{yy} + v_z \tau_{yz} - q_y \end{pmatrix},$$

$$\mathbf{G}_{visc} = \begin{pmatrix} 0 \\ \tau_{zx} \\ \tau_{zy} \\ \tau_{zz} \\ v_x\tau_{zx} + v_y\tau_{zy} + v_z\tau_{zz} - q_z \end{pmatrix},$$

$$e = e_i + \frac{1}{2}|\mathbf{v}|^2, \qquad p = (\gamma - 1)\rho(e - \frac{1}{2}|\mathbf{v}|^2),$$

$$\begin{aligned}
\tau_{xx} &= 2\mu\left(\frac{\partial v_x}{\partial x} - \frac{1}{3}\left(\frac{\partial v_x}{\partial x} + \frac{\partial v_y}{\partial y} + \frac{\partial v_z}{\partial z}\right)\right), \\
\tau_{yy} &= 2\mu\left(\frac{\partial v_y}{\partial y} - \frac{1}{3}\left(\frac{\partial v_x}{\partial x} + \frac{\partial v_y}{\partial y} + \frac{\partial v_z}{\partial z}\right)\right), \\
\tau_{zz} &= 2\mu\left(\frac{\partial v_z}{\partial z} - \frac{1}{3}\left(\frac{\partial v_x}{\partial x} + \frac{\partial v_y}{\partial y} + \frac{\partial v_z}{\partial z}\right)\right), \\
\tau_{xy} &= \tau_{yx} = \mu\left(\frac{\partial v_x}{\partial y} + \frac{\partial v_y}{\partial x}\right), \\
\tau_{xz} &= \tau_{zx} = \mu\left(\frac{\partial v_x}{\partial z} + \frac{\partial v_z}{\partial x}\right), \\
\tau_{yz} &= \tau_{zy} = \mu\left(\frac{\partial v_y}{\partial z} + \frac{\partial v_z}{\partial y}\right).
\end{aligned}$$

5.2 Averaging Process for the Governing Equations

Let us consider the five Eqs. (4.12), (4.13) and (4.19), but now written in tensor notation,[1] which is helpful for the subsequent operations. The external forces and the external heat sources have been omitted for convenience

$$\frac{\partial \rho}{\partial t} + \frac{\partial}{\partial x_j}(\rho v_j) = 0\,, \tag{5.2}$$

$$\frac{\partial \rho v_i}{\partial t} + \frac{\partial}{\partial x_j}(\rho v_i v_j + p\delta_{ij} - \tau_{ij}) = 0, \tag{5.3}$$

$$\frac{\partial \rho e}{\partial t} + \frac{\partial}{\partial x_j}(\rho e v_j + v_j p + q_j - v_i \tau_{ij}) = 0. \tag{5.4}$$

---

[1] The simple rules for multiplication and building derivatives in the frame of the tensor calculus can be found in [1–3].

For adequately high Reynolds numbers, boundary layer flow fields behave chaotic and one speaks about turbulent flow. Generally, it is problematic to resolve such instantaneous flow fields, which would require time steps and grid sizes at least of the order of the time scales and the space scales (grid resolution) of the turbulent flow structures (if numerical methods for solving such flow problems are employed: Direct Numerical Simulation DNS). On the other hand, it turned out that for most of the practical design purposes the consideration of a mean value of the turbulent flow is sufficient. Therefore, the Navier-Stokes equations are averaged statistically. A main requirement for this procedure is that the averaging time t must be sufficiently large compared to the turbulent time scales and sufficiently small compared to the time scales of the mean flow.

We write a dependent variable $\varphi(\bar{x}, t)$, which describes instantaneously a specific aspect of the flow field, as the sum of the mean part $\bar{\varphi}(\bar{x}, t)$ and a fluctuating part $\varphi'(\bar{x}, t)$

$$\varphi(\bar{x}, t) = \bar{\varphi}(\bar{x}, t) + \varphi'(\bar{x}, t),$$

where $\bar{\varphi}(\bar{x}, t)$ is given by the classical time averaging

$$\bar{\varphi} = \frac{1}{T} \int_{t}^{t+T} \varphi(\bar{x}, t) dt \,, \tag{5.5}$$

which is called Reynolds-averaging. Applying this averaging procedure to Eqs. (5.2)–(5.4) leads to a number of extra terms in the averaged equations (open turbulent equations), which must be modeled in order to close the set of equations.

A more appropriate approach is the usage of density weighted time averaging (Favre-averaging), with

$$\varphi(\bar{x}, t) = \hat{\varphi}(\bar{x}, t) + \varphi''(\bar{x}, t),$$

and

$$\hat{\varphi} = \frac{1}{\bar{\rho}} \frac{1}{T} \int_{t}^{t+T} \rho(\bar{x}, t) \varphi(\bar{x}, t) dt, \tag{5.6}$$

which produces fewer extra terms compared with the Reynolds-averaging.

Favre-averaging of the continuity equation (5.2) does not change the formal structure of the equation and yields

$$\frac{\partial \bar{\rho}}{\partial t} + \frac{\partial}{\partial x_j} (\bar{\rho} \hat{v}_j) = 0, \tag{5.7}$$

where the density ρ is Reynolds-averaged and the velocity components v_j are Favre-averaged. Equation (5.7) has a closed form, where no extra term has to be modeled. This behavior changes when the momentum Eq. (5.3) and the energy Eq. (5.4) are Favre-averaged. The density $\bar{\rho}$ and the pressure $\bar{p}$ are Reynolds-averaged, whereas all other dependent variables are Favre-averaged.

The momentum equations then read

$$\frac{\partial \bar{\rho}\hat{v}_i}{\partial t} + \frac{\partial}{\partial x_j}(\bar{\rho}\hat{v}_i\hat{v}_j + \bar{p}\delta_{ij} - \hat{\tau}_{ij}) = -\frac{\partial}{\partial x_j}\overline{\rho v_i'' v_j''}, \tag{5.8}$$

with

$$\hat{\tau}_{ij} = \mu\left(\frac{\partial \hat{v}_j}{\partial x_i} + \frac{\partial \hat{v}_i}{\partial x_j} - \frac{2}{3}\frac{\partial \hat{v}_k}{\partial x_k}\delta_{ij}\right) \tag{5.9}$$

being the Favre-averaged viscous stress tensor,[2] where the term τ_{ij}'' with ($\tau_{ij} = \hat{\tau}_{ij} + \tau_{ij}''$) is neglected. The term $\overline{\rho v_i'' v_j''}$ on the right hand side of Eq. (5.8) is called the Reynolds stress tensor, which can be modeled, when the Boussinesq hypothesis[3] is taken as a basis, by

$$\overline{\rho v_i'' v_j''} \approx \hat{\tau}_{ij}^{turb} = \mu_t\left(\frac{\partial \hat{v}_j}{\partial x_i} + \frac{\partial \hat{v}_i}{\partial x_j} - \frac{2}{3}\frac{\partial \hat{v}_k}{\partial x_k}\delta_{ij}\right) - \frac{2}{3}\bar{\rho}k\delta_{ij}, \tag{5.10}$$

where $k = \frac{\widehat{v_i'' v_i''}}{2}$ is the turbulent kinetic energy. The last term in Eq. (5.10) guarantees that the trace of the Reynolds stress tensor is equal to $-2\bar{\rho}k$.

The Favre-averaged energy equation is obtained by

$$\begin{aligned}
&\frac{\partial \bar{\rho}\hat{e}}{\partial t} + \frac{\partial}{\partial x_j}\left(\bar{\rho}\hat{e}\hat{v}_j + \hat{v}_j\bar{p} - c_p\frac{\mu}{Pr}\frac{\partial \hat{T}}{\partial x_j} - \hat{v}_i\hat{\tau}_{ij}\right) = \\
&= \underbrace{\frac{\partial}{\partial x_j}\left(-\hat{v}_i\overline{\rho v_i'' v_j''} - c_p\overline{\rho v_j'' T''}\right)}_{(1)} + \underbrace{\frac{\partial}{\partial x_j}\left(-\frac{1}{2}\overline{\rho v_j'' v_i'' v_i''} + \overline{v_i''\tau_{ij}}\right)}_{(2)} + \\
&+ \underbrace{\frac{\partial}{\partial x_j}\left(c_p\frac{\mu}{Pr}\frac{\partial \overline{T''}}{\partial x_j} + \hat{v}_i\overline{\tau_{ij}''}\right)}_{(3)}.
\end{aligned} \tag{5.11}$$

[2] Note, no fluctuations of the molecular viscosity are considered.

[3] Newtons hypothesis connects the molecular viscosity μ with the shear components to the molecular stress. Similarly: Boussinesq hypothesis connects the turbulent viscosity (or eddy viscosity) μ_t with the shear components to the Reynolds stress.

The two terms of (1) are essential for the determination of turbulent flow and must be modeled. The two terms of (2) correspond to turbulent transport and molecular diffusion of turbulent energy k and can be neglected, if the turbulent energy is small compared to the enthalpy. On the other hand modeling these terms improves the quality of the solution. The two terms of (3) can be neglected if $\left|\frac{\partial^2 \hat{T}}{\partial x_j^2}\right| >> \left|\frac{\partial^2 \overline{T''}}{\partial x_j^2}\right|$ and $|\hat{\tau}_{ij}| >> |\overline{\tau''_{ij}}|$, which is true for more or less all practical flows. The total energy $\hat{e}$ is given by

$$\hat{e} = \hat{e}_i + \frac{\hat{v}_i \hat{v}_i}{2} + \underbrace{\frac{\widehat{v''_i v''_i}}{2}}_{k},$$

and in the case of a perfect gas flow we find for the pressure

$$\bar{p} = (\gamma - 1)\bar{\rho}\left(\hat{e} - \frac{\hat{v}_i \hat{v}_i}{2} - \frac{\widehat{v''_i v''_i}}{2}\right).$$

The first term in (1) is modeled by the Reynolds stress tensor in the form of Eq. (5.10). The second term is modeled by a Reynolds analogy for the turbulent heat flux, namely

$$c_p \overline{\rho v''_j T''} \approx -\frac{c_p \mu_t}{Pr_t}\frac{\partial \hat{T}}{\partial x_j}, \tag{5.12}$$

with Pr_t being a turbulent Prandtl number. The turbulent Prandtl number is often considered to be a constant.

A widely accepted modeling of the two terms in (2), which corresponds to turbulent transport and molecular diffusion of turbulent energy, is given by

$$-\frac{1}{2}\overline{\rho v''_j v''_i v''_i} + \overline{v''_i \tau_{ij}} \approx \underbrace{\left(\mu + \frac{\mu_t}{\sigma_k}\right)}_{\mu^*}\frac{\partial k}{\partial x_j}, \tag{5.13}$$

where σ_k is a constant of the model.

We now combine the laminar part of the stress tensor $\hat{\tau}_{ij}^{lam}$, Eq. (5.9), with the turbulent part $\hat{\tau}_{ij}^{turb}$, Eq. (5.10), and obtain

$$\hat{\tau}_{ij}^{tot} = \hat{\tau}_{ij}^{lam} + \hat{\tau}_{ij}^{turb} = \check{\mu}\left(\frac{\partial \hat{v}_j}{\partial x_i} + \frac{\partial \hat{v}_i}{\partial x_j} - \frac{2}{3}\frac{\partial \hat{v}_k}{\partial x_k}\delta_{ij}\right) - \frac{2}{3}\bar{\rho}k\delta_{ij}, \tag{5.14}$$

with $\check{\mu} = \mu + \mu_t$.

In the same way, we find for the total heat flux vector

$$\hat{q}_j^{tot} = \hat{q}_j^{lam} + \hat{q}_j^{turb} = -c_p \frac{\mu}{Pr}\frac{\partial \hat{T}}{\partial x_j} - c_p \frac{\mu_t}{Pr_t}\frac{\partial \hat{T}}{\partial x_j} = -\breve{\kappa}\left(\frac{\partial \hat{T}}{\partial x_j}\right), \tag{5.15}$$

with $\breve{\kappa} = \kappa\left(1 + \frac{Pr}{Pr_t}\frac{\mu_t}{\mu}\right)$ and $\kappa = \frac{c_p \mu}{Pr}$ the thermal conductivity.

The closure problem of the energy equation is completed with Eqs. (5.13), (5.14) and (5.15). For more details regarding the averaging process, in particular the estimation of some of the correlations, the reader is referred to [4, 5].

Note:

For flows in chemical or thermal <u>and</u> chemical non-equilibrium the mass diffusion due to the various species plays an important role (see Chap. 9). The diffusion mass flux vector is part of the species continuity equations, Eq. (9.1), and the energy equations, Eqs. (9.5), (9.6), and reads $\mathbf{j}_i = -\rho D_i \mathrm{grad}\omega_i$ with D_i the multicomponent mass diffusion coefficient. The averaging process of such flows leads to the following diffusion coefficient for turbulent flows, [5, 6]

$$\breve{D}_i = D_i + \frac{1}{Sc_t}\frac{\mu_t}{\rho} \tag{5.16}$$

with Sc_t being the turbulent Schmidt number, which is constant and has an approximate value of 0.8.

5.3 The Reynolds-Averaged-Navier-Stokes Equations (RANS), Perfect Gas Flow

In the following passage, we expand the Navier-Stokes equations in the formal vector notation to the so-called Reynolds-Averaged-Navier-Stokes (RANS) equations for turbulent flows. There, all the information is included of Eqs. (5.7), (5.8) and (5.11) and of the relations modeling the correlation terms, Eqs. (5.13),(5.14) and (5.15), which close the set of turbulent equations (The components of the shear stress tensor are those of Eq. (5.14)).

$$\frac{\partial \hat{\mathbf{Q}}}{\partial t} + \frac{\partial \hat{\mathbf{E}}}{\partial x} + \frac{\partial \hat{\mathbf{F}}}{\partial y} + \frac{\partial \hat{\mathbf{G}}}{\partial z} = \frac{\partial \hat{\mathbf{E}}_{visc}}{\partial x} + \frac{\partial \hat{\mathbf{F}}_{visc}}{\partial y} + \frac{\partial \hat{\mathbf{G}}_{visc}}{\partial z}, \tag{5.17}$$

with

$$\hat{\mathbf{Q}} = \begin{pmatrix} \bar{\rho} \\ \bar{\rho}\hat{v}_x \\ \bar{\rho}\hat{v}_y \\ \bar{\rho}\hat{v}_z \\ \bar{\rho}\hat{e} \end{pmatrix},$$

$$\hat{\mathbf{E}} = \begin{pmatrix} \bar{\rho}\hat{v}_x \\ \bar{\rho}\hat{v}_x^2 + \bar{p} \\ \bar{\rho}\hat{v}_x\hat{v}_y \\ \bar{\rho}\hat{v}_x\hat{v}_z \\ (\bar{\rho}\hat{e} + \bar{p})\hat{v}_x \end{pmatrix}, \hat{\mathbf{F}} = \begin{pmatrix} \bar{\rho}\hat{v}_y \\ \bar{\rho}\hat{v}_y\hat{v}_x \\ \bar{\rho}\hat{v}_y^2 + \bar{p} \\ \bar{\rho}\hat{v}_y\hat{v}_z \\ (\bar{\rho}\hat{e} + \bar{p})\hat{v}_y \end{pmatrix}, \hat{\mathbf{G}} = \begin{pmatrix} \bar{\rho}\hat{v}_z \\ \bar{\rho}\hat{v}_z\hat{v}_x \\ \bar{\rho}\hat{v}_z\hat{v}_y \\ \bar{\rho}\hat{v}_z^2 + \bar{p} \\ (\bar{\rho}\hat{e} + \bar{p})\hat{v}_z \end{pmatrix},$$

$$\hat{\mathbf{E}}_{visc} = \begin{pmatrix} 0 \\ \hat{\tau}_{xx} \\ \hat{\tau}_{xy} \\ \hat{\tau}_{xz} \\ \hat{v}_x\hat{\tau}_{xx} + \hat{v}_y\hat{\tau}_{xy} + \hat{v}_z\hat{\tau}_{xz} + \check{\kappa}\frac{\partial \hat{T}}{\partial x} + \mu^*\frac{\partial k}{\partial x} \end{pmatrix},$$

$$\hat{\mathbf{F}}_{visc} = \begin{pmatrix} 0 \\ \hat{\tau}_{yx} \\ \hat{\tau}_{yy} \\ \hat{\tau}_{yz} \\ \hat{v}_x\hat{\tau}_{yx} + \hat{v}_y\hat{\tau}_{yy} + \hat{v}_z\hat{\tau}_{yz} + \check{\kappa}\frac{\partial \hat{T}}{\partial y} + \mu^*\frac{\partial k}{\partial y} \end{pmatrix},$$

$$\hat{\mathbf{G}}_{visc} = \begin{pmatrix} 0 \\ \hat{\tau}_{zx} \\ \hat{\tau}_{zy} \\ \hat{\tau}_{zz} \\ \hat{v}_x\hat{\tau}_{zx} + \hat{v}_y\hat{\tau}_{zy} + \hat{v}_z\hat{\tau}_{zz} + \check{\kappa}\frac{\partial \hat{T}}{\partial z} + \mu^*\frac{\partial k}{\partial z} \end{pmatrix},$$

with

$$\begin{aligned} \check{\mu} &= \mu + \mu_t\,, \\ \mu^* &= \mu + \frac{\mu_t}{\sigma_k}\,, \\ \check{\kappa} &= \kappa\left(1 + \frac{Pr}{Pr_t}\frac{\mu_t}{\mu}\right). \end{aligned}$$

5.4 The Reynolds-Averaged Navier-Stokes Equations (RANS), Non-perfect Gas Flow

5.4.1 Chemical Non-Equilibrium State

The set of equations for turbulent flows with gases in chemical non-equilibrium consists of an adjusted Eq. (5.17), where only the viscous vectors $\hat{\mathbf{E}}_{visc}$, $\hat{\mathbf{F}}_{visc}$, $\hat{\mathbf{G}}_{visc}$ are expanded by the mass diffusion term, which then read

$$\hat{\mathbf{E}}_{visc} = \begin{pmatrix} 0 \\ \hat{\tau}_{xx} \\ \hat{\tau}_{xy} \\ \hat{\tau}_{xz} \\ \hat{v}_x \hat{\tau}_{xx} + \hat{v}_y \hat{\tau}_{xy} + \hat{v}_z \hat{\tau}_{xz} + \check{\kappa} \frac{\partial \hat{T}}{\partial x} + \sum_i \bar{\rho} \check{D}_i \hat{h}_i \frac{\partial \hat{\omega}_i}{\partial x} + \mu^* \frac{\partial k}{\partial x} \end{pmatrix},$$
$$\hat{\mathbf{F}}_{visc} = \begin{pmatrix} 0 \\ \hat{\tau}_{yx} \\ \hat{\tau}_{yy} \\ \hat{\tau}_{yz} \\ \hat{v}_x \hat{\tau}_{yx} + \hat{v}_y \hat{\tau}_{yy} + \hat{v}_z \hat{\tau}_{yz} + \check{\kappa} \frac{\partial \hat{T}}{\partial y} + \sum_i \bar{\rho} \check{D}_i \hat{h}_i \frac{\partial \hat{\omega}_i}{\partial y} + \mu^* \frac{\partial k}{\partial y} \end{pmatrix}, \quad (5.18)$$
$$\hat{\mathbf{G}}_{visc} = \begin{pmatrix} 0 \\ \hat{\tau}_{zx} \\ \hat{\tau}_{zy} \\ \hat{\tau}_{zz} \\ \hat{v}_x \hat{\tau}_{zx} + \hat{v}_y \hat{\tau}_{zy} + \hat{v}_z \hat{\tau}_{zz} + \check{\kappa} \frac{\partial \hat{T}}{\partial z} + \sum_i \bar{\rho} \check{D}_i \hat{h}_i \frac{\partial \hat{\omega}_i}{\partial z} + \mu^* \frac{\partial k}{\partial z} \end{pmatrix}.$$

This set of equations has to be completed by a turbulence model (e.g.: Eq. (8.5)) and the averaged species density Eqs. (9.3), (9.4), both written in a formal vector notation.

5.4.2 Complete Thermodynamic Non-Equilibrium State

The following presents the RANS equations in vector formation for thermal and chemical non-equilibrium states, which also incorporate electron modes. This is a compilation of the Eqs. (5.17), (8.5), (9.3), (9.6), (9.8), where all the variables of the Eqs. (9.3), (9.6), (9.8) are averaged in the sense of Eq. (9.4).

$$\frac{\partial \hat{\mathbf{Q}}_C}{\partial t} + \frac{\partial \hat{\mathbf{E}}_C}{\partial x} + \frac{\partial \hat{\mathbf{F}}_C}{\partial y} + \frac{\partial \hat{\mathbf{G}}_C}{\partial z} = \frac{\partial \hat{\mathbf{E}}_{C,visc}}{\partial x} + \frac{\partial \hat{\mathbf{F}}_{C,visc}}{\partial y} + \frac{\partial \hat{\mathbf{G}}_{C,visc}}{\partial z} + \hat{\mathbf{S}}_C, \quad (5.19)$$

$$
\hat{\mathbf{Q}}_C = \begin{pmatrix} \bar{\rho} \\ \bar{\rho}\hat{v}_x \\ \bar{\rho}\hat{v}_y \\ \bar{\rho}\hat{v}_z \\ \bar{\rho}\hat{e} \\ \bar{\rho}_1 \\ \bar{\rho}_2 \\ \vdots \\ \bar{\rho}_{1'}\hat{e}_{v,1} \\ \bar{\rho}_{2'}\hat{e}_{v,2} \\ \vdots \\ \bar{\rho}\hat{e}_e \end{pmatrix}, \hat{\mathbf{S}}_C = \begin{pmatrix} 0 \\ 0 \\ 0 \\ 0 \\ 0 \\ \hat{\alpha}_1 \\ \hat{\alpha}_2 \\ \vdots \\ Q_{v,1} \\ Q_{v,2} \\ \vdots \\ Q_e \end{pmatrix},
$$

$$
\hat{\mathbf{E}}_C = \begin{pmatrix} \bar{\rho}\hat{v}_x \\ \bar{\rho}\hat{v}_x^2 + \bar{p} \\ \bar{\rho}\hat{v}_x\hat{v}_y \\ \bar{\rho}\hat{v}_x\hat{v}_z \\ (\bar{\rho}\hat{e} + \bar{p})\hat{v}_x \\ \bar{\rho}_1\hat{v}_x \\ \bar{\rho}_2\hat{v}_x \\ \vdots \\ \bar{\rho}_1\hat{e}_{v,1}\hat{v}_x \\ \bar{\rho}_2\hat{e}_{v,2}\hat{v}_x \\ \vdots \\ (\bar{\rho}\hat{e}_e + \bar{p}_e)\hat{v}_x \end{pmatrix}, \hat{\mathbf{F}}_C = \begin{pmatrix} \bar{\rho}\hat{v}_y \\ \bar{\rho}\hat{v}_y\hat{v}_x \\ \bar{\rho}\hat{v}_y^2 + \bar{p} \\ \bar{\rho}\hat{v}_y\hat{v}_z \\ (\bar{\rho}\hat{e} + \bar{p})\hat{v}_y \\ \bar{\rho}_1\hat{v}_y \\ \bar{\rho}_2\hat{v}_y \\ \vdots \\ \bar{\rho}_1\hat{e}_{v,1}\hat{v}_y \\ \bar{\rho}_2\hat{e}_{v,2}\hat{v}_y \\ \vdots \\ (\bar{\rho}\hat{e}_e + \bar{p}_e)\hat{v}_y \end{pmatrix}, \hat{\mathbf{G}}_C = \begin{pmatrix} \bar{\rho}\hat{v}_z \\ \bar{\rho}\hat{v}_z\hat{v}_x \\ \bar{\rho}\hat{v}_z\hat{v}_y \\ \bar{\rho}\hat{v}_z^2 + \bar{p} \\ (\bar{\rho}\hat{e} + \bar{p})\hat{v}_z \\ \bar{\rho}_1\hat{v}_z \\ \bar{\rho}_2\hat{v}_z \\ \vdots \\ \bar{\rho}_1\hat{e}_{v,1}\hat{v}_z \\ \bar{\rho}_2\hat{e}_{v,2}\hat{v}_z \\ \vdots \\ (\bar{\rho}\hat{e}_e + \bar{p}_e)\hat{v}_z \end{pmatrix},
$$

$$
\hat{\mathbf{E}}_{C,visc} = \begin{pmatrix} 0 \\ \hat{\tau}_{xx} \\ \hat{\tau}_{xy} \\ \hat{\tau}_{xz} \\ \hat{v}_x\hat{\tau}_{xx} + \hat{v}_y\hat{\tau}_{xy} + \hat{v}_z\hat{\tau}_{xz} + \hat{q}_x + \mu^*\frac{\partial k}{\partial x} \\ \bar{\rho}\check{D}_1\frac{\partial\hat{\omega}_1}{\partial x} \\ \bar{\rho}\check{D}_2\frac{\partial\hat{\omega}_2}{\partial x} \\ \vdots \\ \kappa_{v,1}\frac{\partial\hat{T}_{v,1}}{\partial x} + \bar{\rho}\hat{e}_{v,1}\check{D}_1\frac{\partial\hat{\omega}_1}{\partial x} \\ \kappa_{v,2}\frac{\partial\hat{T}_{v,2}}{\partial x} + \bar{\rho}\hat{e}_{v,2}\check{D}_2\frac{\partial\hat{\omega}_2}{\partial x} \\ \vdots \\ \kappa_e\frac{\partial\hat{T}_e}{\partial x} + \bar{\rho}\sum_{i=1}^{N}\hat{e}_{e,i}\check{D}_i\frac{\partial\hat{\omega}_i}{\partial x} \end{pmatrix},
$$

$$\hat{\mathbf{F}}_{C,visc} = \begin{pmatrix} 0 \\ \hat{\tau}_{yx} \\ \hat{\tau}_{yy} \\ \hat{\tau}_{yz} \\ \hat{v}_x\hat{\tau}_{yx} + \hat{v}_y\hat{\tau}_{yy} + \hat{v}_z\hat{\tau}_{yz} + \hat{q}_y + \mu^*\frac{\partial k}{\partial y} \\ \bar{\rho}\check{D}_1\frac{\partial\hat{\omega}_1}{\partial y} \\ \bar{\rho}\check{D}_2\frac{\partial\hat{\omega}_2}{\partial y} \\ \vdots \\ \kappa_{v,1}\frac{\partial\hat{T}_{v,1}}{\partial y} + \bar{\rho}\hat{e}_{v,1}\check{D}_1\frac{\partial\hat{\omega}_1}{\partial y} \\ \kappa_{v,2}\frac{\partial\hat{T}_{v,2}}{\partial y} + \bar{\rho}\hat{e}_{v,2}\check{D}_2\frac{\partial\hat{\omega}_2}{\partial y} \\ \vdots \\ \kappa_e\frac{\partial\hat{T}_e}{\partial y} + \bar{\rho}\sum_{i=1}^{N}\hat{e}_{e,i}\check{D}_i\frac{\partial\hat{\omega}_i}{\partial y} \end{pmatrix},$$

$$\hat{\mathbf{G}}_{C,visc} = \begin{pmatrix} 0 \\ \hat{\tau}_{zx} \\ \hat{\tau}_{zy} \\ \hat{\tau}_{zz} \\ \hat{v}_x\hat{\tau}_{zx} + \hat{v}_y\hat{\tau}_{zy} + \hat{v}_z\hat{\tau}_{zz} + \hat{q}_z + \mu^*\frac{\partial k}{\partial z} \\ \bar{\rho}\check{D}_1\frac{\partial\hat{\omega}_1}{\partial z} \\ \bar{\rho}\check{D}_2\frac{\partial\hat{\omega}_2}{\partial z} \\ \vdots \\ \kappa_{v,1}\frac{\partial\hat{T}_{v,1}}{\partial z} + \bar{\rho}\hat{e}_{v,1}\check{D}_1\frac{\partial\hat{\omega}_1}{\partial z} \\ \kappa_{v,2}\frac{\partial\hat{T}_{v,2}}{\partial z} + \bar{\rho}\hat{e}_{v,2}\check{D}_2\frac{\partial\hat{\omega}_2}{\partial z} \\ \vdots \\ \kappa_e\frac{\partial\hat{T}_e}{\partial z} + \bar{\rho}\sum_{i=1}^{N}\hat{e}_{e,i}\check{D}_i\frac{\partial\hat{\omega}_i}{\partial z} \end{pmatrix},$$

The quantities and relations regarding the thermodynamic state are (Eq. (5.19)):

- vibrational coupling term[4] of species j
  $Q_{v,j} = Q_{T-v,j} + Q_{e-v,j} + Q_{D_j} + Q_{v-v,j}$

[4] The definitions of the elements of the vibrational coupling terms can be found in Sect. 9.2.

- electron coupling term[5]
  $$Q_e = \hat{v}_k \frac{\partial \bar{p}_e}{\partial x_k} + Q_{T-e} + Q_{e-v} + Q_I + \hat{q}_{rad,g} ,$$
- vector of heat transfer
  $$\hat{\mathbf{q}} = -\check{\kappa}\,\mathrm{grad}\hat{T} - \sum_{j=1}^{N_{mol}} \kappa_{v,j}\,\mathrm{grad}\hat{T}_{v,j} - \kappa_e\,\mathrm{grad}\hat{T}_e - \sum_{i=1}^{N} \bar{\rho}\hat{h}_i \check{D}_i\,\mathrm{grad}\hat{\omega}_i ,$$
- total energy
  $$\hat{e} = \sum_{i=1}^{N} \frac{\bar{\rho}_i \hat{e}_{i,i}}{\bar{\rho}} + 0.5(\hat{v}_x^2 + \hat{v}_y^2 + \hat{v}_z^2) + k ,$$
- internal energy of species i in chemical equilibrium
  $$\hat{e}_{i,i} ,$$

with j being the number of molecular species and i the number of chemical species. Together with the equation for a turbulence model in formal vector notation, Eq. (8.5), the system is closed.

References

1. Klingbeil, E.: Tensorrechnung fuer Ingenieure. B.I. Wissenschaftsverlag, Mannheim (1966)
2. Teichmann, H.: Physikalische Anwendungen der Vektor- und Tensorrechnung. B.I. Wissenschaftsverlag, Mannheim (1973)
3. Vectors, T.: The Basic Equations of Fluid Mechanics. Rutherford Aris Prentice Hall, Englewood Cliffs, UK (1962)
4. Hirsch, C.: Numerical Computation of Internal and External Flows, vol. II. Wiley, New York, Reprint (1997)
5. Wilcox, D.C.: Turbulence Modelling for CFD. DCW Industries, La Canada CA, USA (2006)
6. Hirschel, E.H., Weiland, C.: Selected Aerothermodynamic Design Problems of Hypersonic Flight Vehicles. Springer, Berlin/Heidelberg, and vol. 229, Progress in Astronautics and Aeronautics, AIAA, Reston, Va. (2009)

[5] The definitions of the elements of the electron coupling terms can be found in Sect. 9.3.

Chapter 6
The Navier-Stokes Equations in Arbitrary Coordinates for Laminar and Turbulent Flow

Abstract In this chapter arbitrary coordinates are introduced. The laminar and turbulent Navier-Stokes equations are transformed to these coordinates in a way, that maintains the conservative properties of the set of equations. The presentation is restricted to perfect gas flow. The Navier-Stokes equations for laminar and turbulent flows including non-equilibrium thermodynamic states are given in Chap. 7.

The Variables and Quantities Used in this Chapter

x, y, z	Cartesian coordinates
ξ, η, ζ	arbitrary coordinates
J	Jacobian of coordinate transformation $x, y, z \Longleftrightarrow \xi, \eta, \zeta$
$\mathbf{E}, \mathbf{F}, \mathbf{G}$	vector: inviscid part
$\mathbf{E}_{visc}, \mathbf{F}_{visc}, \mathbf{G}_{visc}$	vector: viscous part
$\mathbf{Q}$	vector: conservative variables
$\hat{\mathbf{E}}, \hat{\mathbf{F}}, \hat{\mathbf{G}}$	vector: inviscid part of RANS equations
$\hat{\mathbf{E}}_{visc}, \hat{\mathbf{F}}_{visc}, \hat{\mathbf{G}}_{visc}$	vector: viscous part of RANS equations
$\hat{\mathbf{Q}}$	vector: conservative variables of RANS equations
e	total energy per unit mass
ρ	density
p	pressure
v_x, v_y, v_z	components of $\mathbf{v}$ in Cartesian coordinates
V_x, V_y, V_z	components of contravariant velocity vector
$\tau_{xx}, \tau_{yy}, \tau_{zz}$	elements of viscous stress tensor $\bar{\bar{\tau}}$ in Cartesian
$\tau_{xy}, \tau_{xz}, \tau_{yz}$	coordinates
μ	dynamic viscosity
μ_t	turbulent viscosity
$\check{\mu}$	$= \mu + \mu_t$ total viscosity
μ^*	$= \mu + \mu_t/\sigma_k$ abbreviation
σ_k	coefficient
$\beta_x, \beta_y, \beta_z$	abbreviations (laminar flow)
$\hat{\beta}_x, \hat{\beta}_y, \hat{\beta}_z$	abbreviations (turbulent flow)

C. Weiland, *The Hierarchy of Fluid Dynamic Equations*,
https://doi.org/10.1007/978-3-031-71198-5_6

k	turbulent kinetic energy
$\hat{e}_i$	internal energy in turbulent flow
$\hat{e}$	$= \hat{e}_i + \frac{1}{2}\|\hat{\mathbf{v}}\|^2 + k$ total energy in turbulent flow
$\bar{p}$	$= (\gamma - 1)\bar{\rho}(\hat{e} - \frac{1}{2}\|\hat{\mathbf{v}}\|^2 - k)$ pressure in turbulent flow, perfect gas

6.1 Arbitrary Coordinates

The description of flow around complex configurations like aircrafts or space planes requires arbitrary coordinate systems.[1] Usually the surface contour $S(x, y, z, t)$ of such a configuration is a constant coordinate surface. Let the set of arbitrary coordinates be chosen by ξ, η, ζ and the time (for unsteady flows or moving grids) be denoted by τ, then we have

$$\xi = \xi(x, y, z, t), \quad \eta = \eta(x, y, z, t), \quad \zeta = \zeta(x, y, z, t), \quad \tau = t.$$

In order to transform the Navier-Stokes equations Eqs. (5.1) or (5.17) to these arbitrary coordinates, the metric terms must be known. These terms can be evaluated by applying the chain rule and the appropriate laws of linear algebra, and one obtains

$$\begin{aligned}
\xi_x &= J(y_\eta z_\zeta - z_\eta y_\zeta), & \eta_x &= J(z_\xi y_\zeta - y_\xi z_\zeta), & \zeta_x &= J(y_\xi z_\eta - z_\xi y_\eta),\\
\xi_y &= J(z_\eta x_\zeta - x_\eta z_\zeta), & \eta_y &= J(x_\xi z_\zeta - z_\xi x_\zeta), & \zeta_y &= J(z_\xi x_\eta - x_\xi z_\eta),\\
\xi_z &= J(x_\eta y_\zeta - y_\eta x_\zeta), & \eta_z &= J(y_\xi x_\zeta - x_\xi y_\zeta), & \zeta_z &= J(x_\xi y_\eta - y_\xi x_\eta),
\end{aligned}$$

$$\begin{aligned}
\xi_t &= -x_\tau \xi_x - y_\tau \xi_y - z_\tau \xi_z,\\
\eta_t &= -x_\tau \eta_x - y_\tau \eta_y - z_\tau \eta_z,\\
\zeta_t &= -x_\tau \zeta_x - y_\tau \zeta_y - z_\tau \zeta_z,
\end{aligned} \tag{6.1}$$

where the Jacobian J reads

$$J = \left\{ \frac{\partial(\xi, \eta, \zeta)}{\partial(x, y, z)} \right\} = \left\{ \frac{\partial(x, y, z)}{\partial(\xi, \eta, \zeta)} \right\}^{-1}.$$

The determinant of the Jacobian is

$$J^{-1} = x_\xi y_\eta z_\zeta + x_\zeta y_\xi z_\eta + x_\eta y_\zeta z_\xi - x_\xi y_\zeta z_\eta - x_\eta y_\xi z_\zeta - x_\zeta y_\eta z_\xi. \tag{6.2}$$

[1] Arbitrary coordinates means non-orthogonal curvilinear coordinates.

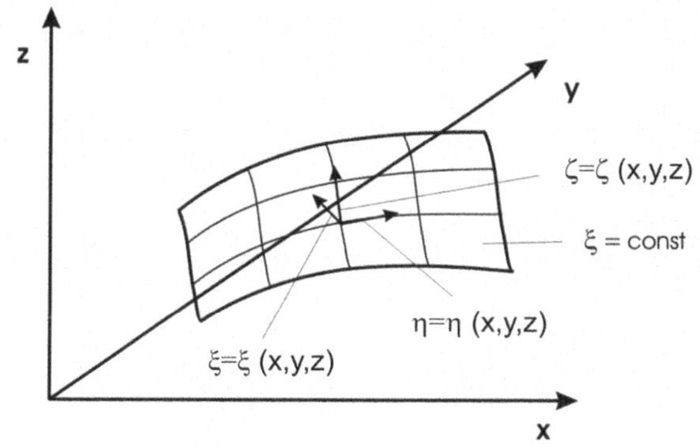

Fig. 6.1 Arbitrary steady coordinates ξ, η, ζ for geometrically fixed systems with an iso-surface ξ= const defining the surface of a configuration

In Fig. 6.1 the definition of the ξ, η, ζ coordinates is presented, where usually the η, ζ coordinates are located in the surface plane of the configuration with $S(x, y, z) = \xi(x, y, z) =$ const whereas ξ points outward. Some examples of surface planes, which are defined by $\xi(x, y, z) =$ const, are given in Fig. 6.2 [1–3].

So far, we have considered boundaries and coordinate systems, that are not time-dependent. In the case that the boundaries of a fluid flow regime are time-dependent, such as the contour of an aircraft or space plane, which undergoes changes due to structural forces [4], or the outer boundary of a supersonic or hypersonic flow field when it alignes with a bow shock (bow shock fitting approach [5, 6]), the arbitrary coordinates are time-dependent. Therefore, boundary surfaces like a vehicle contour are defined by $\xi(x, y, z, t) =$ const (Fig. 6.3).

6.2 The Navier-Stokes Equations for Laminar, Perfect Gas Flow

The systems of differential equations (5.1) and (5.17) are conservatively formulated, which means that by applying appropriate numerical approximations (conservative discretisation) these equations can describe flow fields containing discontinuities like shock waves, shear layers and contact discontinuities in contrast to the sets of differential equations in non-conservative formulation. For more details see Chap. 16. Normally, if coordinate transformations are carried out, the components of vector field functions (e.g. the velocity field) are transformed, too. In the case of the Navier-Stokes equations, this would dramatically increase the number of terms and a formulation with conservative properties is not attainable. Therefore, Viviand [7], and Vinokur [8], proposed a formulation in which the components of the vector field functions of the transformed equations are a combination of the basic and the transformed (contravariant) components. The astonishing result was that the trans-

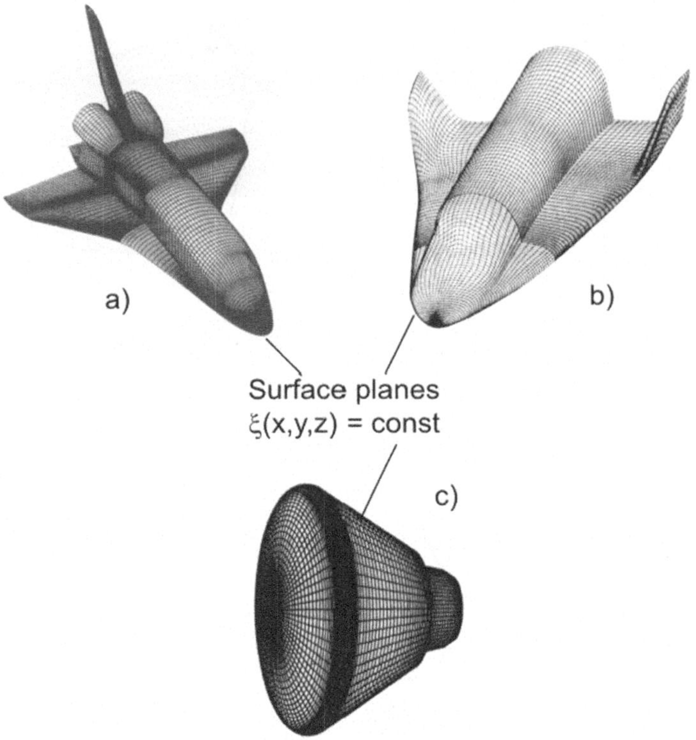

Fig. 6.2 Examples of surface planes $\xi(x, y, z)$ = const: **a** the Space Shuttle Orbiter [2], **b** the Hermes Space Plane [1] and **c** the Aerothermodynamic Reentry Demonstrator (ARD) [3]

formed set of equations differs not very much from the Cartesian one and features full conservative properties. This is confirmed by the Navier-Stokes equations shown below, where Eq. (5.1) is transformed to arbitrary coordinates. The external forces are dropped in this version.

$$J^{-1}\frac{\partial \mathbf{Q}}{\partial \tau}+\frac{\partial \mathbf{E}}{\partial \xi}+\frac{\partial \mathbf{F}}{\partial \eta}+\frac{\partial \mathbf{G}}{\partial \zeta}=\frac{\partial \mathbf{E}_{visc}}{\partial \xi}+\frac{\partial \mathbf{F}_{visc}}{\partial \eta}+\frac{\partial \mathbf{G}_{visc}}{\partial \zeta}, \tag{6.3}$$

$$\mathbf{Q}=\begin{pmatrix}\rho\\ \rho v_x\\ \rho v_y\\ \rho v_z\\ \rho e\end{pmatrix},\ \mathbf{E}=J^{-1}\begin{pmatrix}\rho V_x\\ \rho v_x V_x+\xi_x p\\ \rho v_y V_x+\xi_y p\\ \rho v_z V_x+\xi_z p\\ (\rho e+p)V_x-\xi_t p\end{pmatrix},$$

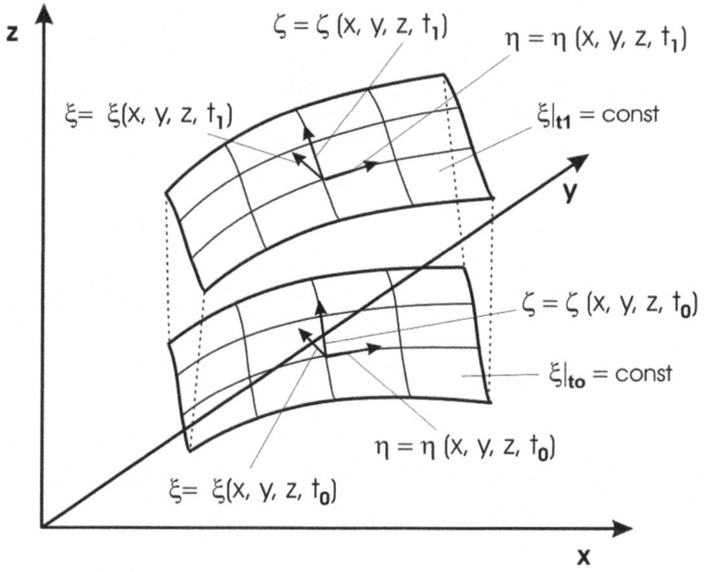

Fig. 6.3 Arbitrary time-dependent coordinates with iso-surfaces $\xi(x, y, z, t)$ = const defining the moving surface of the contour of a configuration or an outer boundary of a flow field at times t_0 and t_1

$$\mathbf{F} = J^{-1} \begin{pmatrix} \rho V_y \\ \rho v_x V_y + \eta_x p \\ \rho v_y V_y + \eta_y p \\ \rho v_z V_y + \eta_z p \\ (\rho e + p) V_y - \eta_t p \end{pmatrix}, \quad \mathbf{G} = J^{-1} \begin{pmatrix} \rho V_z \\ \rho v_x V_z + \zeta_x p \\ \rho v_y V_z + \zeta_y p \\ \rho v_z V_z + \zeta_z p \\ (\rho e + p) V_z - \zeta_t p \end{pmatrix},$$

$$\mathbf{E}_{visc} = J^{-1} \begin{pmatrix} 0 \\ \xi_x \tau_{xx} + \xi_y \tau_{xy} + \xi_z \tau_{xz} \\ \xi_x \tau_{yx} + \xi_y \tau_{yy} + \xi_z \tau_{yz} \\ \xi_x \tau_{zx} + \xi_y \tau_{zy} + \xi_z \tau_{zz} \\ \xi_x \beta_x + \xi_y \beta_y + \xi_z \beta_z \end{pmatrix},$$

$$\mathbf{F}_{visc} = J^{-1} \begin{pmatrix} 0 \\ \eta_x \tau_{xx} + \eta_y \tau_{xy} + \eta_z \tau_{xz} \\ \eta_x \tau_{yx} + \eta_y \tau_{yy} + \eta_z \tau_{yz} \\ \eta_x \tau_{zx} + \eta_y \tau_{zy} + \eta_z \tau_{zz} \\ \eta_x \beta_x + \eta_y \beta_y + \eta_z \beta_z \end{pmatrix},$$

$$\mathbf{G}_{visc} = J^{-1} \begin{pmatrix} 0 \\ \zeta_x \tau_{xx} + \zeta_y \tau_{xy} + \zeta_z \tau_{xz} \\ \zeta_x \tau_{yx} + \zeta_y \tau_{yy} + \zeta_z \tau_{yz} \\ \zeta_x \tau_{zx} + \zeta_y \tau_{zy} + \zeta_z \tau_{zz} \\ \zeta_x \beta_x + \zeta_y \beta_y + \zeta_z \beta_z \end{pmatrix},$$

with

$$\begin{aligned} V_x &= \xi_t + \xi_x v_x + \xi_y v_y + \xi_z v_z, \\ V_y &= \eta_t + \eta_x v_x + \eta_y v_y + \eta_z v_z, \\ V_z &= \zeta_t + \zeta_x v_x + \zeta_y v_y + \zeta_z v_z, \end{aligned} \tag{6.4}$$

$$\begin{aligned} \tau_{xx} &= 2\mu \left(\frac{\partial v_x}{\partial x} - \frac{1}{3} \left(\frac{\partial v_x}{\partial x} + \frac{\partial v_y}{\partial y} + \frac{\partial v_z}{\partial z} \right) \right), \\ \tau_{yy} &= 2\mu \left(\frac{\partial v_y}{\partial y} - \frac{1}{3} \left(\frac{\partial v_x}{\partial x} + \frac{\partial v_y}{\partial y} + \frac{\partial v_z}{\partial z} \right) \right), \\ \tau_{zz} &= 2\mu \left(\frac{\partial v_z}{\partial z} - \frac{1}{3} \left(\frac{\partial v_x}{\partial x} + \frac{\partial v_y}{\partial y} + \frac{\partial v_z}{\partial z} \right) \right), \end{aligned}$$

$$\begin{aligned} \tau_{xy} = \tau_{yx} &= \mu \left(\frac{\partial v_x}{\partial y} + \frac{\partial v_y}{\partial x} \right), \\ \tau_{xz} = \tau_{zx} &= \mu \left(\frac{\partial v_x}{\partial z} + \frac{\partial v_z}{\partial x} \right), \\ \tau_{yz} = \tau_{zy} &= \mu \left(\frac{\partial v_y}{\partial z} + \frac{\partial v_z}{\partial y} \right), \end{aligned} \tag{6.5}$$

$$\begin{aligned} \beta_x &= \kappa \frac{\partial T}{\partial x} + v_x \tau_{xx} + v_y \tau_{xy} + v_z \tau_{xz}, \\ \beta_y &= \kappa \frac{\partial T}{\partial y} + v_x \tau_{yx} + v_y \tau_{yy} + v_z \tau_{yz}, \\ \beta_z &= \kappa \frac{\partial T}{\partial z} + v_x \tau_{zx} + v_y \tau_{zy} + v_z \tau_{zz}. \end{aligned}$$

The Cartesian derivative, for example, of the velocity component v_x is given by the chain rule

$$\frac{\partial v_x}{\partial x} = \xi_x \frac{\partial v_x}{\partial \xi} + \eta_x \frac{\partial v_x}{\partial \eta} + \zeta_x \frac{\partial v_x}{\partial \zeta}, \tag{6.6}$$

which is valid for all the other variables and Cartesian coordinates.

Other possible forms of Eq. (6.4) are, for example,

$$V_x = \xi_x(u - x_\tau) + \xi_y(v - y_\tau) + \xi_z(w - z_\tau), \tag{6.7}$$

and similarly for V_y and V_z, where x_τ, y_τ, z_τ are the Cartesian components of the grid velocity. Considering Eq. (6.3) and applying Eq. (6.7), one obtains after some mathematical operations

$$\begin{aligned} \frac{\partial J^{-1}\mathbf{Q}}{\partial t} &+ J^{-1}\left(\frac{\partial(\mathbf{E} - x_\tau \mathbf{Q})}{\partial x} + \frac{\partial(\mathbf{F} - y_\tau \mathbf{Q})}{\partial y} + \frac{\partial(\mathbf{G} - z_\tau \mathbf{Q})}{\partial z}\right) \\ &= J^{-1}\left(\frac{\partial \mathbf{E}_{visc}}{\partial x} + \frac{\partial \mathbf{F}_{visc}}{\partial y} + \frac{\partial \mathbf{G}_{visc}}{\partial z}\right), \end{aligned} \tag{6.8}$$

which is identical to Eq. (6.3). The definitions of the vectors $\mathbf{E}$, $\mathbf{F}$, $\mathbf{G}$, $\mathbf{E}_{visc}$, $\mathbf{F}_{visc}$, $\mathbf{G}_{visc}$ are those of Eq. (5.1). Equation (6.8) is named the Arbitrary Lagrange-Euler formulation (ALE), which is often used for fluid-structure interaction problems [4]. It describes problems, where grid points move with velocities other than the velocities of the local fluid elements. Generally, if grid points move with the same speed as the fluid elements, the equation is called a Lagrange formulation, whereas if grid points are stationary while the fluid elements travel through the grid, we speak about an Euler formulation.

6.3 The Navier-Stokes Equations for Turbulent (RANS), Perfect Gas Flow

For completeness, we are also providing the non-orthogonal curvilinear Reynolds averaged Navier-Stokes equations for turbulent flows (see Eq. (5.17)).

$$J^{-1}\frac{\partial \hat{\mathbf{Q}}}{\partial \tau} + \frac{\partial \hat{\mathbf{E}}}{\partial \xi} + \frac{\partial \hat{\mathbf{F}}}{\partial \eta} + \frac{\partial \hat{\mathbf{G}}}{\partial \zeta} = \frac{\partial \hat{\mathbf{E}}_{visc}}{\partial \xi} + \frac{\partial \hat{\mathbf{F}}_{visc}}{\partial \eta} + \frac{\partial \hat{\mathbf{G}}_{visc}}{\partial \zeta}, \tag{6.9}$$

$$\hat{\mathbf{Q}} = \begin{pmatrix} \bar{\rho} \\ \bar{\rho}\hat{v}_x \\ \bar{\rho}\hat{v}_y \\ \bar{\rho}\hat{v}_z \\ \bar{\rho}\hat{e} \end{pmatrix}, \quad \hat{\mathbf{E}} = J^{-1}\begin{pmatrix} \bar{\rho}\hat{V}_x \\ \bar{\rho}\hat{v}_x\hat{V}_x + \xi_x\bar{p} \\ \bar{\rho}\hat{v}_y\hat{V}_x + \xi_y\bar{p} \\ \bar{\rho}\hat{v}_z\hat{V}_x + \xi_z\bar{p} \\ (\bar{\rho}\hat{e} + \bar{p})\hat{V}_x - \xi_t\bar{p} \end{pmatrix},$$

$$\hat{\mathbf{F}} = J^{-1} \begin{pmatrix} \bar{\rho}\hat{V}_y \\ \bar{\rho}\hat{v}_x\hat{V}_y + \eta_x\bar{p} \\ \bar{\rho}\hat{v}_y\hat{V}_y + \eta_y\bar{p} \\ \bar{\rho}\hat{v}_z\hat{V}_y + \eta_z\bar{p} \\ (\bar{\rho}\hat{e} + \bar{p})\hat{V}_y - \eta_t\bar{p} \end{pmatrix}, \quad \hat{\mathbf{G}} = J^{-1} \begin{pmatrix} \bar{\rho}\hat{V}_z \\ \bar{\rho}\hat{v}_x\hat{V}_z + \zeta_x\bar{p} \\ \bar{\rho}\hat{v}_y\hat{V}_z + \zeta_y\bar{p} \\ \bar{\rho}\hat{v}_z\hat{V}_z + \zeta_z\bar{p} \\ (\bar{\rho}\hat{e} + \bar{p})\hat{V}_z - \zeta_t\bar{p} \end{pmatrix},$$

$$\hat{\mathbf{E}}_{visc} = J^{-1} \begin{pmatrix} 0 \\ \xi_x\hat{\tau}_{xx} + \xi_y\hat{\tau}_{xy} + \xi_z\hat{\tau}_{xz} \\ \xi_x\hat{\tau}_{yx} + \xi_y\hat{\tau}_{yy} + \xi_z\hat{\tau}_{yz} \\ \xi_x\hat{\tau}_{zx} + \xi_y\hat{\tau}_{zy} + \xi_z\hat{\tau}_{zz} \\ \xi_x\hat{\beta}_x + \xi_y\hat{\beta}_y + \xi_z\hat{\beta}_z \end{pmatrix},$$

$$\hat{\mathbf{F}}_{visc} = J^{-1} \begin{pmatrix} 0 \\ \eta_x\hat{\tau}_{xx} + \eta_y\hat{\tau}_{xy} + \eta_z\hat{\tau}_{xz} \\ \eta_x\hat{\tau}_{yx} + \eta_y\hat{\tau}_{yy} + \eta_z\hat{\tau}_{yz} \\ \eta_x\hat{\tau}_{zx} + \eta_y\hat{\tau}_{zy} + \eta_z\hat{\tau}_{zz} \\ \eta_x\hat{\beta}_x + \eta_y\hat{\beta}_y + \eta_z\hat{\beta}_z \end{pmatrix},$$

$$\hat{\mathbf{G}}_{visc} = J^{-1} \begin{pmatrix} 0 \\ \zeta_x\hat{\tau}_{xx} + \zeta_y\hat{\tau}_{xy} + \zeta_z\hat{\tau}_{xz} \\ \zeta_x\hat{\tau}_{yx} + \zeta_y\hat{\tau}_{yy} + \zeta_z\hat{\tau}_{yz} \\ \zeta_x\hat{\tau}_{zx} + \zeta_y\hat{\tau}_{zy} + \zeta_z\hat{\tau}_{zz} \\ \zeta_x\hat{\beta}_x + \zeta_y\hat{\beta}_y + \zeta_z\hat{\beta}_z \end{pmatrix},$$

with

$$\begin{aligned} \hat{V}_x &= \xi_t + \xi_x\hat{v}_x + \xi_y\hat{v}_y + \xi_z\hat{v}_z\,, \\ \hat{V}_y &= \eta_t + \eta_x\hat{v}_x + \eta_y\hat{v}_y + \eta_z\hat{v}_z\,, \\ \hat{V}_z &= \zeta_t + \zeta_x\hat{v}_x + \zeta_y\hat{v}_y + \zeta_z\hat{v}_z, \end{aligned}$$

$$\begin{aligned} \hat{\tau}_{xx} &= 2\breve{\mu}\left(\frac{\partial\hat{v}_x}{\partial x} - \frac{1}{3}\left(\frac{\partial\hat{v}_x}{\partial x} + \frac{\partial\hat{v}_y}{\partial y} + \frac{\partial\hat{v}_z}{\partial z}\right)\right) - \frac{2}{3}\bar{\rho}k, \\ \hat{\tau}_{yy} &= 2\breve{\mu}\left(\frac{\partial\hat{v}_y}{\partial y} - \frac{1}{3}\left(\frac{\partial\hat{v}_x}{\partial x} + \frac{\partial\hat{v}_y}{\partial y} + \frac{\partial\hat{v}_z}{\partial z}\right)\right) - \frac{2}{3}\bar{\rho}k, \\ \hat{\tau}_{zz} &= 2\breve{\mu}\left(\frac{\partial\hat{v}_z}{\partial z} - \frac{1}{3}\left(\frac{\partial\hat{v}_x}{\partial x} + \frac{\partial\hat{v}_y}{\partial y} + \frac{\partial\hat{v}_z}{\partial z}\right)\right) - \frac{2}{3}\bar{\rho}k, \end{aligned}$$

$$\hat{\tau}_{xy} = \hat{\tau}_{yx} = \check{\mu}\left(\frac{\partial \hat{v}_x}{\partial y} + \frac{\partial \hat{v}_y}{\partial x}\right),$$

$$\hat{\tau}_{xz} = \hat{\tau}_{zx} = \check{\mu}\left(\frac{\partial \hat{v}_x}{\partial z} + \frac{\partial \hat{v}_z}{\partial x}\right),$$

$$\hat{\tau}_{yz} = \hat{\tau}_{zy} = \check{\mu}\left(\frac{\partial \hat{v}_y}{\partial z} + \frac{\partial \hat{v}_z}{\partial y}\right),$$

$$\hat{\beta}_x = \check{\kappa}\frac{\partial \hat{T}}{\partial x} + \hat{v}_x\hat{\tau}_{xx} + \hat{v}_y\hat{\tau}_{xy} + \hat{v}_z\hat{\tau}_{xz} + \mu^*\frac{\partial k}{\partial x},$$

$$\hat{\beta}_y = \check{\kappa}\frac{\partial \hat{T}}{\partial y} + \hat{v}_x\hat{\tau}_{yx} + \hat{v}_y\hat{\tau}_{yy} + \hat{v}_z\hat{\tau}_{yz} + \mu^*\frac{\partial k}{\partial y},$$

$$\hat{\beta}_z = \check{\kappa}\frac{\partial \hat{T}}{\partial z} + \hat{v}_x\hat{\tau}_{zx} + \hat{v}_y\hat{\tau}_{zy} + \hat{v}_z\hat{\tau}_{zz} + \mu^*\frac{\partial k}{\partial z},$$

and with the Cartesian derivatives

$$\frac{\partial \hat{v}_x}{\partial x} = \xi_x\frac{\partial \hat{v}_x}{\partial \xi} + \eta_x\frac{\partial \hat{v}_x}{\partial \eta} + \zeta_x\frac{\partial \hat{v}_x}{\partial \zeta} \qquad \text{etc.}. \tag{6.10}$$

The total energy $\hat{e}$, the pressure $\bar{p}$ and the turbulent kinetic energy k are

$$\hat{e} = \hat{e}_i + \frac{1}{2}(\hat{v}_x^2 + \hat{v}_y^2 + \hat{v}_z^2) + k\,,$$

$$\bar{p} = (\gamma - 1)\bar{\rho}(\hat{e} - \frac{1}{2}(\hat{v}_x^2 + \hat{v}_y^2 + \hat{v}_z^2) - k)\,,$$

$$k = \frac{1}{2}((\widehat{v_x''})^2 + (\widehat{v_y''})^2 + (\widehat{v_z''})^2)\,.$$

Note: The "bar" means Reynolds-averaged ($\bar{\varphi}$) and the "hat" Favre-averaged ($\hat{\varphi}$), see Sect. 5.2.

References

1. Menne, S., Weiland, C., Pfitzner, M.: Computation of 3-D hypersonic flows in chemical non-equilibrium including transport phenomena. AIAA Paper 92-2876, 27th Thermophysics Conference, Nashville USA (1992)
2. Hartmann, G., Menne, S., Schröder, W.: Gittergenerierung und Strömungsfelduntersuchungen für komplexe Wiedereintrittskörper. DGLR Jahrestagung, Erlangen Germany (1994)
3. Haidinger, F.A.: Numerical investigation of the heat transfer on the front shield of a capsule. DGLR Jahrestagung, Paper DGLR-JT 96-0137 (1996)

4. Hirschel, E.H., Weiland, C.: Selected Aerothermodynamic Design Problems of Hypersonic Flight Vehicles. Springer Verlag, Berlin/Heidelberg and AIAA (Progress in Astronautics and Aeronautics), **229**, Reston, USA (2009)
5. Pfitzner, M., Weiland, C.: 3-D Euler solutions for hypersonic freestream Mach numbers. AGARD-CP-428, Paper 22, Bristol, UK (1987)
6. Weiland, C., Pfitzner, M.: 3-D and 2-D solutions of the quasi-conservative Euler equations. In: Notes in Physics, vol. 264. Springer, Berlin (1986)
7. Viviand, H.: Conservative forms of gas dynamic equations. In: La Recherche Aerospatiale, No. 1, pp. 65–68 (1974)
8. Vinokur, M.J.: Conservative equations of gas dynamics in curvilinear coordinate systems. J. Comp. Phys. **14**, 105–125 (1974)

Chapter 7
Dimensionless Navier-Stokes Equations

Abstract The solution of any system of governing equations either Potential, Euler or Navier-Stokes equations, either for laminar or turbulent flow or either with perfect gas, equilibrium real gas or any kind of thermodynamic non-equilibrium gas assumption, is performed today in an approximative way by the application of numerical methods. This usually requires a dimensionless variant of the system of equations. There are different ways for making the variables dimensionless (see for example [1]). The non-dimensionality procedure is demonstrated for the RANS equations, with the equation for the turbulence model and the equation for a gas in chemical and thermal non-equilibrium.

The Variables and Quantities Used in this Chapter

$\mathbf{E}, \mathbf{F}, \mathbf{G}$	vector: inviscid part of RANS equations
$\mathbf{E}_{visc}, \mathbf{F}_{visc}, \mathbf{G}_{visc}$	vector: viscous part of RANS equations
$\mathbf{Q}$	vector: conservative variables of RANS equations
$\mathbf{E}_T, \mathbf{F}_T, \mathbf{G}_T$	vector: "inviscid part" of turbulence model
$\mathbf{E}_{T,visc}, \mathbf{F}_{T,visc}, \mathbf{G}_{T,visc}$	vector: "viscous part" of turbulence model
$\mathbf{Q}_T$	vector: variables of turbulence model
$\mathbf{S}_T$	vector: source terms of turbulence model
$\mathbf{E}_S, \ldots$	vector: "inviscid part" of species conservation
$\mathbf{E}_{S,visc}, \ldots$	vector: "viscous part" of species conservation
$\mathbf{Q}_S$	vector: species densities
$\mathbf{S}_S$	vector: source terms of species conservation
$\mathbf{E}_{vib}, \ldots$	vector: "inviscid part" of vibrational energy
$\mathbf{E}_{vib,visc}, \ldots$	vector: "viscous part" of vibrational energy
$\mathbf{Q}_{vib}$	vector: vibrational energies
$\mathbf{S}_{vib}$	vector: source terms of vibrational energies
e	total energy per unit mass (RANS equations)
ρ	density (RANS equations)
p	pressure (RANS equations)
T	temperature (RANS equations)

C. Weiland, *The Hierarchy of Fluid Dynamic Equations*,
https://doi.org/10.1007/978-3-031-71198-5_7

v_x, v_y, v_z	components of $\mathbf{v}$ in Cartesian coordinates
$\tau_{xx}, \tau_{yy}, \tau_{zz}$	elements of viscous stress tensor $\bar{\bar{\tau}}$ in Cartesian
$\tau_{xy}, \tau_{xz}, \tau_{yz}$	coordinates (RANS equations)
τ_{ij}	components of $\bar{\bar{\tau}}$ in tensor notation
μ	dynamic viscosity
μ_t	turbulent viscosity
$\breve{\mu}$	$= \mu + \mu_t$ total viscosity
μ^*	$= \mu + \mu_t/\sigma_k$ abbreviation
σ_k	coefficient, constant of turbulence model
κ	thermal conductivity
$\breve{\kappa}$	$= \kappa(1 + Pr\mu_t/Pr_t\mu)$ total thermal conductivity
D_i	multicomponent mass diffusion coefficient of species i
$\bar{D}$	reference value of mass diffusion coefficient
$\breve{D}_i$	$= D_i + \frac{1}{Sc_t}\frac{\mu_t}{\rho}$ total mass diffusion coefficient for turbulent flow
h_i	enthalpy of species i
α_i	source term of species conservation
ω_i	mass fraction of species i
ω	specific dissipation rate
k	turbulent kinetic energy
Re	Reynolds number
Pr	Prandtl number
Pr_t	turbulent Prandtl number
Sc	Schmidt number
Sc_t	turbulent Schmidt number
$\beta, \beta*$	constants of turbulence model
C_k	$= \frac{\kappa_\infty T_\infty}{\mu_\infty v_\infty^2}$ abbreviation
$e_{v,j}$	vibrational energy of molecular species j
$Q_{T-v,j}$	energy exchange between transitional and vibrational modes of species j
$Q_{e-v,j}$	energy exchange between electron and vibrational modes of species j
Q_{Dj}	energy removed at dissociation or gained at recombination of species j
$Q_{v-v,j}$	energy exchange between different vibrational modes

Definition of a set of non-dimensional variables:[1]

$$
\begin{aligned}
&x'_k = \frac{x_k}{\bar{l}}, \quad t' = \frac{\bar{v}}{\bar{l}}t, \quad v'_k = \frac{v_k}{\bar{v}}, \quad \rho' = \frac{\rho}{\bar{\rho}}, \quad \rho'_i = \frac{\rho_i}{\bar{\rho}}, \\
&p' = \frac{p}{\bar{\rho}\bar{v}^2}, \quad T' = \frac{T}{\bar{T}}, \quad e' = \frac{e}{\bar{v}^2}, \quad h'_i = \frac{h_i}{\bar{v}^2}, \quad \kappa' = \frac{\kappa}{\bar{\kappa}}, \\
&\mu' = \frac{\mu}{\bar{\mu}}, \quad k' = \frac{k}{\bar{v}^2}, \quad \omega' = \frac{\omega\bar{\mu}}{\bar{\rho}\bar{v}^2}, \quad D'_i = \frac{D_i}{\bar{D}},
\end{aligned} \tag{7.1}
$$

The clean variables have dimensions, the ones with a prime are dimensionless and the ones with a bar are the reference values; k components of coordinates, i number of species .

7.1 RANS Equations in Cartesian Coordinates for Chemical Non-Equilibrium Flow

Numerical approximation methods of flow fields with real gas effects, for example those at hypersonic flight of space vehicles or flow in rocket motors (combustion chamber and nozzle extension flows), often require the dimensionless turbulent Navier-Stokes equations formulated for flow in chemical non-equilibrium. Of course, a space vehicle, which re-enters from an Earth orbit or a moon mission, experiences due to the high Mach numbers and the low densities during the first part of the re-entry trajectory, that the air passing around its configuration is in a chemical non-equilibrium state. On the other hand the low density of the air provides that the flow essentially behaves laminar. Nevertheless, turbulent flow fields are the more general ones. We therefore firstly present in this section the dimensionless form of the RANS equations, the equation of the turbulence model and the species continuity equations for chemical non-equilibrium flows in Cartesian coordinates (see Eqs. (5.17), (5.18), (8.5), (9.3)).

The essential mathematical manipulation is shown by performing the procedure on the energy equation with only the derivatives of the x-coordinate.

From Eqs. (5.17), (5.18) we extract

$$
\begin{aligned}
\frac{\partial \rho e}{\partial t} + \frac{\partial(\rho e + p)v_x}{\partial x} = \frac{\partial}{\partial x}\Bigg(& v_x\tau_{xx} + v_y\tau_{xy} + v_z\tau_{xz} + \check{\kappa}\frac{\partial T}{\partial x} + \\
&+ \sum_i \rho D_i h_i \frac{\partial \omega_i}{\partial x} + \mu^*\frac{\partial k}{\partial x}\Bigg),
\end{aligned} \tag{7.2}
$$

[1] For convenience in all the equations considered, the “bars” and “hats”, denoting the averaging process, on top of the variables are dropped.

where we apply the relations Eq. (7.1) and use the definition of the multicomponent mass diffusion component for turbulent flow $\check{D}_i$, Eq. (5.16). One obtains

$$\left\{\frac{\bar{\rho}\bar{v}^3}{\bar{l}}\right\}\frac{\partial \rho' e'}{\partial t'} + \left\{\frac{\bar{\rho}\bar{v}^3}{\bar{l}}\right\}\frac{\partial(\rho' e' + p')v'_x}{\partial x'} =$$
$$\frac{1}{\bar{l}}\frac{\partial}{\partial x'}\left[\left\{\frac{\bar{\mu}\bar{v}^2}{\bar{l}}\right\}(v'_x\tau'_{xx} + v'_y\tau'_{xy} + v'_z\tau'_{xz}) + \left\{\bar{\kappa}\frac{\bar{T}}{\bar{l}}\right\}\kappa'\left(1 + \frac{Pr\,\mu'_t\bar{\mu}}{Pr_t\,\mu'\bar{\mu}}\right)\frac{\partial T'}{\partial x'} + \right.$$
$$\left.\left\{\frac{\bar{\rho}\bar{v}^2}{\bar{l}}\right\}\sum_i \rho' h'_i \bar{D}\left(D'_i + \frac{1}{Sc_t}\frac{\mu'_t\bar{\mu}}{\bar{D}\rho'\bar{\rho}}\right)\frac{\partial \omega_i}{\partial x'} + \left\{\frac{\bar{\mu}\bar{v}^2}{\bar{l}}\right\}\mu'^*\frac{\partial k'}{\partial x'}\right]. \tag{7.3}$$

This results in

$$\frac{\partial \rho' e'}{\partial t'} + \frac{\partial(\rho' e' + p')v'_x}{\partial x'} = \frac{1}{Re}\left[v'_x\tau'_{xx} + v'_y\tau'_{xy} + v'_z\tau'_{xz} + \right.$$
$$\left. + \frac{\bar{\kappa}\bar{T}}{\bar{\mu}\bar{v}^2}\kappa'\left(1 + \frac{Pr\,\mu'_t}{Pr_t\,\mu'}\right)\frac{\partial T'}{\partial x'} + \frac{1}{Sc}\sum_i \rho' h'_i\left(D'_i + \frac{Sc}{Sc_t}\frac{\mu'_t}{\rho'}\right)\frac{\partial \omega_i}{\partial x'} + \mu'^*\frac{\partial k'}{\partial x'}\right], \tag{7.4}$$

where the elements of the turbulence stress tensor are (see Eq. (5.14))

$$\tau'_{ij} = \check{\mu}'\left(\frac{\partial v'_i}{\partial x_j} + \frac{\partial v'_j}{\partial x_i} - \frac{2}{3}\frac{\partial v'_k}{\partial x_k}\delta_{ij}\right) - \frac{2}{3}Re\,\rho' k'\delta_{ij}$$

with $\check{\mu}' = \mu' + \mu'_t$ and $Sc = \bar{\mu}/\bar{D}\bar{\rho}$ being the material based Schmidt number, $Pr = \bar{\mu}\bar{c}_p/\bar{\kappa}$ the material based Prandtl number as well as Sc_t the turbulent Schmidt number and Pr_t the turbulent Prandtl number.

In the same sense as above we obtain for the dimensionless species continuity equations (for chemical non-equilibrium flow) Eq. (9.4)

$$\frac{\partial \rho'_i}{\partial t'} + \mathrm{div}(\rho'_i v') - \underbrace{\frac{1}{Re\,Sc}}_{\frac{\bar{D}}{\bar{v}\bar{l}}}\mathrm{div}\left(\rho'\left(D'_i + \frac{Sc}{Sc_t}\frac{\mu'_t}{\rho'}\right)\mathrm{grad}\omega_i\right) = \frac{\bar{l}}{\bar{\rho}\bar{v}}\alpha_i\,, \tag{7.5}$$

and the dimensionless form of the two equation $k-\omega$ turbulence model, Eq. (8.4)

$$\frac{\partial \rho' k'}{\partial t'} + \frac{(\rho' v'_j k')}{\partial x'_j} = \frac{1}{Re}\tau'^{turb}_{ij}\frac{\partial v'_i}{\partial x'_j} - Re\,\beta^*\rho'\omega' k' + \frac{1}{Re}\frac{\partial}{\partial x'_j}\left[\left(\mu' + \frac{\mu'_t}{\sigma_k}\right)\frac{\partial k'}{\partial x'_j}\right],$$
$$\frac{\partial \rho' \omega'}{\partial t'} + \frac{(\rho' v'_j \omega')}{\partial x'_j} = \frac{1}{Re}\alpha\frac{\omega'}{k'}\tau'^{turb}_{ij}\frac{\partial v'_i}{\partial x'_j} - Re\,\beta\rho'\omega'^2 + \frac{1}{Re}\frac{\partial}{\partial x'_j}\left[\left(\mu' + \frac{\mu'_t}{\sigma_\omega}\right)\frac{\partial \omega'}{\partial x'_j}\right]. \tag{7.6}$$

Below the dimensionless RANS equations in formal vector notation for flows with gas in chemical non-equilibrium based on Eqs. (5.17), (5.18), (6.9) are presented in total.[2]

$$\frac{\partial \mathbf{Q}}{\partial t} + \frac{\partial \mathbf{E}}{\partial x} + \frac{\partial \mathbf{F}}{\partial y} + \frac{\partial \mathbf{G}}{\partial z} = \frac{1}{Re}\left(\frac{\partial \mathbf{E}_{visc}}{\partial x} + \frac{\partial \mathbf{F}_{visc}}{\partial y} + \frac{\partial \mathbf{G}_{visc}}{\partial z}\right), \tag{7.7}$$

with

$$\mathbf{Q} = \begin{pmatrix} \rho \\ \rho v_x \\ \rho v_y \\ \rho v_z \\ \rho e \end{pmatrix},$$

$$\mathbf{E} = \begin{pmatrix} \rho v_x \\ \rho v_x^2 + p \\ \rho v_x v_y \\ \rho v_x v_z \\ (\rho e + p) v_x \end{pmatrix}, \quad \mathbf{F} = \begin{pmatrix} \rho v_y \\ \rho v_y v_x \\ \rho v_y^2 + p \\ \rho v_y v_z \\ (\rho e + p) v_y \end{pmatrix}, \quad \mathbf{G} = \begin{pmatrix} \rho v_z \\ \rho v_z v_x \\ \rho v_z v_y \\ \rho v_z^2 + p \\ (\rho e + p) v_z \end{pmatrix},$$

$$\mathbf{E}_{visc} = \begin{pmatrix} 0 \\ \tau_{xx} \\ \tau_{xy} \\ \tau_{xz} \\ v_x \tau_{xx} + v_y \tau_{xy} + v_z \tau_{xz} + C\kappa \left(1 + \frac{Pr}{Pr_t}\frac{\mu_t}{\mu}\right)\frac{\partial T}{\partial x} + \cdots \\ \cdots + \frac{1}{Sc}\sum_i \rho h_i \left(D_i + \frac{Sc}{Sc_t}\frac{\mu_t}{\rho}\right)\frac{\partial \omega_i}{\partial x} + (\mu + \mu_t/\sigma_k)\frac{\partial k}{\partial x} \end{pmatrix},$$

$$\mathbf{F}_{visc} = \begin{pmatrix} 0 \\ \tau_{yx} \\ \tau_{yy} \\ \tau_{yz} \\ v_x \tau_{yx} + v_y \tau_{yy} + v_z \tau_{yz} + C\kappa \left(1 + \frac{Pr}{Pr_t}\frac{\mu_t}{\mu}\right)\frac{\partial T}{\partial y} + \cdots \\ \cdots + \frac{1}{Sc}\sum_i \rho h_i \left(D_i + \frac{Sc}{Sc_t}\frac{\mu_t}{\rho}\right)\frac{\partial \omega_i}{\partial y} + (\mu + \mu_t/\sigma_k)\frac{\partial k}{\partial y} \end{pmatrix},$$

[2] All "checks", "bars", "hats" and "primes" are dropped.

$$\mathbf{G}_{visc} = \begin{pmatrix} 0 \\ \tau_{zx} \\ \tau_{zy} \\ \tau_{zz} \\ v_x\tau_{zx} + v_y\tau_{zy} + v_z\tau_{zz} + C\kappa\left(1 + \frac{Pr\,\mu_t}{Pr_t\,\mu}\right)\frac{\partial T}{\partial z} + \cdots \\ \cdots + \frac{1}{Sc}\sum_i \rho h_i\left(D_i + \frac{Sc\,\mu_t}{Sc_t\,\rho}\right)\frac{\partial \omega_i}{\partial z} + (\mu + \mu_t/\sigma_k)\frac{\partial k}{\partial z} \end{pmatrix},$$

where $C_K = \frac{\bar{\kappa}\bar{T}}{\bar{\mu}\bar{v}^2} \equiv \frac{\kappa_\infty T_\infty}{\mu_\infty v_\infty^2}$. The system of equations is completed with Eqs. (7.5), (7.6).

7.2 RANS Equations in Arbitrary Coordinates for Chemical Non-Equilibrium Flow

Practical flow problems require the transformation of the set of governing equations into arbitrary coordinates. Therefore, the RANS equations, the equation for the turbulence model and the equations for the species continuity for chemical non-equilibrium flow, given in the above section, are written for arbitrary coordinates taking into consideration the Eqs. (5.18), (6.9),(7.7), (8.6), (9.14).

For laminar flow the dimensionless form of the vibrational energy equations for flow in thermal non-equilibrium in arbitrary coordinates is also added, Eq. (9.6).

$$J^{-1}\frac{\partial \mathbf{Q}}{\partial \tau} + \frac{\partial \mathbf{E}}{\partial \xi} + \frac{\partial \mathbf{F}}{\partial \eta} + \frac{\partial \mathbf{G}}{\partial \zeta} = \frac{1}{Re}\left(\frac{\partial \mathbf{E}_{visc}}{\partial \xi} + \frac{\partial \mathbf{F}_{visc}}{\partial \eta} + \frac{\partial \mathbf{G}_{visc}}{\partial \zeta}\right), \tag{7.8}$$

$$\mathbf{Q} = \begin{pmatrix} \rho \\ \rho v_x \\ \rho v_y \\ \rho v_z \\ \rho e \end{pmatrix}, \quad \mathbf{E} = J^{-1}\begin{pmatrix} \rho V_x \\ \rho v_x V_x + \xi_x p \\ \rho v_y V_x + \xi_y p \\ \rho v_z V_x + \xi_z p \\ (\rho e + p)V_x - \xi_t p \end{pmatrix},$$

$$\mathbf{F} = J^{-1}\begin{pmatrix} \rho V_y \\ \rho v_x V_y + \eta_x p \\ \rho v_y V_y + \eta_y p \\ \rho v_z V_y + \eta_z p \\ (\rho e + p)V_y - \eta_t p \end{pmatrix}, \quad \mathbf{G} = J^{-1}\begin{pmatrix} \rho V_z \\ \rho v_x V_z + \zeta_x p \\ \rho v_y V_z + \zeta_y p \\ \rho v_z V_z + \zeta_z p \\ (\rho e + p)V_x - \zeta_t p \end{pmatrix},$$

$$
\mathbf{E}_{visc} = J^{-1} \begin{pmatrix} 0 \\ \xi_x \tau_{xx} + \xi_y \tau_{xy} + \xi_z \tau_{xz} \\ \xi_x \tau_{yx} + \xi_y \tau_{yy} + \xi_z \tau_{yz} \\ \xi_x \tau_{zx} + \xi_y \tau_{zy} + \xi_z \tau_{zz} \\ v_x(\xi_x \tau_{xx} + \xi_y \tau_{xy} + \xi_z \tau_{xz}) + v_y(\xi_x \tau_{yx} + \xi_y \tau_{yy} + \xi_z \tau_{yz}) + \cdots \\ \cdots v_z(\xi_x \tau_{zx}) + \xi_y \tau_{zy} \xi_z \tau_{zz}) + \cdots \\ \cdots C\kappa \left(1 + \frac{Pr\,\mu_t}{Pr_t\,\mu}\right)\left(\xi_x \frac{\partial T}{\partial x} + \xi_y \frac{\partial T}{\partial y} + \xi_z \frac{\partial T}{\partial z}\right) + \cdots \\ \cdots \frac{1}{Sc} \sum_i \rho h_i \left(D_i + \frac{Sc}{Sc_t}\frac{\mu_t}{\rho}\right)\left(\xi_x \frac{\partial \omega_i}{\partial x} + \xi_y \frac{\partial \omega_i}{\partial y} + \xi_z \frac{\partial \omega_i}{\partial z}\right) + \cdots \\ \cdots (\mu + \mu_t/\sigma_k)\left(\xi_x \frac{\partial k}{\partial x} + \xi_y \frac{\partial k}{\partial y} + \xi_z \frac{\partial k}{\partial z}\right) \end{pmatrix},
$$

$$
\mathbf{F}_{visc} = J^{-1} \begin{pmatrix} 0 \\ \eta_x \tau_{xx} + \eta_y \tau_{xy} + \eta_z \tau_{xz} \\ \eta_x \tau_{yx} + \eta_y \tau_{yy} + \eta_z \tau_{yz} \\ \eta_x \tau_{zx} + \eta_y \tau_{zy} + \eta_z \tau_{zz} \\ v_x(\eta_x \tau_{xx} + \eta_y \tau_{xy} + \eta_z \tau_{xz}) + v_y(\eta_x \tau_{yx} + \eta_y \tau_{yy} + \eta_z \tau_{yz}) + \cdots \\ \cdots v_z(\eta_x \tau_{zx}) + \eta_y \tau_{zy} \eta_z \tau_{zz}) + \cdots \\ \cdots C\kappa \left(1 + \frac{Pr\,\mu_t}{Pr_t\,\mu}\right)\left(\eta_x \frac{\partial T}{\partial x} + \eta_y \frac{\partial T}{\partial y} + \eta_z \frac{\partial T}{\partial z}\right) + \cdots \\ \cdots \frac{1}{Sc} \sum_i \rho h_i \left(D_i + \frac{Sc}{Sc_t}\frac{\mu_t}{\rho}\right)\left(\eta_x \frac{\partial \omega_i}{\partial x} + \eta_y \frac{\partial \omega_i}{\partial y} + \eta_z \frac{\partial \omega_i}{\partial z}\right) + \cdots \\ \cdots (\mu + \mu_t/\sigma_k)\left(\eta_x \frac{\partial k}{\partial x} + \eta_y \frac{\partial k}{\partial y} + \eta_z \frac{\partial k}{\partial z}\right) \end{pmatrix},
$$

$$
\mathbf{G}_{visc} = J^{-1} \begin{pmatrix} 0 \\ \zeta_x \tau_{xx} + \zeta_y \tau_{xy} + \zeta_z \tau_{xz} \\ \zeta_x \tau_{yx} + \zeta_y \tau_{yy} + \zeta_z \tau_{yz} \\ \zeta_x \tau_{zx} + \zeta_y \tau_{zy} + \zeta_z \tau_{zz} \\ v_x(\zeta_x \tau_{xx} + \zeta_y \tau_{xy} + \zeta_z \tau_{xz}) + v_y(\zeta_x \tau_{yx} + \zeta_y \tau_{yy} + \zeta_z \tau_{yz}) + \cdots \\ \cdots v_z(\zeta_x \tau_{zx}) + \zeta_y \tau_{zy} \zeta_z \tau_{zz}) + \cdots \\ \cdots C\kappa \left(1 + \frac{Pr\,\mu_t}{Pr_t\,\mu}\right)\left(\zeta_x \frac{\partial T}{\partial x} + \zeta_y \frac{\partial T}{\partial y} + \zeta_z \frac{\partial T}{\partial z}\right) + \cdots \\ \cdots \frac{1}{Sc} \sum_i \rho h_i \left(D_i + \frac{Sc}{Sc_t}\frac{\mu_t}{\rho}\right)\left(\zeta_x \frac{\partial \omega_i}{\partial x} + \zeta_y \frac{\partial \omega_i}{\partial y} + \zeta_z \frac{\partial \omega_i}{\partial z}\right) + \cdots \\ \cdots (\mu + \mu_t/\sigma_k)\left(\zeta_x \frac{\partial k}{\partial x} + \zeta_y \frac{\partial k}{\partial y} + \zeta_z \frac{\partial k}{\partial z}\right) \end{pmatrix}.
$$

where all the inner derivatives with respect to the Cartesian coordinates must be transformed by the chain rule, see Eq. (6.10).

The equation describing the $k - \omega$ turbulence model is given in dimensionless form by Eqs. (8.4), (8.5), (8.6)

$$J^{-1}\frac{\partial \mathbf{Q}_T}{\partial t} + \frac{\partial \mathbf{E}_T}{\partial \xi} + \frac{\partial \mathbf{F}_T}{\partial \eta} + \frac{\partial \mathbf{G}_T}{\partial \zeta} = \frac{1}{Re}\left(\frac{\partial \mathbf{E}_{T,visc}}{\partial \xi} + \frac{\partial \mathbf{F}_{T,visc}}{\partial \eta} + \frac{\partial \mathbf{G}_{T,visc}}{\partial \zeta}\right) + S_T\,, \tag{7.9}$$

$$\mathbf{Q}_T = \begin{pmatrix} \rho k \\ \rho\omega \end{pmatrix}, \mathbf{E}_T = J^{-1}\begin{pmatrix} \rho V_x k \\ \rho V_x \omega \end{pmatrix}, \mathbf{F}_T = J^{-1}\begin{pmatrix} \rho V_y k \\ \rho V_y \omega \end{pmatrix}, \mathbf{G}_T = J^{-1}\begin{pmatrix} \rho V_z k \\ \rho V_z \omega \end{pmatrix},$$

$$\mathbf{E}_{T,visc} = J^{-1}\mu^* \begin{pmatrix} \xi_x \dfrac{\partial k}{\partial x} + \xi_y \dfrac{\partial k}{\partial y} + \xi_z \dfrac{\partial k}{\partial z} \\ \xi_x \dfrac{\partial \omega}{\partial x} + \xi_y \dfrac{\partial \omega}{\partial y} + \xi_z \dfrac{\partial \omega}{\partial z} \end{pmatrix},$$

$$\mathbf{F}_{T,visc} = J^{-1}\mu^* \begin{pmatrix} \eta_x \dfrac{\partial k}{\partial x} + \eta_y \dfrac{\partial k}{\partial y} + \eta_z \dfrac{\partial k}{\partial z} \\ \eta_x \dfrac{\partial \omega}{\partial x} + \eta_y \dfrac{\partial \omega}{\partial y} + \eta_z \dfrac{\partial \omega}{\partial z} \end{pmatrix},$$

$$\mathbf{G}_{T,visc} = J^{-1}\mu^* \begin{pmatrix} \zeta_x \dfrac{\partial k}{\partial x} + \zeta_y \dfrac{\partial k}{\partial y} + \zeta_z \dfrac{\partial k}{\partial z} \\ \zeta_x \dfrac{\partial \omega}{\partial x} + \zeta_y \dfrac{\partial \omega}{\partial y} + \zeta_z \dfrac{\partial \omega}{\partial z} \end{pmatrix},$$

$$\mathbf{S}_T = \begin{pmatrix} \dfrac{1}{Re}\tau_{ij}^{turb}\dfrac{\partial \hat{v}_i}{\partial x_j} - Re\beta^* \bar{\rho}\omega k \\ \dfrac{1}{Re}\alpha \dfrac{\omega}{k}\tau_{ij}^{turb}\dfrac{\partial \hat{v}_i}{\partial x_j} - Re\beta\bar{\rho}\omega^2 \end{pmatrix}.$$

The species continuity equations complete the system (see Eqs. (7.4), (9.3), (9.14)):

$$J^{-1}\frac{\partial \mathbf{Q}_S}{\partial t} + \frac{\partial \mathbf{E}_S}{\partial \xi} + \frac{\partial \mathbf{F}_S}{\partial \eta} + \frac{\partial \mathbf{G}_S}{\partial \zeta} = \frac{1}{Re}\left(\frac{\partial \mathbf{E}_{S,visc}}{\partial \xi} + \frac{\partial \mathbf{F}_{S,visc}}{\partial \eta} + \frac{\partial \mathbf{G}_{S,visc}}{\partial \zeta}\right) + \mathbf{S}_S, \tag{7.10}$$

$$\mathbf{Q}_S = \begin{pmatrix} \rho_1 \\ \rho_2 \\ \rho_3 \\ \vdots \end{pmatrix}, \mathbf{S}_S = \frac{l_{ref}}{\rho_\infty v_\infty} \begin{pmatrix} \alpha_1 \\ \alpha_2 \\ \alpha_3 \\ \vdots \end{pmatrix},$$

$$\mathbf{E}_S = J^{-1} \begin{pmatrix} \rho_1 V_x \\ \rho_2 V_x \\ \rho_3 V_x \\ \vdots \end{pmatrix}, \mathbf{F}_S = J^{-1} \begin{pmatrix} \rho_1 V_y \\ \rho_2 V_y \\ \rho_3 V_y \\ \vdots \end{pmatrix}, \mathbf{G}_S = J^{-1} \begin{pmatrix} \rho_1 V_z \\ \rho_2 V_z \\ \rho_3 V_z \\ \vdots \end{pmatrix},$$

$$\mathbf{E}_{S,visc} = J^{-1} \frac{\rho}{Sc} \begin{pmatrix} \left(D_1 + \frac{Sc}{Sc_t} \frac{\mu_t}{\rho}\right) \left(\xi_x \frac{\partial \omega_1}{\partial x} + \xi_y \frac{\partial \omega_1}{\partial y} + \xi_z \frac{\partial \omega_1}{\partial z}\right) \\ \left(D_2 + \frac{Sc}{Sc_t} \frac{\mu_t}{\rho}\right) \left(\xi_x \frac{\partial \omega_2}{\partial x} + \xi_y \frac{\partial \omega_2}{\partial y} + \xi_z \frac{\partial \omega_2}{\partial z}\right) \\ \left(D_3 + \frac{Sc}{Sc_t} \frac{\mu_t}{\rho}\right) \left(\xi_x \frac{\partial \omega_3}{\partial x} + \xi_y \frac{\partial \omega_3}{\partial y} + \xi_z \frac{\partial \omega_3}{\partial z}\right) \\ \vdots \end{pmatrix},$$

$$\mathbf{F}_{S,visc} = J^{-1} \frac{\rho}{Sc} \begin{pmatrix} \left(D_1 + \frac{Sc}{Sc_t} \frac{\mu_t}{\rho}\right) \left(\eta_x \frac{\partial \omega_1}{\partial x} + \eta_y \frac{\partial \omega_1}{\partial y} + \eta_z \frac{\partial \omega_1}{\partial z}\right) \\ \left(D_2 + \frac{Sc}{Sc_t} \frac{\mu_t}{\rho}\right) \left(\eta_x \frac{\partial \omega_2}{\partial x} + \eta_y \frac{\partial \omega_2}{\partial y} + \eta_z \frac{\partial \omega_2}{\partial z}\right) \\ \left(D_3 + \frac{Sc}{Sc_t} \frac{\mu_t}{\rho}\right) \left(\eta_x \frac{\partial \omega_3}{\partial x} + \eta_y \frac{\partial \omega_3}{\partial y} + \eta_z \frac{\partial \omega_3}{\partial z}\right) \\ \vdots \end{pmatrix},$$

$$\mathbf{G}_{S,visc} = J^{-1} \frac{\rho}{Sc} \begin{pmatrix} \left(D_1 + \frac{Sc}{Sc_t} \frac{\mu_t}{\rho}\right) \left(\zeta_x \frac{\partial \omega_1}{\partial x} + \zeta_y \frac{\partial \omega_1}{\partial y} + \zeta_z \frac{\partial \omega_1}{\partial z}\right) \\ \left(D_2 + \frac{Sc}{Sc_t} \frac{\mu_t}{\rho}\right) \left(\zeta_x \frac{\partial \omega_2}{\partial x} + \zeta_y \frac{\partial \omega_2}{\partial y} + \zeta_z \frac{\partial \omega_2}{\partial z}\right) \\ \left(D_3 + \frac{Sc}{Sc_t} \frac{\mu_t}{\rho}\right) \left(\zeta_x \frac{\partial \omega_3}{\partial x} + \zeta_y \frac{\partial \omega_3}{\partial y} + \zeta_z \frac{\partial \omega_3}{\partial z}\right) \\ \vdots \end{pmatrix}.$$

Today, we are able with great success to integrate with specific numerical methods the various versions of the Navier-Stokes equations. A demonstration of this capability is presented in Fig. 7.1. The DAVIS-VOL code of Daimler-Benz-Aerospace (Dasa) (later EADS and Airbus Group) on the basis of Eqs. (7.8)–(7.10) was employed, [2–4]. The flow around the Hopper-Phoenix configuration is considered, Fig. 7.1a. A radiation adiabatic wall temperature condition is implemented, where the hot wall radiates the heat away with an emission coefficient $\varepsilon = 0.8$, [5]. Due to the high gas temperatures behind the bow shock a real gas approach with equilibrium thermodynamics is applied, Sect. 9.4, [4].

For the freeflight conditions $M_\infty = 14.4, \alpha = 31.3^0$, $p_\infty = 51.9 Pa, T_\infty = 264 K$, $Re = 8.41 \cdot 10^6$ the computed skin-friction lines and the contours of the pressure coefficient at the windward side of the body surface are plotted in Fig. 7.1b, where the influence of the deflected flap ($\eta_{BF} = 20^0$) on the pressure can impressively be observed and also the separation line along the wing tip. Wall temperatures at the windward side are displayed in Fig. 7.1c, where the expected temperature increase along the wing tip line and on the deflected body flap can be monitored. Figure 7.1d shows at the windward side the wall streamlines and the pressure coefficient contours on the wing part for another freeflight condition ($M_\infty = 3.2, \alpha = 15^0$, $p_\infty = 591 Pa, T_\infty = 242 K, Re = 2.31 \cdot 10^7$) with a body flap deflection angle $\eta_{BF} = 40^0$. A pronounced separation bubble can be observed in the body flap regime. The details of the separation and reattachment are very distinct, which can also be seen on the pressure levels on the body flap.

7.3 The Thermal Non-Equilibrium State in Arbitrary Coordinates for Laminar Flow

The dimensionless form in arbitrary coordinates of the vibrational energy equations describing the thermal non-equilibrium state for laminar flows reads (see Sect. 9.2)

$$J^{-1}\frac{\partial \mathbf{Q}_{vib}}{\partial t} + \frac{\partial \mathbf{E}_{vib}}{\partial \xi} + \frac{\partial \mathbf{F}_{vib}}{\partial \eta} + \frac{\partial \mathbf{G}_{vib}}{\partial \zeta} =$$
$$= \frac{1}{Re}\left(\frac{\partial \mathbf{E}_{vib,visc}}{\partial \xi} + \frac{\partial \mathbf{F}_{vib,visc}}{\partial \eta} + \frac{\partial \mathbf{G}_{vib,visc}}{\partial \zeta}\right) + \mathbf{S}_{vib} \qquad (7.11)$$

$$\mathbf{Q}_{vib} = \begin{pmatrix} \rho_1 e_{v,1} \\ \rho_2 e_{v,2} \\ \rho_3 e_{v,3} \\ \vdots \end{pmatrix}, \mathbf{S}_{vib} = \frac{l_{ref}}{\rho_\infty v_\infty^3}\begin{pmatrix} Q_{T-v,1} + Q_{e-v,1} + Q_{D_1} + Q_{v-v,1} \\ Q_{T-v,2} + Q_{e-v,2} + Q_{D_2} + Q_{v-v,2} \\ Q_{T-v,3} + Q_{e-v,3} + Q_{D_3} + Q_{v-v,3} \\ \vdots \end{pmatrix},$$

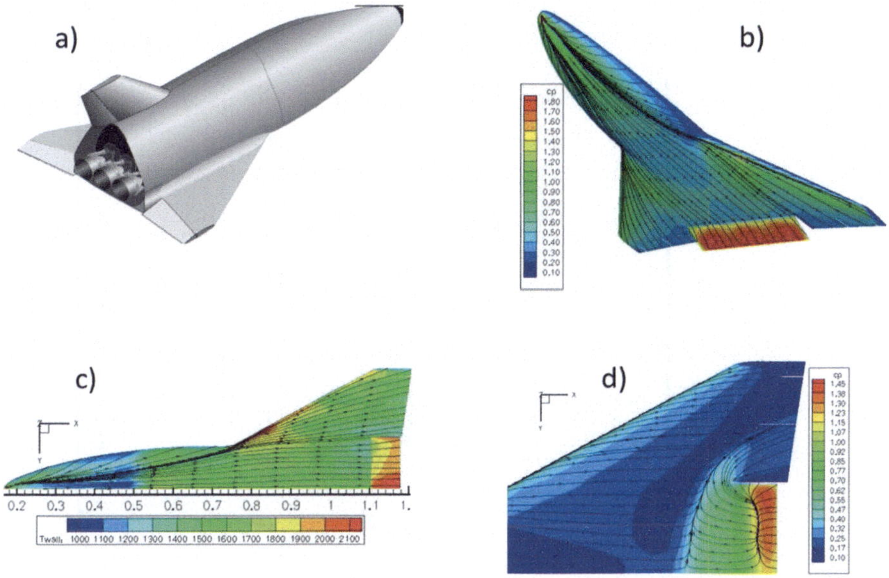

Fig. 7.1 Solution of RANS equations in thermodynamic equilibrium, [4] **a** Hopper-Phoenix configuration. **b** Skin-friction lines and contours of the pressure coefficient at windward side, $M_\infty = 14.4$, $\alpha = 31.3^0$, $p_\infty = 51.9 Pa$, $T_\infty = 264K$, $Re = 8.41 \cdot 10^6$, body flap deflected by $\eta_{BF} = 20^0$ and **c** skin-friction lines and wall temperature contours at windward side. **d** Skin-friction lines and pressure coefficient contours at windward side on wing part with a deflected body flap of $\eta_{BF} = 40^0$ for $M_\infty = 3.2$, $\alpha = 15^0$, $p_\infty = 591 Pa$, $T_\infty = 242K$, $Re = 2.31 \cdot 10^7$

$$\mathbf{E}_{vib} = J^{-1} \begin{pmatrix} \rho_1 e_{v,1} V_x \\ \rho_2 e_{v,2} V_x \\ \rho_3 e_{v,3} V_x \\ \vdots \end{pmatrix}, \mathbf{F}_{vib} = J^{-1} \begin{pmatrix} \rho_1 e_{v,1} V_y \\ \rho_2 e_{v,2} V_y \\ \rho_3 e_{v,3} V_y \\ \vdots \end{pmatrix}, \mathbf{G}_{vib} = J^{-1} \begin{pmatrix} \rho_1 e_{v,1} V_z \\ \rho_2 e_{v,2} V_z \\ \rho_3 e_{v,3} V_z \\ \vdots \end{pmatrix},$$

$$\mathbf{E}_{vib,visc} = J^{-1} \begin{pmatrix} \frac{1}{Sc} \rho e_{v,1} D_1 \left(\xi_x \frac{\partial \omega_1}{\partial x} + \xi_y \frac{\partial \omega_1}{\partial y} + \xi_z \frac{\partial \omega_1}{\partial z} \right) + \cdots \\ \cdots + \frac{\kappa_{v_\infty} T_\infty}{\mu_\infty v_\infty^2} \kappa_{v,1} \left(\xi_x \frac{\partial T_{v,1}}{\partial x} + \xi_y \frac{\partial T_{v,1}}{\partial y} + \xi_z \frac{\partial T_{v,1}}{\partial z} \right) \\ \frac{1}{Sc} \rho e_{v,2} D_2 \left(\xi_x \frac{\partial \omega_2}{\partial x} + \xi_y \frac{\partial \omega_2}{\partial y} + \xi_z \frac{\partial \omega_2}{\partial z} \right) + \cdots \\ \cdots + \frac{\kappa_{v_\infty} T_\infty}{\mu_\infty v_\infty^2} \kappa_{v,2} \left(\xi_x \frac{\partial T_{v,2}}{\partial x} + \xi_y \frac{\partial T_{v,2}}{\partial y} + \xi_z \frac{\partial T_{v,2}}{\partial z} \right) \\ \frac{1}{Sc} \rho e_{v,3} D_3 \left(\xi_x \frac{\partial \omega_3}{\partial x} + \xi_y \frac{\partial \omega_3}{\partial y} + \xi_z \frac{\partial \omega_3}{\partial z} \right) + \cdots \\ \cdots + \frac{\kappa_{v_\infty} T_\infty}{\mu_\infty v_\infty^2} \kappa_{v,3} \left(\xi_x \frac{\partial T_{v,3}}{\partial x} + \xi_y \frac{\partial T_{v,3}}{\partial y} + \xi_z \frac{\partial T_{v,3}}{\partial z} \right) \\ \vdots \end{pmatrix},$$

$$\mathbf{F}_{vib,visc} = J^{-1} \begin{pmatrix} \frac{1}{Sc}\rho e_{v,1} D_1 \left(\eta_x \frac{\partial \omega_1}{\partial x} + \eta_y \frac{\partial \omega_1}{\partial y} + \eta_z \frac{\partial \omega_1}{\partial z} \right) + \cdots \\ \cdots + \frac{\kappa_{v_\infty} T_\infty}{\mu_\infty v_\infty^2} \kappa_{v,1} \left(\eta_x \frac{\partial T_{v,1}}{\partial x} + \eta_y \frac{\partial T_{v,1}}{\partial y} + \eta_z \frac{\partial T_{v,1}}{\partial z} \right) \\ \frac{1}{Sc}\rho e_{v,2} D_2 \left(\eta_x \frac{\partial \omega_2}{\partial x} + \eta_y \frac{\partial \omega_2}{\partial y} + \eta_z \frac{\partial \omega_2}{\partial z} \right) + \cdots \\ \cdots + \frac{\kappa_{v_\infty} T_\infty}{\mu_\infty v_\infty^2} \kappa_{v,2} \left(\eta_x \frac{\partial T_{v,2}}{\partial x} + \eta_y \frac{\partial T_{v,2}}{\partial y} + \eta_z \frac{\partial T_{v,2}}{\partial z} \right) \\ \frac{1}{Sc}\rho e_{v,3} D_3 \left(\eta_x \frac{\partial \omega_3}{\partial x} + \eta_y \frac{\partial \omega_3}{\partial y} + \eta_z \frac{\partial \omega_3}{\partial z} \right) + \cdots \\ \cdots + \frac{\kappa_{v_\infty} T_\infty}{\mu_\infty v_\infty^2} \kappa_{v,3} \left(\eta_x \frac{\partial T_{v,3}}{\partial x} + \eta_y \frac{\partial T_{v,3}}{\partial y} + \eta_z \frac{\partial T_{v,3}}{\partial z} \right) \\ \vdots \end{pmatrix},$$

$$\mathbf{G}_{vib,visc} = J^{-1} \begin{pmatrix} \frac{1}{Sc}\rho e_{v,1} D_1 \left(\zeta_x \frac{\partial \omega_1}{\partial x} + \zeta_y \frac{\partial \omega_1}{\partial y} + \zeta_z \frac{\partial \omega_1}{\partial z} \right) + \cdots \\ \cdots + \frac{\kappa_{v_\infty} T_\infty}{\mu_\infty v_\infty^2} \kappa_{v,1} \left(\zeta_x \frac{\partial T_{v,1}}{\partial x} + \zeta_y \frac{\partial T_{v,1}}{\partial y} + \zeta_z \frac{\partial T_{v,1}}{\partial z} \right) \\ \frac{1}{Sc}\rho e_{v,2} D_2 \left(\zeta_x \frac{\partial \omega_2}{\partial x} + \zeta_y \frac{\partial \omega_2}{\partial y} + \zeta_z \frac{\partial \omega_2}{\partial z} \right) + \cdots \\ \cdots + \frac{\kappa_{v_\infty} T_\infty}{\mu_\infty v_\infty^2} \kappa_{v,2} \left(\zeta_x \frac{\partial T_{v,2}}{\partial x} + \zeta_y \frac{\partial T_{v,2}}{\partial y} + \zeta_z \frac{\partial T_{v,2}}{\partial z} \right) \\ \frac{1}{Sc}\rho e_{v,3} D_3 \left(\zeta_x \frac{\partial \omega_3}{\partial x} + \zeta_y \frac{\partial \omega_3}{\partial y} + \zeta_z \frac{\partial \omega_3}{\partial z} \right) + \cdots \\ \cdots + \frac{\kappa_{v_\infty} T_\infty}{\mu_\infty v_\infty^2} \kappa_{v,3} \left(\zeta_x \frac{\partial T_{v,3}}{\partial x} + \zeta_y \frac{\partial T_{v,3}}{\partial y} + \zeta_z \frac{\partial T_{v,3}}{\partial z} \right) \\ \vdots \end{pmatrix}.$$

There were some test cases for numerical methods and physical approaches of thermodynamic non-equilibrium effects (chemical and thermal (vibrational) non-equilibrium real gas effects) in the frame of ESA's[3] Manned Space Transportation Program (MSTP) activities, performed during the second half of the 1990s. One of these test cases was the axisymmetric flow past a hemisphere-cylinder-flare configuration (flare angle $\delta = 20^0$) with the free-stream conditions $M_\infty = 20$, $p_\infty = 22.5\,Pa$, $T_\infty = 254\,K$, $H = 60\,km$, $T_{wall} = 1500\,K$, $Re/m = 1.1 \cdot 10^5$. The nose radius of the hemisphere was $R = 0.5\,m$.

[3] ESA $\Longrightarrow$ European Space Agency.

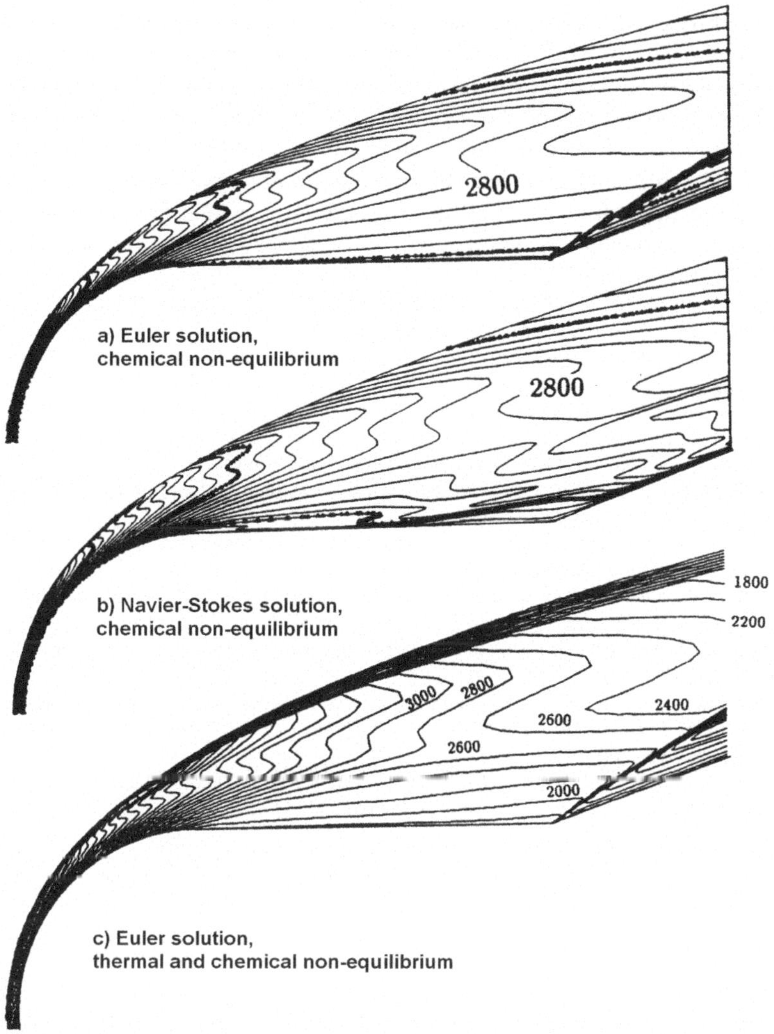

Fig. 7.2 Hypersonic flow past a hemisphere-cylinder-flare configuration, [6]. Temperature contours for three approximation cases. $M_\infty = 20$, $p_\infty = 22.5 Pa$, $T_\infty = 254 K$, $H = 60 km$, $T_{wall} = 1500 K$, $Re/m = 1.1 \cdot 10^5$

In Fig. 7.2 three solutions with different physical approximation levels are presented[4], where the temperature contours are plotted, [6]. An Euler solution with chemical non-equilibrium thermodynamics is given in Fig. 7.2a, while in Fig. 7.2b an extension to the Navier-Stokes equations with the same thermodynamics is shown. The main difference between both solutions consists in the junction regime between

[4] For the details of the approximation levels see Sects. (6.2), (9.1), (9.2).

cylinder and flare, where the necessary flow recompression by the Euler solution is achieved with a sharp "ramp-shock", whereas by considering the viscous effects, a boundary layer exists, which produces a recirculation regime in the cylinder-flare corner. This results in the production of a weak separation shock. Finally in Fig. 7.2c an Euler solution with thermal **and** chemical non-equilibrium thermodynamics is presented. The difference between the two Euler solutions consists in the temperature levels in the cylinder part of the configuration, which are somewhat lower for the thermal and chemical non-equilibrium case compared to the pure chemical non-equilibrium flow.

References

1. Weiland, C.: Mechanics of Flow Similarities. Springer, Berlin (2020)
2. Weber, C., Behr, R., Weiland, C.: Investigation of hypersonic turbulent flow over the X-38 crew return vehicle. AIAA Paper 2000–2601 (2000)
3. Schröder, W., Hartmann, G.: Analysis of inviscid and viscous hypersonic flows past a two-stage spacecraft. J. Spacecraft Rockets **30**, 8–13 (1993)
4. Haeberle, J.: Einfluss heisser Oberflächen auf aerothermodynamische Flugeigenschaften von Hopper-Phoenix (Influence of Hot Surfaces on Aerothermodynamic Flight Properties of Hopper-Phoenix). Diploma Thesis, Institut fuer Aerodynamik und Gasdynamik, Universität Stuttgart, Germany (2004)
5. Hirschel, E.H., Weiland, C.: Selected Aerothermodynamic Design Problems of Hypersonic Flight Vehicles. Springer, Berlin/Heidelberg, and **229**, Progress in Astronautics and Aeronautics, AIAA, Reston, Va (2009)
6. Menne, S., Weiland, C., d'Ambrosio, D., Pandolfi, M.: Comparison of real gas simulations using different numerical methods. Comp. Fluids **24**, 189–208 (1995)

Chapter 8
Turbulence Models

Abstract Since the seminal work of Osborne Reynolds regarding the observation of turbulent flow in a pipe (1883) and Ludwig Prandtl with the introduction of the boundary layer concept (1904) and later his presentation of the mixing length idea for the description of turbulent boundary layer flow (1925) [1, 2], a huge number of researchers and scientists has tried to understand the physics of turbulence and were engaged to find theoretical approaches and mathematical descriptions of the turbulence phenomenon. But it is a matter of fact, that the physics of turbulence are still not completely understood. The consequence is that based on empirical experience various models were generated. One class of such models consists of the Reynolds-averaged statistical approach using the Boussinesq eddy viscosity approximation, given for example by the algebraic mixing length idea, the one equation models as well as the two equation models. This group of models has delivered solutions accurate enough for many practical industrial applications. Therefore we focus in this chapter our interest on this group. Of course, there are other approaches for the computational simulation of turbulence like the "Large Eddy Simulation" (LES), the "Detached Eddy Simulation" (DES) or the "Direct Numerical Simulation" (DNS) [3]. All these methods try to resolve the turbulent eddies partly (LES,DES) or completely (DNS) by fine or very fine numerical grids, while the remaining eddies (LES,DES) are still modeled.

The Variables and Quantities Used in this Chapter

x, y, z	Cartesian coordinates
ξ, η, ζ	arbitrary coordinates
$\hat{\tau}_{ij}^{turb}$	components of Reynolds stress tensor in tensor notation
$\hat{\tau}_{xy}^{turb}$	xy component of Reynolds stress tensor, 2-D flow
ν	kinematic eddy viscosity
$\nu_{t,i}$	$= l_{mix}^2 \lvert\omega\rvert$ kinematic eddy viscosity, inner layer Baldwin-Lomax model
$\nu_{t,o}$	$= \alpha C_{cp} F_{wake} F_{Kleb}$ kinematic eddy viscosity, outer layer Baldwin-Lomax model

C. Weiland, *The Hierarchy of Fluid Dynamic Equations*,
https://doi.org/10.1007/978-3-031-71198-5_8

ν_t	$= \tilde{\nu} f_{v1}$ kinematic eddy viscosity, Spalart-Allmaras model
$\tilde{\nu}$	variable of transport equation, Spalart-Allmaras model
$\lvert\omega\rvert$	$= \lvert\text{rot } \hat{\mathbf{v}}\rvert$
u_t	$= \sqrt{\tau_{wall}/\rho}$
y^+	$= \frac{u_t y}{\nu}$
$\bar{\rho}$	Reynolds-averaged density
k	turbulent kinetic energy
ω	specific dissipation rate
μ	dynamic viscosity
μ_t	turbulent dynamic viscosity, Wilcox $k - \omega$ model, $\mu_t = \alpha \frac{\bar{\rho} k}{\omega}$
$\mu*$	$= \mu + \frac{\mu_t}{\sigma}$ abbreviation
σ_k, σ_ω	coefficient of $k - \omega$ turbulence model
$\hat{v}_x, \hat{v}_y, \hat{v}_z$	Favre-averaged Cartesian velocity components of $\mathbf{v}$
$\hat{v}_i, \hat{v}_j, \hat{v}_k$	Favre-averaged velocity components in tensor notation
$\hat{V}_x, \hat{V}_y, \hat{V}_z$	Favre-averaged components of contravariant velocity vector
$\hat{\mathbf{E}}_T, \ldots$	vector: "inviscid part" of $k - \omega$ turbulence model
$\hat{\mathbf{E}}_{T,visc}, \ldots$	vector: "viscous part of" $k - \omega$ turbulence modelm
$\hat{\mathbf{Q}}_T$	vector: variables of $k - \omega$ turbulence model
$\hat{\mathbf{S}}_T$	vector: source terms of $k - \omega$ turbulence model

8.1 Turbulence Models in Cartesian Coordinates

There exist some groups of statistical turbulence models to be usable in the Reynolds-Averaged Navier-Stokes (RANS) equations.[1] The background of the statistical turbulence models is the experience that an analogy consists between the shear stress resulting from the molecular transport of momentum (shear stress tensor) and the shear emanating from the turbulent transport of momentum (Reynolds stress tensor).

The first group are the algebraic turbulence models. These models introduce for the definition of the eddy viscosity the mixing length hypothesis of L. Prandtl, which is in analogy to the free mean path defining the molecular viscosity in the frame of the kinetic theory of gases, [3, 5]. The component of the Reynolds stress tensor for a two dimensional flow reads[2]

$$\hat{\tau}_{xy}^{turb} = \nu_t \frac{d\hat{v}_x}{dy} \quad \text{with} \quad \nu_t = l_{mix}^2 \left| \frac{d\hat{v}_x}{dy} \right| .$$

[1] In [4] one can find more information regarding the physics of the statistical turbulence modeling of super/hypersonic flows.

[2] In three dimensions one has $\nu_t = l_{mix}^2 \sqrt{2 S_{ij} S_{ij}},\ S_{ij} = \frac{1}{2}\left(\frac{\partial \hat{v}_i}{\partial x_j} + \frac{\partial \hat{v}_j}{\partial x_i}\right)$.

This is the basis for all mixing length turbulence models including the one of Cebeci-Smith, [6] and the most prominent one for industrial applications the model of Baldwin-Lomax [7]. These models are tuned regarding "optimal" applicability with respect to specific types of flow. Cebeci-Smith and Baldwin-Lomax have formulated two-layer models, where the boundary layer is divided into an inner and an outer layer. We present the Baldwin-Lomax model.

Inner layer:

$$\nu_{t,i} = l_{mix}^2 |\omega|, \quad l_{mix} = \kappa y \left(1 - e^{-\frac{y^+}{A_0^+}}\right). \tag{8.1}$$

Outer layer:

$$\nu_{t,o} = \alpha C_{cp} F_{wake} F_{Kleb}(y, y_{max}/C_{Kleb}) \tag{8.2}$$

with

$$F_{wake} = \min(y_{max} F_{max}, C_{wk} y_{max} v_{dif}^2 / F_{max}),$$

$$F_{max} = \frac{1}{\kappa} \max_y (l_{mix} |\omega|),$$

$$|\omega| = |\mathrm{rot}\,\hat{\mathbf{v}}| = \left[\left(\frac{\partial \hat{v}_y}{\partial x} - \frac{\partial \hat{v}_x}{\partial y}\right)^2 + \left(\frac{\partial \hat{v}_z}{\partial y} - \frac{\partial \hat{v}_y}{\partial z}\right)^2 + \left(\frac{\partial \hat{v}_x}{\partial z} - \frac{\partial \hat{v}_z}{\partial x}\right)^2\right]^{\frac{1}{2}},$$

$$v_{dif} = \left.\sqrt{\hat{v}_x^2 + \hat{v}_y^2 + \hat{v}_z^2}\right|_{max} - \left.\sqrt{\hat{v}_x^2 + \hat{v}_y^2 + \hat{v}_z^2}\right|_{y=y_{max}},$$

$$F_{Kleb} = \left(1 + 5.5\left(\frac{y}{y_{max}/C_{Kleb}}\right)^6\right)^{-1}, \quad y^+ = \frac{u_t y}{\nu}, \quad u_t = \sqrt{\tau_{wall}/\rho},$$

where y_{max} denotes the value of y at which $l_{mix}|\omega|$ has its maximum. F_{Kleb} is the Klebanoff intermittency function. The closure coefficients are[3]

$$\kappa = 0.40, \ \alpha = 0.0168, \ A_o^+ = 26, \ C_{cp} = 1.6, \ C_{Kleb} = 0.3, \ C_{wk} = 1.$$

Other algebraic models are those based on the Renormalization Group Theory (RNG) proposed by Yakhot and Orszag, [8, 9].

The second group encompasses the one-equation models. Traditionally, a transport equation for the kinetic energy of turbulence k is formulated as basis for the establishment of one and two equation turbulence models. However, experience has shown that in the case of a one-equation model a transport equation of the kinematic eddy viscosity ν_t provides more appropriate results in a wide range of flow cases.

[3] More details can be found in [3, 7].

The models of Baldwin-Barth and Spalart-Allmaras belong to this class, [3]. For example the Spalart-Allmaras model reads, [10, 11]

$$\underbrace{\frac{\partial \tilde{\nu}}{\partial t} + \hat{v}_j \frac{\partial \tilde{\nu}}{\partial x_j} = c_{b1} \tilde{S} \tilde{\nu} + \frac{1}{\sigma} \frac{\partial}{\partial x_k} \left[(\nu + \tilde{\nu}) \frac{\partial \tilde{\nu}}{\partial x_k} + c_{b2} \left(\frac{\partial \tilde{\nu}}{\partial x_k} \right)^2 \right] - c_{w1} f_w \left(\frac{\tilde{\nu}}{d} \right)^2}_{transport\ equation\ for\ eddy\ viscosity},$$

$$\underbrace{\nu_t = \tilde{\nu} f_{v1}}_{kinematic\ eddy\ viscosity} \quad , \tag{8.3}$$

where $c_{b1}, c_{b2}, c_{v1}, \sigma, c_{w1}, c_{w2}, c_{w3}, \kappa$ are coefficients and $f_{v1}, f_{v2}, f_w, \chi, g, r, \tilde{S}$, S, Ω_{ij} are functions with

$$f_{v1} = \frac{\chi^3}{\chi^3 + c_{v1}^3}, \quad f_{v2} = 1 - \frac{\chi}{1 + \chi f_{v1}}, \quad f_w = g \left(\frac{1 + c_{w3}^6}{g^6 + c_{w3}^6} \right)^{\frac{1}{6}},$$

$$\chi = \frac{\tilde{\nu}}{\nu}, \quad g = r + c_{w2}(r^6 - r), \quad r = \frac{\tilde{\nu}}{\tilde{S} \kappa^2 d^2}, \quad \tilde{S} = S + \frac{\tilde{\nu}}{\kappa^2 d^2} f_{v2},$$

$$S = \sqrt{2 \Omega_{ij} \Omega_{ij}}, \quad \Omega_{ij} = \frac{1}{2} \left(\frac{\partial \hat{v}_i}{\partial x_j} - \frac{\partial \hat{v}_j}{\partial x_i} \right).$$

$$c_{b1} = 0.1355, \quad c_{b2} = 0.622, \quad c_{v1} = 7.1, \quad \sigma = \frac{2}{3},$$

$$c_{w1} = \frac{c_{b1}}{\kappa^2} + \frac{1 + c_{b2}}{\sigma}, \quad c_{w2} = 0.3, \quad c_{w3} = 2, \quad \kappa = 0.41 .$$

The distance from the closest surface is denoted by d. Further information can be found in [3, 10].

The third group comprises the so-called two equation turbulence models, which are based essentially upon the transport equation for the turbulent energy k. Most popular two equations models are the $k - \varepsilon$ and the $k - \omega$ approaches. Since one of the purposes of this book is to illustrate, how the governing equations can handle turbulence models, the decision was taken to demonstrate this by the standard $k - \omega$ model of Wilcox, [3], with k being the turbulent kinetic energy and ω the specific dissipation rate.

$$\underbrace{\frac{\partial \bar{\rho} k}{\partial t}+\frac{\partial \bar{\rho}\hat{v}_j k}{\partial x_j}=\hat{\tau}_{ij}^{turb}\frac{\partial \hat{v}_i}{\partial x_j}-\beta^*\bar{\rho}\omega k+\frac{\partial}{\partial x_j}\left[\left(\mu+\frac{\mu_t}{\sigma_k}\right)\frac{\partial k}{\partial x_j}\right]}_{transport\ equation\ for\ turbulent\ kinetic\ energy},$$

$$\underbrace{\frac{\partial \bar{\rho}\omega}{\partial t}+\frac{\partial \bar{\rho}\hat{v}_j\omega}{\partial x_j}=\alpha\frac{\omega}{k}\hat{\tau}_{ij}^{turb}\frac{\partial \hat{v}_i}{\partial x_j}-\beta\bar{\rho}\omega^2+\frac{\partial}{\partial x_j}\left[\left(\mu+\frac{\mu_t}{\sigma_\omega}\right)\frac{\partial \omega}{\partial x_j}\right]}_{transport\ equation\ for\ specific\ dissipation\ rate} \tag{8.4}$$

with

$$\hat{\tau}_{ij}^{turb}=\mu_t\left(\frac{\partial \hat{v}_j}{\partial x_i}+\frac{\partial \hat{v}_i}{\partial x_j}-\frac{2}{3}\frac{\partial \hat{v}_k}{\partial x_k}\delta_{ij}\right)-\frac{2}{3}\bar{\rho}k\delta_{ij}.$$

The eddy viscosity is calculated by $\mu_t=\alpha^*\frac{\bar{\rho}k}{\omega}$. The constants α, α^*, β, β^*, σ_k, σ_ω can be found in [3, 12]. For compressible flow the model must be corrected in order to prevent that the normal components of $\hat{\tau}_{ij}^{turb}$ are going to be smaller than zero. In the literature one can find a lot of proposals in this regard, e.g. [13–15].

The two equation turbulence models can also be written in formal vector notation (see Eq. (5.17)). We demonstrate this with the $k-\omega$ turbulence model, Eq. (8.4).

$$\frac{\partial \hat{\mathbf{Q}}_T}{\partial t}+\frac{\partial \hat{\mathbf{E}}_T}{\partial x}+\frac{\partial \hat{\mathbf{F}}_T}{\partial y}+\frac{\partial \hat{\mathbf{G}}_T}{\partial z}=\frac{\partial \hat{\mathbf{E}}_{T,visc}}{\partial x}+\frac{\partial \hat{\mathbf{F}}_{T,visc}}{\partial y}+\frac{\partial \hat{\mathbf{G}}_{T,visc}}{\partial z}+\hat{\mathbf{S}}_T, \tag{8.5}$$

$$\hat{\mathbf{Q}}_T=\begin{pmatrix}\bar{\rho}k\\ \bar{\rho}\omega\end{pmatrix},\ \hat{\mathbf{E}}_T=\begin{pmatrix}\bar{\rho}\hat{v}_x k\\ \bar{\rho}\hat{v}_x\omega\end{pmatrix},\ \hat{\mathbf{F}}_T=\begin{pmatrix}\bar{\rho}\hat{v}_y k\\ \bar{\rho}\hat{v}_y\omega\end{pmatrix},\ \hat{\mathbf{G}}_T=\begin{pmatrix}\bar{\rho}\hat{v}_z k\\ \bar{\rho}\hat{v}_z\omega\end{pmatrix},$$

$$\hat{\mathbf{E}}_{T,visc}=\mu^*\begin{pmatrix}\frac{\partial k}{\partial x}\\ \frac{\partial \omega}{\partial x}\end{pmatrix},\ \hat{\mathbf{F}}_{T,visc}=\mu^*\begin{pmatrix}\frac{\partial k}{\partial y}\\ \frac{\partial \omega}{\partial y}\end{pmatrix},\ \hat{\mathbf{G}}_{T,visc}=\mu^*\begin{pmatrix}\frac{\partial k}{\partial z}\\ \frac{\partial \omega}{\partial z}\end{pmatrix},$$

$$\hat{\mathbf{S}}_T=\begin{pmatrix}\hat{\tau}_{ij}^{turb}\frac{\partial \hat{v}_i}{\partial x_j}-\beta^*\bar{\rho}\omega k\\ \alpha\frac{\omega}{k}\hat{\tau}_{ij}^{turb}\frac{\partial \hat{v}_i}{\partial x_j}-\beta\bar{\rho}\omega^2\end{pmatrix},$$

with $\mu^*=\mu+\mu_t/\sigma$ and $\sigma=\sigma_k=\sigma_\omega$.

In recent years flow field predictions in aircraft as well as in space industry did require the application of advanced numerical methods. Flow past spacecraft configuration and through the elements of modern rocket motors, for example the

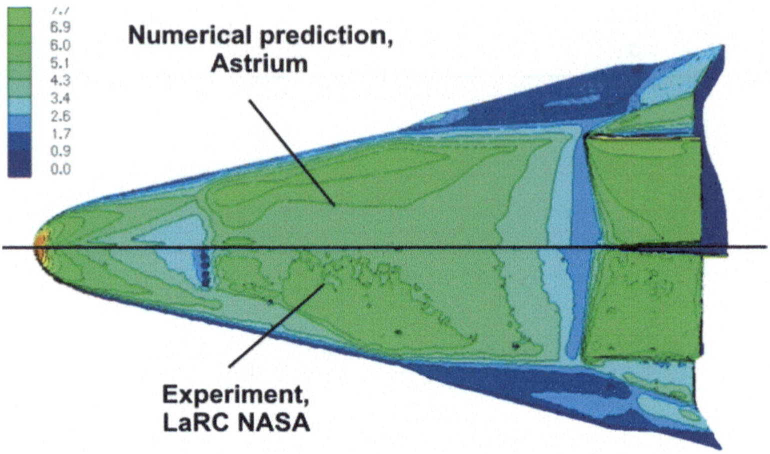

Fig. 8.1 X-38 Space Vehicle. $M_\infty = 6.0$, $\alpha = 40^0$, $Re = 3.571 \cdot 10^6$, $T_{wall} = 300K$. Wall heat flux distribution at the windward side. Comparison of numerical solution applying a $k - \omega$ model with compressible corrections, [16], and experimental data of LaRC NASA, [17]

combustion chamber and the nozzle extension part, can no longer be determined by simple potential equation (panel methods), Newton method or other handbook methods due to today's strong demand on the efficiency and performance of modern space transportation systems. Therefore, the Navier-Stokes equations as high level description of all kinds of flow fields are the method of choice for the prediction of the above-mentioned flow problems.

Because depending on the Reynolds number these flows are often turbulent, we present in Figs. 8.1 and 8.2 two results of Navier-Stokes solutions for turbulent flows, where the turbulence is modeled by the two equation $k - \omega$ model, Eq. (8.4), and with various compressibility corrections.

The first example shows the flow past the X-38 space vehicle with $M_\infty = 6.0$, $\alpha = 40^0$, $Re = 3.571 \cdot 10^6$, $T_{wall} = 300K$, Fig. 8.1, where the wall heat fluxes at the windward side are plotted, [16, 17]. Comparison of the predicted data with the experimental ones exhibits very good agreement. The second example shows the turbulent flow field in a rocket motor nozzle, [18]. The goal of this investigation was to predict the thrust and the specific impulse (Isp) as well as the wall heat flux as precisely as possible, Fig. 8.2. The quality of these solutions in a broad range is sufficient for industrial purposes.

8.2 A Turbulence Model in Arbitrary Coordinates

With the RANS equations of Sect. 5.3 and the two equation $k - \omega$ turbulence model of Sect. 8.1 we did demonstrate exemplarily how a complete set of governing equations

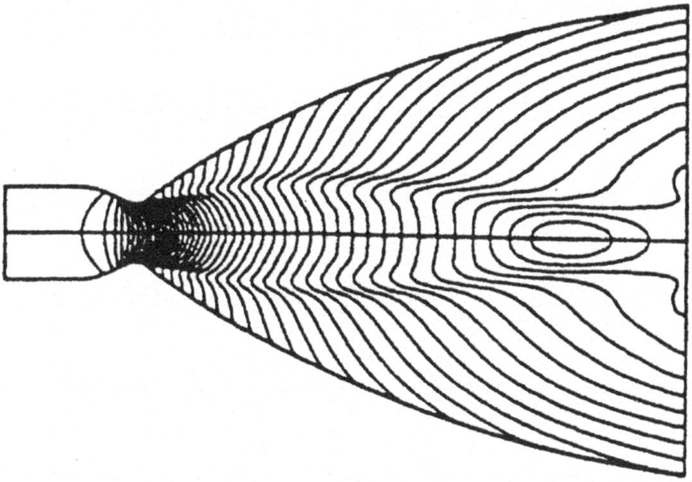

Fig. 8.2 Rocket motor nozzle flow. Chamber pressure $p_c = 117$ bar, mixing ratio of the fuel (O/F) of H_2 to O_2 amounts to 7.2. Adiabatic wall condition. Equilibrium real gas Navier-Stokes (RANS) solution with turbulent modelling on $k - \omega$ two equation basis with compressibility corrections, [18]. Mach number distribution

looks like for the computation of turbulent flow fields with perfect gases provided that this flow field can be described with sufficient accuracy by the mean values of the dependent variables. In Sect. 6.3 we presented the RANS equations in arbitrary coordinates and now transform the two equation turbulence model Eq. (8.5) also to arbitrary coordinates.

$$J^{-1}\frac{\partial \hat{\mathbf{Q}}_T}{\partial t} + \frac{\partial \hat{\mathbf{E}}_T}{\partial \xi} + \frac{\partial \hat{\mathbf{F}}_T}{\partial \eta} + \frac{\partial \hat{\mathbf{G}}_T}{\partial \zeta} = \frac{\partial \hat{\mathbf{E}}_{T,visc}}{\partial \xi} + \frac{\partial \hat{\mathbf{F}}_{T,visc}}{\partial \eta} + \frac{\partial \hat{\mathbf{G}}_{T,visc}}{\partial \zeta} + \hat{\mathbf{S}}_T\,, \tag{8.6}$$

$$\hat{\mathbf{Q}}_T = \begin{pmatrix} \bar{\rho} k \\ \bar{\rho}\omega \end{pmatrix}, \; \hat{\mathbf{E}}_T = J^{-1}\begin{pmatrix} \bar{\rho}\hat{V}_x k \\ \bar{\rho}\hat{V}_x\omega \end{pmatrix}, \; \hat{\mathbf{F}}_T = J^{-1}\begin{pmatrix} \bar{\rho}\hat{V}_y k \\ \bar{\rho}\hat{V}_y\omega \end{pmatrix}, \; \hat{\mathbf{G}}_T = J^{-1}\begin{pmatrix} \bar{\rho}\hat{V}_z k \\ \bar{\rho}\hat{V}_z\omega \end{pmatrix},$$

$$\hat{\mathbf{E}}_{T,visc} = J^{-1}\mu^* \begin{pmatrix} \xi_x \dfrac{\partial k}{\partial x} + \xi_y \dfrac{\partial k}{\partial y} + \xi_z \dfrac{\partial k}{\partial z} \\[2ex] \xi_x \dfrac{\partial \omega}{\partial x} + \xi_y \dfrac{\partial \omega}{\partial y} + \xi_z \dfrac{\partial \omega}{\partial z} \end{pmatrix},$$

$$\hat{\mathbf{F}}_{T,visc} = J^{-1}\mu^* \begin{pmatrix} \eta_x \frac{\partial k}{\partial x} + \eta_y \frac{\partial k}{\partial y} + \eta_z \frac{\partial k}{\partial z} \\ \eta_x \frac{\partial \omega}{\partial x} + \eta_y \frac{\partial \omega}{\partial y} + \eta_z \frac{\partial \omega}{\partial z} \end{pmatrix},$$

$$\hat{\mathbf{G}}_{T,visc} = J^{-1}\mu^* \begin{pmatrix} \zeta_x \frac{\partial k}{\partial x} + \zeta_y \frac{\partial k}{\partial y} + \zeta_z \frac{\partial k}{\partial z} \\ \zeta_x \frac{\partial \omega}{\partial x} + \zeta_y \frac{\partial \omega}{\partial y} + \zeta_z \frac{\partial \omega}{\partial z} \end{pmatrix},$$

$$\hat{\mathbf{S}}_T = \begin{pmatrix} \hat{\tau}_{ij}^{turb} \frac{\partial \hat{v}_i}{\partial x_j} - \beta^* \bar{\rho} \omega k \\ \alpha \frac{\omega}{k} \hat{\tau}_{ij}^{turb} \frac{\partial \hat{v}_i}{\partial x_j} - \beta \bar{\rho} \omega^2 \end{pmatrix}.$$

The term $\hat{\tau}_{ij}^{turb} \frac{\partial \hat{v}_i}{\partial x_j}$ in $\hat{\mathbf{S}}_T$ is expanded and reads

$$\begin{aligned} \hat{\tau}_{ij}^{turb} \frac{\partial \hat{v}_i}{\partial x_j} = \; & \hat{\tau}_{xx} \frac{\partial \hat{v}_x}{\partial x} + \hat{\tau}_{xy} \frac{\partial \hat{v}_x}{\partial y} + \hat{\tau}_{xz} \frac{\partial \hat{v}_x}{\partial z} + \\ & \hat{\tau}_{yx} \frac{\partial \hat{v}_y}{\partial x} + \hat{\tau}_{yy} \frac{\partial \hat{v}_y}{\partial y} + \hat{\tau}_{yz} \frac{\partial \hat{v}_y}{\partial z} + \\ & \hat{\tau}_{zx} \frac{\partial \hat{v}_x}{\partial z} + \hat{\tau}_{zy} \frac{\partial \hat{v}_z}{\partial y} + \hat{\tau}_{zz} \frac{\partial \hat{v}_z}{\partial z}, \end{aligned}$$

where the inner derivatives with respect to the Cartesian coordinates (of $\hat{\tau}_{ij}^{turb}$ and $\frac{\partial \hat{v}_i}{\partial x_j}$) must be transformed by the chain rule, see Eq. (6.10).

References

1. Eckert, M.: The Dawn of Fluid Dynamics. Wiley-VCH, Weinheim (2006)
2. Rotta, J.C.: Die Aerodynamische Versuchsanstalt in Göttingen, ein Werk Ludwig Prandtls. Vandenhoek & Ruprecht, Göttingen (1990)
3. Wilcox, D.C.: Turbulence Modelling for CFD. DCW Industries, La Canada CA, USA (2006)
4. Hirschel, E.H.: Basics of Aerothermodynamics, 2nd edn. Springer, Berlin (2015)
5. Hänel, D.: Molekulare Gasdynamik. Springer, Berlin (2004)
6. Cebeci, T., Smith, A.M.O.: Analysis of Turbulent Boundary Layer. Acacdemic Press, New York (1978)
7. Baldwin, B.S., Lomax, H.: Thin layer approximation and algebraic model for separated turbulent flows. AIAA Paper 78-0257 (1978)
8. Yakhot, V., Orszag, S.A.: Renormalisation group analysis of turbulence. J. Sci. Comput. **1**, 3–51 (1986)

9. Kirtley, K.R.: Renormalisation group based algebraic turbulence model for three-dimensional turbomachinery flows. AIAA J. **30**, 1500–1506 (1992)
10. Spalart, P.R., Allmaras, S.R.: A one-equation turbulence model for aerodynamic flows. La Recherche Aerospatiale **1**, 5–21 (1994)
11. Spalart, P.R., Allmaras, S.R.: A one-equation turbulence model for aerodynamic flows. AIAA Paper-92-0439 (1992)
12. Gross, A.: Numerische Untersuchung abgelöster Düsenströmungen (Numerical Investigations of Separated Nozzle Flows). Doctoral Thesis, Department of Combustion and Propulsion, RWTH Aachen, Germany (2000)
13. Wilcox, D.C.: Simulation of transition with a two equation turbulence model. AIAA J. **32**, 247–255 (1994)
14. Sarkar, S., Erlebacher, G., Hussaini, M.Y., Kreiss, H.O.: The analysis and modelling of dilatational terms in compressible turbulence. J. Fluid Mech. **227** (1991)
15. Zeman, O.: A new model for super/hypersonic turbulent boundary layers. AIAA Paper 93-0897 (1993)
16. Weber, C., Behr, R., Weiland, C.: Investigation of hypersonic turbulent flow over the X-38 crew return vehicle. AIAA Paper 2000-2601 (2000)
17. Horvath, T.J., Berry, S.A., Fitzgerald, S.M.: X-38 Experimental aerothermodynamics. AIAA Paper 2000-2685 (2000)
18. Haidinger, F.A.: Influence of turbulence modeling on the performance prediction for rocket engine nozzles. J. Propulsion Power **15**(4) (1999)

Chapter 9
Thermodynamics for Real Gas Flow

Abstract Atmospheric flight of space vehicles during take-off and re-entry from space (either from a planetary mission or from any Earth orbit) normally takes place with free-stream Mach numbers up to approximately 30. For such high flight Mach numbers and low air densities the gases around the space vehicles are in thermodynamic non-equilibrium. The related phenomena are:

- chemical reactions, dissoziations, recombinations
  $\Longrightarrow$ chemical non-equilibrium,
- excitation of molecular vibrations
  $\Longrightarrow$ thermal non-equilibrium,
- electron modes
  $\Longrightarrow$ ionization.

The following three sections describe the thermodynamic non-equilibrium due to chemical reactions, excitation of molecular vibrations as well as ionization. Section 9.4 addresses gases in thermodynamic equilibrium. The species continuity equations in arbitrary coordinates is given in Sect. 9.5.

The Variables and Quantities Used in this Chapter

$\mathbf{E}_S, \mathbf{F}_S, \mathbf{G}_S$	vector: "inviscid part" of species conservation
$\mathbf{E}_{S,visc}, \ldots$	vector: "viscous part" of species conservation
$\mathbf{Q}_S$	vector: species density
$\mathbf{S}_S$	vector: source term of species conservation
$\hat{\mathbf{E}}_S, \hat{\mathbf{F}}_S, \hat{\mathbf{G}}_S$	vector: "inviscid part" of species conservation in turbulent flow and arbitrary coordinates
$\hat{\mathbf{E}}_{S,visc}, \ldots$	vector: "viscous part" of species conservation in turbulent flow and arbitrary coordinates
$\hat{\mathbf{Q}}_S$	vector: species density in turbulent flow
$\hat{\mathbf{S}}_S$	vector: source term of species conservation in turbulent flow
v_x, v_y, v_z	Cartesian coordinates
$\hat{V}_x, \hat{V}_y, \hat{V}_z$	Favre-averaged components of contravariant velocity vector
k	Arrhenius formula

C. Weiland, *The Hierarchy of Fluid Dynamic Equations*,
https://doi.org/10.1007/978-3-031-71198-5_9

k_c	equilibrium constant of Arrhenius formula
$\check{D}_i$	$= D_i + \frac{1}{Sc_t}\frac{\mu_t}{\rho}$ mass diffusion coefficient for turbulent flow
h_i	enthalpy of species i
$e_{i,i}$	internal energy of species i
R_i	specific gas constant of species i
p_i	pressure of species i
μ	dynamic viscosity
κ	heat conduction coefficient

9.1 Flow with Gases in Chemical Non-Equilibrium

In this section the equations are presented, which describe the chemical non-equilibrium gas[1], [2–6]. The equation of species conservation is given by

$$\underbrace{\rho\frac{\partial\omega_i}{\partial t} + \rho\mathbf{v}\,\mathrm{grad}\,\omega_i}_{\frac{\partial\rho_i}{\partial t}+\mathrm{div}\rho_i\mathbf{v}} + \mathrm{div}\,\underbrace{\rho_i\mathbf{U}_i^*}_{\mathbf{j}_i} = \alpha_i, \tag{9.1}$$

$$\alpha_i = M_i\frac{\partial[X_i]}{\partial t} = M_i\sum_r(\nu''_{i,r} - \nu'_{i,r})\left\{k_{f,r}\prod_j[X_j]^{\nu'_{j,r}} - k_{b,r}\prod_k[X_k]^{\nu''_{k,r}}\right\}, \tag{9.2}$$

with

$\mathbf{U}_i^* = -D_i\frac{\mathrm{grad}\,\omega_i}{\omega_i}$	Fick's law, velocity of diffusion
$\mathbf{j}_i = \rho_i\mathbf{U}_i^* = -\rho D_i\,\mathrm{grad}\,\omega_i$	diffusion mass flux vector
$[X_i] = \frac{\rho_i}{M_i}$	molar concentration
M_i	mol mass
D_i	multicomponent mass diffusion coefficient
ρ_i	species density
$\omega_i = \frac{\rho_i}{\rho}$	mass fraction

[1] Real gas effects occur, when temperatures and pressures (or densities) are beyond the limits of the range of thermally and calorically perfect gases. For high pressures and low temperatures van der Waals effects come into play, which must be respected [1]. Also in the case of high pressures and high temperatures van der Waals effects exist, but they are small [2]. Therefore for the flow situation in hypersonic flight van der Waals effects can be neglected. The remaining real gas effects are called high temperature real gas effects and are the object of this chapter.

α_i	rate of change equation, source term
$k_{f,r}$	forward reaction rate of reaction r
$k_{b,r}$	backward reaction rate of reaction r
$\nu'_{i,r}$	stoichiometric coefficient of forward reaction r
$\nu''_{i,r}$	stoichiometric coefficient of backward reaction r
r	all reactions where species i take place
j	number of partners of reaction r
k	number of reaction products of reaction r
$i = 1...N$	number of species i

Usually, the reaction rates $k_{f,r}$ and $k_{b,r}$ are calculated by an Arrhenius approximation, which has the form

$$k = CT^s e^{-\frac{A}{T}},$$

where the constants C, s, A are determined experimentally. Both reaction rates are connected by the equilibrium constant

$$k_c = \frac{k_f}{k_b}.$$

Equation (9.1) can be written in formal vector notation, too

$$\frac{\partial \mathbf{Q}_S}{\partial t} + \frac{\partial \mathbf{E}_S}{\partial x} + \frac{\partial \mathbf{F}_S}{\partial y} + \frac{\partial \mathbf{G}_S}{\partial z} = \frac{\partial \mathbf{E}_{S,visc}}{\partial x} + \frac{\partial \mathbf{F}_{S,visc}}{\partial y} + \frac{\partial \mathbf{G}_{S,visc}}{\partial z} + \mathbf{S}_S, \tag{9.3}$$

$$\mathbf{Q}_S = \begin{pmatrix} \rho_1 \\ \rho_2 \\ \rho_3 \\ \vdots \end{pmatrix}, \quad \mathbf{S}_S = \begin{pmatrix} \alpha_1 \\ \alpha_2 \\ \alpha_3 \\ \vdots \end{pmatrix},$$

$$\mathbf{E}_S = \begin{pmatrix} \rho_1 v_x \\ \rho_2 v_x \\ \rho_3 v_x \\ \vdots \end{pmatrix}, \quad \mathbf{F}_S = \begin{pmatrix} \rho_1 v_y \\ \rho_2 v_y \\ \rho_3 v_y \\ \vdots \end{pmatrix}, \quad \mathbf{G}_S = \begin{pmatrix} \rho_1 v_z \\ \rho_2 v_z \\ \rho_3 v_z \\ \vdots \end{pmatrix},$$

$$\mathbf{E}_{S,visc} = \begin{pmatrix} \rho D_1 \dfrac{\partial \omega_1}{\partial x} \\ \rho D_2 \dfrac{\partial \omega_2}{\partial x} \\ \rho D_3 \dfrac{\partial \omega_3}{\partial x} \\ \vdots \end{pmatrix}, \ \mathbf{F}_{S,visc} = \begin{pmatrix} \rho D_1 \dfrac{\partial \omega_1}{\partial y} \\ \rho D_2 \dfrac{\partial \omega_2}{\partial y} \\ \rho D_3 \dfrac{\partial \omega_3}{\partial y} \\ \vdots \end{pmatrix}, \ \mathbf{G}_{S,visc} = \begin{pmatrix} \rho D_1 \dfrac{\partial \omega_1}{\partial z} \\ \rho D_2 \dfrac{\partial \omega_2}{\partial z} \\ \rho D_3 \dfrac{\partial \omega_3}{\partial z} \\ \vdots \end{pmatrix}.$$

For the application of Eq. (9.1) in turbulent flows (Navier-Stokes Eqs. (5.17)) an averaging has to be carried out

$$\frac{\partial \bar{\rho}_i}{\partial t} + \text{div}\,(\bar{\rho}_i \hat{\mathbf{v}}) - \text{div}\,(\bar{\rho}\, \check{D}_i\, \text{grad}\, \hat{\omega}_i) = \hat{\alpha}_i, \tag{9.4}$$

with $\check{D}_i = D_i + \dfrac{1}{Sc_t}\dfrac{\mu_t}{\rho}$, see Eq. (5.16).

The components of the gases can be deemed as thermally perfect. For the species enthalpy h_i, the species internal energy $e_{i,i}$ and the binary diffusion coefficients D_{ij} usually polynomials with respect to the temperature are available. Exemplarily, for the air chemistry (oxygen-nitrogen) or the oxygen-hydrogen chemistry (for rocket motors) the functions given in [7] for h_i and $e_{i,i}$ can be used

$$\frac{h_i}{R_i} = a_{i6} + a_{i1}T + a_{i2}\frac{T^2}{2} + a_{i3}\frac{T^3}{3} + a_{i4}\frac{T^4}{4} + a_{i5}\frac{T^5}{5},$$

$$e_{i,i} = h_i - R_i T = R_i\left(a_{i6} + (a_{i1} - 1)T + a_{i2}\frac{T^2}{2} + a_{i3}\frac{T^3}{3} + a_{i4}\frac{T^4}{4} + a_{i5}\frac{T^5}{5}\right),$$

whereas for the binary diffusion coefficients the approximative functions of [8] are a good choice

$$D_{ij} = \frac{1}{p} exp\,(d_{1jk} + d_{2jk} lnT + d_{3jk}(lnT)^2 + d_{4jk}(lnT)^3).$$

The enthalpy h, the internal energy e_i and the gas constant R are summed up with the mass fractions to give $h = \sum_i \omega_i h_i$, $e_i = \sum_i \omega_i e_{i,i}$, $R = \sum_i \omega_i R_i$. The perfect gas equation is valid for the species pressure $p_i = \rho_i R_i T$ and the mixture pressure $p = \rho R T$. The pressure and the density are found by summing up the species pressures p_i by $p = \sum_i p_i$, as well as the species densities ρ_i by $\rho = \sum_i \rho_i$.

Finally, the mol fractions are $X_i = \frac{M}{M_i}\omega_i$ with $\sum_i X_i = 1$ and $M = \frac{1}{\sum_i \frac{\omega_i}{M_i}}$. The multicomponent diffusion coefficients D_i are determined by the binary diffusion components, [4], via

$$D_i = \frac{1 - X_i}{\sum_j \frac{X_j}{D_{ij}}}$$

The molecular viscosity μ and the heat conduction coefficient κ are then[2]

$$\mu = \frac{1}{2}\left(\sum_i X_i Z_i + \frac{1}{\sum_i X_i/Z_i}\right),$$

$$\kappa = \frac{1}{2}\left(\sum_i X_i Z_i + \frac{1}{\sum_i X_i/Z_i}\right),$$

$$Z_i = exp\,(b_{1i} + b_{2i} lnT + b_{3i}(lnT)^2 + b_{4i}(lnT)^3),$$

where the coefficients $b_{1i}, b_{2i}, b_{3i}, b_{4i}$ are different for μ and κ and are listed in tables, [8].

For completion of these considerations we mention that for flows with gases in chemical non-equilibrium the transport of heat $\mathbf{q}$ is given by Eq. (3.10)

$$\mathbf{q} = -\kappa \text{ grad } T - \sum_i \rho D_i h_i \text{ grad } \omega_i \qquad \text{(laminar flow)},$$

$$\hat{\mathbf{q}} = -\check{\kappa} \text{ grad } \tilde{T} - \sum_i \bar{\rho} \check{D}_i \hat{h}_i \text{ grad } \hat{\omega}_i \qquad \text{(turbulent flow)}. \qquad (9.5)$$

9.2 Flow with Gases in Thermal Non-Equilibrium State

The vibrational energy equation of each molecular species j describes the thermal non-equilibrium state, [9–11].

[2] The formulas for μ and κ have the same form, but differs in the magnitude of the values of Z_i.

$$\frac{\partial}{\partial t}(\rho_j e_{v,j}) + \frac{\partial}{\partial x_k}(\rho_j e_{v,j} v_k) = -\frac{\partial}{\partial x_k}\left(\rho_j e_{v,j} U^*_{j,k}\right) - \frac{\partial q_{v,j,k}}{\partial x_k} + \\ + Q_{T-v,j} + Q_{e-v,j} + Q_{D_j} + Q_{v-v,j}, \quad (9.6)$$

where the meaning of the individual terms with respect to the species j is as follows

$\mathbf{U}^*_j$	velocity of mass diffusion
$\mathbf{j}_j = \rho_j \mathbf{U}^*_j = -\rho D_j \operatorname{grad} \omega_j$	diffusion mass flux vector
D_j	multicomponent mass diffusion coefficient
ρ_j	species density
$\omega_j = \frac{\rho_j}{\rho}$	mass fraction
$\mathbf{q}_{v,j} = -\kappa_{v,j} \operatorname{grad} T_{v,j}$	conduction of vibrational energy due to vibrational temperature gradients
$e_{v,j}$	vibrational energy
$T_{v,j}$	vibrational temperature
$\kappa_{v,j}$	frozen thermal conductivity coefficient of vibrational energy
$Q_{T-v,j} = \rho_j \frac{\bar{e}_{v,j}(T) - e_{v,j}}{<\tau_j>}$	energy exchange between translational and vibrational modes, T translational-vibrational temperature, $\bar{e}_{v,j}(T)$ vibrational energy at temperature T, $<\tau_j>$ characteristic relaxation time for translational-vibrational energy exchange
$Q_{e-v,j} = \rho_j \frac{\bar{\bar{e}}_{v,j}(T_e) - e_{v,j}}{<\tau_{e,j}>}$	energy exchange between electronic and vibrational modes, T_e electronic-vibrational temperature, $\bar{\bar{e}}_{v,j}(T_e)$ vibrational energy at temperature T_e, $<\tau_{e,j}>$ characteristic relaxation time for electronic-vibrational energy exchange
$Q_{D_j} = \alpha_j \hat{e}_{v,j}$	energy removed due to dissociation or gained due to recombination, α_j mass rate of formation $\hat{e}_{v,j}$ Vibrational energy generated or lost with mass rate α_j ($\hat{e}_{v,j}$ can be higher than the vibrational energy $e_{v,j}$)
$Q_{v-v,j}$	energy exchange between different vibrational modes
j	$= 1, ...N_{mol}$ number of molecular species j

To reduce the computational effort it is sometimes sufficient to apply a vibrational model with one averaged temperature T_v. The energy equation on the basis of Eq. (9.6) then reads

$$\frac{\partial}{\partial t}(\rho e_v) + \frac{\partial}{\partial x_k}(\rho e_v v_k) = \frac{\partial}{\partial x_k}\left(\rho \sum_{j=1}^{N_{mol}} e_{v,j} D_j \frac{\partial \omega_j}{\partial x_k}\right) - \frac{\partial q_{v,k}}{\partial x_k} + \\ + Q_{T-v} + Q_{e-v} + Q_D + Q_{v-v}, \tag{9.7}$$

now with the total quantities (see also the above list)

$\rho e_v = \sum_{j=1}^{N_{mol}} \rho_j e_{v,j}$	vibrational energy per unit volume
$\mathbf{q_v}, = -\kappa_v \text{grad}\, T_v$	conduction of vibrational energy due to vibrational temperature gradients
$Q_{T-v} = \sum_{j=1}^{N_{mol}} \rho_j \frac{\bar{e}_{v,j}(T) - e_{v,j}}{<\tau_j>}$	energy exchange between translational and vibrational modes
$Q_{e-v} = \sum_{j=1}^{N_{mol}} \rho_j \frac{\bar{\bar{e}}_{v,j}(T_e) - e_{v,j}}{<\tau_{e,j}>}$	energy exchange between electronic and vibrational modes
$Q_D = \sum_{j=1}^{N_{mol}} \alpha_j \hat{e}_{v,j}$	energy removed at dissociation or gained at recombination
$Q_{v-v} = \sum_{j=1}^{N_{mol}} Q_{v-v_j}$	energy exchange between different vibrational modes.

For turbulent flow Eqs. (9.6), (9.7) have to be averaged in the sense of Eq. (9.4).

9.3 Flow with Gases Including Excited Electron Modes

The energy equation for electron modes basically has the same appearance as Eq. (9.6) for the vibration of the molecular species, [9, 10, 12, 13].

$$\frac{\partial}{\partial t}(\rho e_e) + \frac{\partial}{\partial x_k}((\rho e_e + p_e) v_k) = \frac{\partial}{\partial x_k}\left(\rho \sum_i e_{e,i} D_i \frac{\partial \omega_i}{\partial x_k}\right) - \frac{\partial q_{e,k}}{\partial x_k} + v_k \frac{\partial p_e}{\partial x_k} \\ + Q_{T-e} + Q_{e-v} + Q_I + q_{rad,g}. \tag{9.8}$$

The meaning of the individual terms is

$\rho e_e = \sum_{i=1}^{N} \rho_i e_{e,i}$	electron energy per unit volume
$q_{e,k} = -\kappa_e \dfrac{\partial T_e}{\partial x_k}$	conduction of electron energy due to electron temperature gradient
$q_{rad,g}$	energy amount due to gas radiation caused by electron transition
κ_e	frozen thermal conductivity for electron energy
p_e	pressure of electrons
ρ_e	density of electrons
T_e	temperature of electrons
$\mathcal{R}$	universal gas constant
ν_e	collision frequency for electrons and heavy particles
M_i	mol mass of species i
$\dot{n}_{e,i''}$	number density of species i'' produced by electron impact ionization
$I_{i''}$	ionization potential of species i''
$Q_{T-e} = 3\rho\mathcal{R}(T - T_e)\sum_{i=1}^{N-1} \dfrac{\nu_{e,i}}{M_i}$	energy exchange due to elastic collisions between electrons and heavy particles
$Q_{e-v} = \sum_{i=1}^{N} Q_{e-v,i}$	energy exchange between electronic and vibrational modes
$Q_I = \sum_{i''=1}^{N_{ions}} \dot{n}_{e,i''} I_{i''}$	energy amount due to ionization by electron impact

with i being the number of electronic modes and i'' the number of ions. For turbulent flows Eq. (9.8) has to be averaged in the sense of Eq. (9.4).

9.4 Flow with Gases in Thermodynamic Equilibrium

The equations for gas flows in thermodynamic equilibrium are much simpler than the equations for gas flows in chemical and thermal non-equilibrium Eqs. (9.1), (9.3) and (9.6). First, the species continuity equations can be dropped and second, the term of the mass diffusion $\sum_i \rho D_i h_i \operatorname{grad} \omega_i$ in the molecular transport of heat $\mathbf{q}$ can be cancelled.

The flow field is characterized by the six dependent variables v_x, v_y, v_z, ρ, p and T or e_i. The five Eqs. (5.1) and the combination of the two equations for the thermal and caloric state, describing either the temperature or the internal energy, close the system. For a perfect gas the state equations are

$$\begin{aligned} p &= \rho R T && \text{thermal equation of state,} \\ e_i &= c_v T && \text{caloric equation of state,} \end{aligned} \tag{9.9}$$

with R being the specific gas constant and c_v the specific heat at constant volume.

Generally for flows with real gases in equilibrium the thermal and caloric equations of state read

$$\begin{aligned} p &= \rho R T Z(\rho, T) && \text{thermal equation of state,} \\ e_i &= e_i(\rho, T) && \text{caloric equation of state,} \end{aligned} \tag{9.10}$$

with the compressibility factor $Z = Z(\rho, T)$.

From thermodynamics it is known that every thermodynamic variable of a single-phase pure substance in equilibrium is completely described by two other thermodynamic variables. Thus it is convenient to use the density ρ and the internal energy e_i as independent variables, which results in

$$\begin{aligned} p &= \rho R T Z(\rho, e_i) = \rho R T(\rho, e_i), \\ T &= T(\rho, e_i) \end{aligned} \tag{9.11}$$

Further, we also have $\mu = \mu(\rho, e_i)$ and $\kappa = \kappa(\rho, e_i)$. The speed of sound, classically defined by

$$c^2 = \left(\frac{\partial p}{\partial \rho}\right)_s = \frac{\left(\frac{\partial h}{\partial \rho}\right)_p}{\frac{1}{\rho} - \left(\frac{\partial h}{\partial p}\right)_\rho},$$

often used in the solution process for the convective part of the equations, has to be defined in terms of the independent thermodynamic variables ρ and e_i, too.

From the relations

$$\begin{aligned} T ds &= de_i + p d\frac{1}{\rho}, \\ dp &= \left(\frac{\partial p}{\partial \rho}\right)_s d\rho + \left(\frac{\partial p}{\partial s}\right)_\rho ds && \text{for } p = p(\rho, s), \\ dp &= \left(\frac{\partial p}{\partial \rho}\right)_{e_i} d\rho + \left(\frac{\partial p}{\partial e_i}\right)_\rho de_i && \text{for } p = p(\rho, e_i), \end{aligned} \tag{9.12}$$

we obtain with $ds = 0$

$$\frac{de_i}{d\rho} = \frac{p}{\rho^2}, \quad \frac{dp}{d\rho} = \left(\frac{\partial p}{\partial \rho}\right)_s ,$$

$$c^2(\rho, e_i) = \frac{p}{\rho^2}\left(\frac{\partial p}{\partial e_i}\right)_\rho + \left(\frac{\partial p}{\partial \rho}\right)_{e_i} . \tag{9.13}$$

All the variables $p = p(\rho, e_i)$, $T = T(\rho, e_i)$, $c^2 = c^2(\rho, e_i)$, $\mu = \mu(\rho, e_i)$, $\kappa = \kappa(\rho, e_i)$, are determined, for example, either by the computer program of [7] or by a table look-up [14, 15].

9.5 Species Continuity Equations in Arbitrary Coordinates

The species continuity equations necessary for the description of flow fields with gases in chemical non-equilibrium are presented for laminar flow in vector notation by Eq. (9.3) and for turbulent flow in formal notation by Eq. (9.4). The species continuity equations for turbulent flow are now transformed to arbitrary coordinates (in vector notation).

$$J^{-1}\frac{\partial \hat{\mathbf{Q}}_S}{\partial t} + \frac{\partial \hat{\mathbf{E}}_S}{\partial \xi} + \frac{\partial \hat{\mathbf{F}}_S}{\partial \eta} + \frac{\partial \hat{\mathbf{G}}_S}{\partial \zeta} = \frac{\partial \hat{\mathbf{E}}_{S,visc}}{\partial \xi} + \frac{\partial \hat{\mathbf{F}}_{S,visc}}{\partial \eta} + \frac{\partial \hat{\mathbf{G}}_{S,visc}}{\partial \zeta} + \hat{\mathbf{S}}_S \tag{9.14}$$

$$\hat{\mathbf{Q}}_S = \begin{pmatrix} \bar{\rho}_1 \\ \bar{\rho}_2 \\ \bar{\rho}_3 \\ \vdots \end{pmatrix}, \quad \hat{\mathbf{S}}_S = \begin{pmatrix} \hat{\alpha}_1 \\ \hat{\alpha}_2 \\ \hat{\alpha}_3 \\ \vdots \end{pmatrix},$$

$$\hat{\mathbf{E}}_S = J^{-1}\begin{pmatrix} \bar{\rho}_1 \hat{V}_x \\ \bar{\rho}_2 \hat{V}_x \\ \bar{\rho}_3 \hat{V}_x \\ \vdots \end{pmatrix}, \quad \hat{\mathbf{F}}_S = J^{-1}\begin{pmatrix} \bar{\rho}_1 \hat{V}_y \\ \bar{\rho}_2 \hat{V}_y \\ \bar{\rho}_3 \hat{V}_y \\ \vdots \end{pmatrix}, \quad \hat{\mathbf{G}}_S = J^{-1}\begin{pmatrix} \bar{\rho}_1 \hat{V}_z \\ \bar{\rho}_2 \hat{V}_z \\ \bar{\rho}_3 \hat{V}_z \\ \vdots \end{pmatrix},$$

$$\hat{\mathbf{E}}_{S,visc} = J^{-1}\begin{pmatrix} \bar{\rho}\check{D}_1\left(\xi_x \frac{\partial \hat{\omega}_1}{\partial x} + \xi_y \frac{\partial \hat{\omega}_1}{\partial y} + \xi_z \frac{\partial \hat{\omega}_1}{\partial z}\right) \\ \bar{\rho}\check{D}_2\left(\xi_x \frac{\partial \hat{\omega}_2}{\partial x} + \xi_y \frac{\partial \hat{\omega}_2}{\partial y} + \xi_z \frac{\partial \hat{\omega}_2}{\partial z}\right) \\ \bar{\rho}\check{D}_3\left(\xi_x \frac{\partial \hat{\omega}_3}{\partial x} + \xi_y \frac{\partial \hat{\omega}_3}{\partial y} + \xi_z \frac{\partial \hat{\omega}_3}{\partial z}\right) \\ \vdots \end{pmatrix},$$

$$\hat{\mathbf{F}}_{S,visc} = J^{-1} \begin{pmatrix} \bar{\rho}\check{D}_1 \left(\eta_x \dfrac{\partial \hat{\omega}_1}{\partial x} + \eta_y \dfrac{\partial \hat{\omega}_1}{\partial y} + \eta_z \dfrac{\partial \hat{\omega}_1}{\partial z} \right) \\ \bar{\rho}\check{D}_2 \left(\eta_x \dfrac{\partial \hat{\omega}_2}{\partial x} + \eta_y \dfrac{\partial \hat{\omega}_2}{\partial y} + \eta_z \dfrac{\partial \hat{\omega}_2}{\partial z} \right) \\ \bar{\rho}\check{D}_3 \left(\eta_x \dfrac{\partial \hat{\omega}_3}{\partial x} + \eta_y \dfrac{\partial \hat{\omega}_3}{\partial y} + \eta_z \dfrac{\partial \hat{\omega}_3}{\partial z} \right) \\ \vdots \end{pmatrix},$$

$$\hat{\mathbf{G}}_{S,visc} = J^{-1} \begin{pmatrix} \bar{\rho}\check{D}_1 \left(\zeta_x \dfrac{\partial \hat{\omega}_1}{\partial x} + \zeta_y \dfrac{\partial \hat{\omega}_1}{\partial y} + \zeta_z \dfrac{\partial \hat{\omega}_1}{\partial z} \right) \\ \bar{\rho}\check{D}_2 \left(\zeta_x \dfrac{\partial \hat{\omega}_2}{\partial x} + \zeta_y \dfrac{\partial \hat{\omega}_2}{\partial y} + \zeta_z \dfrac{\partial \hat{\omega}_2}{\partial z} \right) \\ \bar{\rho}\check{D}_3 \left(\zeta_x \dfrac{\partial \hat{\omega}_3}{\partial x} + \zeta_y \dfrac{\partial \hat{\omega}_3}{\partial y} + \zeta_z \dfrac{\partial \hat{\omega}_3}{\partial z} \right) \\ \vdots \end{pmatrix},$$

with $\check{D}_i = D_i + \dfrac{1}{Sc_t} \dfrac{\mu_t}{\rho}$, see Eq. (5.16). Further the inner derivatives with respect to the Cartesian coordinates are transformed by the chain rule, see Eq. (6.10).

References

1. Oertel, H.: Stossrohre (Shock Tubes). Springer, Wien (1966)
2. Hirschel, E.H.: Basics of Aerothermodynamics, 2nd edn. Springer, Berlin (2015)
3. Gross, A.: Numerische Untersuchung abgelöster Düsenströmungen (Numerical Investigations of Separated Nozzle Flows). Doctoral Thesis, Department of Combustion and Propulsion, RWTH Aachen, Germany (2000)
4. Anderson, J.D.: Hypersonic and High Temperature Gas Dynamics. McGraw-Hill Book Company (1989)
5. Vincenti, W.G., Kruger, C.H., Jr.: Introduction to Physical Gas Dynamics. Wiley, New York (1966)
6. Park, C.: On the convergence of computation of chemically reacting flows. AIAA Paper 85-0247 (1985)
7. Gordon, S., McBride, B.J.: Computer Program for Calculation of Complex Chemical Equilibrium Compositions, Rocket Performance, Incident and Reflected Shocks, and Chapman-Jouguet Detonations. NASA TN D-7056, Lewis Research Center (1973)
8. Warnatz, J.: Private Communication (1997)
9. Hirschel, E.H., Weiland, C.: Selected Aerothermodynamic Design Problems of Hypersonic Flight Vehicles. Springer, Berlin/Heidelberg, and vol. 229, Progress in Astronautics and Aeronautics, AIAA, Reston, Va. (2009)
10. Gnoffo, P.A., Gupta, R.N., Shinn, J.L.: Conservation equations and physical models for hypersonic air flows in thermal and chemical nonequilibrium. NASA-TN-2867 (1989)
11. Candler, G.V., MacCormack, R.W.: The computation of hypersonic ionized flows in chemical and thermal nonequilibrium. AIAA-Paper 88-0511 (1988)

12. Otsu, H., Suzuki, K., Fujita, K., Abe, T.: Radiative heating analysis around the MUSES-C reentry capsule at a superorbital speed. AIAA-Paper 98-2447 (1998)
13. Surzhikov, S.T.: Computing system for solving radiative gasdynamic problems of entry and re-entry space vehicles. ESA-SP-533 (2003)
14. Srinivasan, S., Tannehill, J.C., Weilmuenster, K.J.: Simplified Curve Fits for the Thermodynamic Properties of Equilibrium Air. ISU-ERI-Ames-86401, Iowa State University (1986)
15. Mundt, Ch., Keraus, R., Fischer, J.: New, Accurate vectorized approximations of the state surfaces for the thermodynamic and transport properties of equilibrium Air. Zeitschrift für Flugwissenschaften und Weltraumforschung (ZfW) **15**(3), 179–184 (1991)

Chapter 10
The Thin Layer Approximation (TLA) of the Navier-Stokes Equations

Abstract Flow with high Reynolds numbers around aircraft and space vehicles generates viscous layers. The viscous layers are thin as long as they are attached onto the configuration (wings, fuselages, aerodynamic controls, etc.). These flows are mainly governed by the velocity gradient perpendicular to the body contour and can be treated by the so-called Thin Layer Approximation (TLA), [1–3]. The TLA procedure is very similar to the boundary layer approach with the difference that no matching is necessary between inviscid flow solution (e.g. Euler solution) and the boundary layer solution. In contrast to the classical boundary layer theory, the momentum equation normal to the wall is maintained and is not replaced by $\partial p/\partial n \approx \partial p/\partial z = 0$ as in boundary layer theory, which means that the pressure must not be constant normal to the body surface.

The Variables and Quantities Used in this Chapter

$\mathbf{E}, \mathbf{F}, \mathbf{G}$	vector: inviscid part of TLA equations
$\mathbf{H}_{visc}$	vector: viscous part of TLA equations
$\mathbf{Q}$	vector: conservative variables of TLA equations
$\hat{\mathbf{E}}, \hat{\mathbf{F}}, \hat{\mathbf{G}}$	vector: inviscid part of Reynolds-averaged TLA equations
$\hat{\mathbf{H}}_{visc}$	vector: viscous part of Reynolds-averaged TLA equations
$\hat{\mathbf{Q}}$	vector: conservative variables of Reynolds-averaged TLA equations
$\tau_{xz}, \tau_{yz}, \tau_{zz}$	components of viscous shear tensor relevant for TLA equations
v_x, v_y, v_z	velocity components in Cartesian coordinates
μ	dynamic viscosity
μ_t	turbulent viscosity
$\check{\mu}$	$= \mu + \mu_t$ total viscosity
μ^*	$= \mu + \mu_t/\sigma_k$ abbreviation
κ	thermal conductivity
$\check{\kappa}$	$= \kappa(1 + Pr\mu_t/Pr_t\mu)$ total thermal conductivity
σ_k	constant
k	turbulent kinetic energy

C. Weiland, *The Hierarchy of Fluid Dynamic Equations*,
https://doi.org/10.1007/978-3-031-71198-5_10

Re	Reynolds number
Pr	Prandtl number
Pr_t	turbulent Prandtl number
Sc	Schmidt number
Sc_t	turbulent Schmidt number
D_i	multicomponent mass diffusion coefficient of species i
h_i	enthalpy of species i
ξ, η, ζ	arbitrary coordinates
$C_{\xi^2}, C_{\xi \cdot \upsilon_\xi}, C_{\xi \cdot \upsilon}$	abbreviations, laminar TLA equations
$\hat{C}_{\xi \cdot \upsilon_\xi}, \hat{C}_{\xi \cdot \upsilon}, \hat{C}_{\upsilon \cdot \upsilon_\xi}$	abbreviations, turbulent TLA equations

10.1 The TLA Equations for Laminar Perfect Gas Flow in Cartesian Coordinates

Let us consider a flow field in Cartesian coordinates x, y, z, where x, y are lying in the main flow plane and z perpendicular to this (for example a flat plate flow), Fig. 10.1. Then all derivatives of the viscous terms with respect to x and y can be neglected and only the z derivatives are maintained. In flow regimes with strong viscous effects in the x and y directions, for example in separation areas with flow recirculations, wake flows, strong concave corners, etc., the thin-layer approach is not valid.

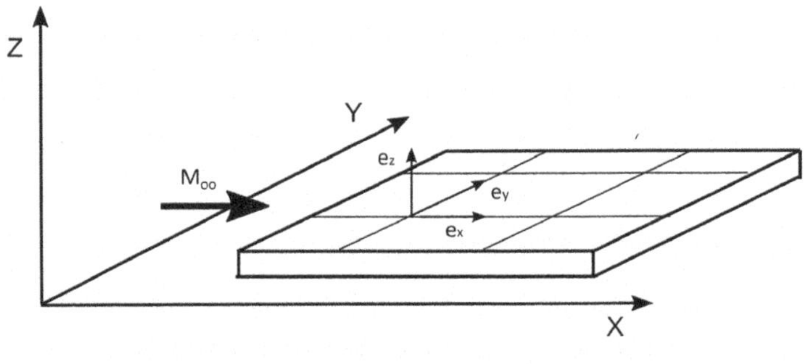

Fig. 10.1 Sketch illustrating the thin-layer approach for flow cases described in Cartesian coordinates with the unit vectors e_x, e_y, e_z

From the vector operation Eq. (4.3) div$\bar{\bar{\vartheta}}$ with the tensor $\bar{\bar{\tau}}$ it remains

$$\operatorname{div} \bar{\bar{\tau}}|_{TLA} = \begin{pmatrix} \dfrac{\partial \tau_{xz}}{\partial z} \\ \dfrac{\partial \tau_{yz}}{\partial z} \\ \dfrac{\partial \tau_{zz}}{\partial z} \end{pmatrix} \tag{10.1}$$

with

$$\tau_{xz} = \mu \frac{\partial v_x}{\partial z}, \quad \tau_{yz} = \mu \frac{\partial v_y}{\partial z}, \quad \tau_{zz} = \frac{4}{3} \mu \frac{\partial v_z}{\partial z},$$

and from the vector operation Eq.(4.5) div$\bar{\bar{\vartheta}} \cdot \mathbf{a}$ with the tensor $\bar{\bar{\tau}}$ and the vector $\mathbf{v}$ it follows

$$\operatorname{div}(\bar{\bar{\tau}} \cdot \mathbf{v})|_{TLA} = \frac{\partial}{\partial z} \mu \left(\frac{\partial v_x}{\partial z} v_x + \frac{\partial v_y}{\partial z} v_y + \frac{4}{3} \frac{\partial v_z}{\partial z} v_z \right).$$

Further the heat conduction term reduces to div$\mathbf{q}\,|_{TLA} = \dfrac{\partial}{\partial z} \kappa \dfrac{\partial T}{\partial z}$. Thus from the viscous vectors $\mathbf{E}_{visc}$, $\mathbf{F}_{visc}$, $\mathbf{G}_{visc}$ of Eq. (5.1) only

$$\mathbf{H}_{visc} = \begin{pmatrix} 0 \\ \mu \frac{\partial v_x}{\partial z} \\ \mu \frac{\partial v_y}{\partial z} \\ \mu \frac{4}{3} \frac{\partial v_z}{\partial z} \\ \mu \left(\frac{\partial v_x}{\partial z} v_x + \frac{\partial v_y}{\partial z} v_y + \frac{4}{3} \frac{\partial v_z}{\partial z} v_z \right) + \kappa \frac{\partial T}{\partial z} \end{pmatrix}$$

is left over and Eq. (5.1) then reads in the TLA version ($\mathbf{S}_{NS}$ is dropped)

$$\frac{\partial \mathbf{Q}}{\partial t} + \frac{\partial \mathbf{E}}{\partial x} + \frac{\partial \mathbf{F}}{\partial y} + \frac{\partial \mathbf{G}}{\partial z} = \frac{\partial \mathbf{H}_{visc}}{\partial z}. \tag{10.2}$$

10.2 The TLA Equations for Turbulent Perfect Gas Flow in Cartesian Coordinates

Turning to the turbulent flow case one obtains with the relations Eqs. (5.14), (5.15) the TLA equations corresponding to Eq. (5.17)

$$\frac{\partial \hat{\mathbf{Q}}}{\partial t} + \frac{\partial \hat{\mathbf{E}}}{\partial x} + \frac{\partial \hat{\mathbf{F}}}{\partial y} + \frac{\partial \hat{\mathbf{G}}}{\partial z} = \frac{\partial \hat{\mathbf{H}}_{visc}}{\partial z}, \tag{10.3}$$

with

$$\hat{\mathbf{H}}_{visc} = \begin{pmatrix} 0 \\ \check{\mu}\frac{\partial \hat{v}_x}{\partial z} \\ \check{\mu}\frac{\partial \hat{v}_y}{\partial z} \\ \check{\mu}\frac{4}{3}\frac{\partial \hat{v}_z}{\partial z} - \frac{2}{3}\bar{\rho}k \\ \check{\mu}\left(\frac{\partial \hat{v}_x}{\partial z}\hat{v}_x + \frac{\partial \hat{v}_y}{\partial z}\hat{v}_y + \frac{4}{3}\frac{\partial \hat{v}_z}{\partial z}\hat{v}_z\right) - \frac{2}{3}\bar{\rho}k\hat{v}_z + \check{\kappa}\frac{\partial \hat{T}}{\partial z} + \mu^*\frac{\partial k}{\partial z} \end{pmatrix}$$

10.3 The TLA Equations for Laminar Perfect Gas Flow in Arbitrary Coordinates

For the TLA equations in arbitrary coordinates for laminar perfect gas flow we refer to Eq. (6.3). Assuming that the η and ζ coordinates are parallel to the body contour and the ξ coordinate is the body surface normal (or close to it), see Fig. 10.2, all derivatives with respect to η, ζ in the viscous term $\mathbf{H}_{visc}$ are neglected. Only the ξ derivatives are retained. This leads to

$$J^{-1}\frac{\partial \mathbf{Q}}{\partial \tau} + \frac{\partial \mathbf{E}}{\partial \xi} + \frac{\partial \mathbf{F}}{\partial \eta} + \frac{\partial \mathbf{G}}{\partial \zeta} = \frac{\partial \mathbf{H}_{visc}}{\partial \xi}, \tag{10.4}$$

and

$$\mathbf{H}_{visc} = J^{-1} \begin{pmatrix} 0 \\ \mu C_{\xi^2}\frac{\partial v_x}{\partial \xi} + \frac{\mu}{3}C_{\xi \cdot v_\xi}\xi_x \\ \mu C_{\xi^2}\frac{\partial v_y}{\partial \xi} + \frac{\mu}{3}C_{\xi \cdot v_\xi}\xi_y \\ \mu C_{\xi^2}\frac{\partial v_z}{\partial \xi} + \frac{\mu}{3}C_{\xi \cdot v_\xi}\xi_z \\ C_{\xi \cdot v}\frac{\mu}{3}C_{\xi \cdot v_\xi} + \cdots \\ \cdots + C_{\xi^2}(\kappa\frac{\partial T}{\partial \xi} + \mu(v_x\frac{\partial v_x}{\partial \xi} + v_y\frac{\partial v_y}{\partial \xi} + v_z\frac{\partial v_z}{\partial \xi})) \end{pmatrix},$$

with

$$C_{\xi^2} = \xi_x^2 + \xi_y^2 + \xi_z^2,$$
$$C_{\xi \cdot v_\xi} = \xi_x\frac{\partial v_x}{\partial \xi} + \xi_y\frac{\partial v_y}{\partial \xi} + \xi_z\frac{\partial v_z}{\partial \xi},$$
$$C_{\xi \cdot v} = \xi_x v_x + \xi_y v_y + \xi_z v_z.$$

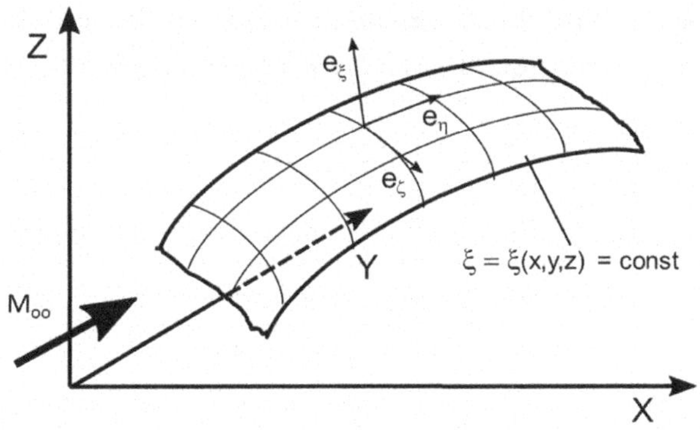

Fig. 10.2 Sketch illustrating the thin-layer approach for flow cases described by arbitrary body oriented coordinates ξ, η, ζ with unit vectors e_ξ, e_η, e_ζ

10.4 The TLA Equations for Turbulent Perfect Gas Flow in Arbitrary Coordinates

As in the subsection before in the viscous part of the equations only the derivatives in ξ direction are maintained, whereas the η and ζ derivatives are dropped. We refer to Eq. (6.9), which is reduced to

$$J^{-1}\frac{\partial \hat{\mathbf{Q}}}{\partial \tau} + \frac{\partial \hat{\mathbf{E}}}{\partial \xi} + \frac{\partial \hat{\mathbf{F}}}{\partial \eta} + \frac{\partial \hat{\mathbf{G}}}{\partial \zeta} = \frac{\partial \hat{\mathbf{H}}_{visc}}{\partial \xi}, \tag{10.5}$$

and

$$\hat{\mathbf{H}}_{visc} = J^{-1}\begin{pmatrix} 0 \\ \check{\mu} C_{\xi^2}\frac{\partial v_x}{\partial \xi} + \frac{\check{\mu}}{3}\hat{C}_{\xi\cdot v_\xi}\xi_x - \frac{2}{3}\xi_x\bar{\rho}k \\ \check{\mu} C_{\xi^2}\frac{\partial \hat{v}_y}{\partial \xi} + \frac{\check{\mu}}{3}\hat{C}_{\xi\cdot v_\xi}\xi_y - \frac{2}{3}\xi_y\bar{\rho}k \\ \check{\mu} C_{\xi^2}\frac{\partial \hat{v}_z}{\partial \xi} + \frac{\check{\mu}}{3}\hat{C}_{\xi\cdot v_\xi}\xi_z - \frac{2}{3}\xi_z\bar{\rho}k \\ \hat{C}_{\xi\cdot\hat{v}}(\frac{\check{\mu}}{3}\hat{C}_{\xi\cdot v_\xi} - \frac{2}{3}\bar{\rho}k) + \cdots \\ \cdots + C_{\xi^2}(\check{\kappa}\frac{\partial \hat{T}}{\partial \xi} + \check{\mu}\hat{C}_{v\cdot v_\xi} + \mu^*\frac{\partial k}{\partial \xi}) \end{pmatrix},$$

with

$$\begin{aligned} C_{\xi^2} &= \xi_x^2 + \xi_y^2 + \xi_z^2, \\ \hat{C}_{\xi\cdot v_\xi} &= \xi_x\frac{\partial \hat{v}_x}{\partial \xi} + \xi_y\frac{\partial \hat{v}_y}{\partial \xi} + \xi_z\frac{\partial \hat{v}_z}{\partial \xi}, \\ \hat{C}_{\xi\cdot v} &= \xi_x\hat{v}_x + \xi_y\hat{v}_y + \xi_z\hat{v}_z, \\ \hat{C}_{v\cdot v_\xi} &= \hat{v}_x\frac{\partial \hat{v}_x}{\partial \xi} + \hat{v}_y\frac{\partial \hat{v}_y}{\partial \xi} + \hat{v}_z\frac{\partial \hat{v}_z}{\partial \xi}, \end{aligned}$$

$$\check{\mu} = \mu + \mu_t, \quad \mu^* = \mu + \mu_t/\sigma_k, \quad \check{\kappa} = \kappa\left(1 + \frac{Pr\mu_t}{Pr_t\mu}\right).$$

10.5 The TLA Equations for Turbulent Chemical Non-equilibrium Flow in Arbitrary Coordinates

In context with Eq. (7.8) the TLA equations are deduced for turbulent flows in chemical non-equilibrium, which we also have made dimensionless using the relations (7.1). (As in Chap. 7 for convenience in this representation all the "hats" and "bars" are dropped.)

$$J^{-1}\frac{\partial \mathbf{Q}}{\partial \tau}+\frac{\partial \mathbf{E}}{\partial \xi}+\frac{\partial \mathbf{F}}{\partial \eta}+\frac{\partial \mathbf{G}}{\partial \zeta}=\frac{1}{Re}\left(\frac{\partial \mathbf{H}_{visc}}{\partial \xi}\right), \tag{10.6}$$

and

$$\mathbf{H}_{visc}=J^{-1}\begin{pmatrix} 0 \\ \check{\mu}C_{\xi^2}\frac{\partial v_x}{\partial \xi}+\frac{\check{\mu}}{3}C_{\xi\cdot v_\xi}\xi_x-\frac{2}{3}\xi_x\,Re\,\rho k \\ \check{\mu}C_{\xi^2}\frac{\partial v_y}{\partial \xi}+\frac{\check{\mu}}{3}C_{\xi\cdot v_\xi}\xi_y-\frac{2}{3}\xi_y\,Re\,\rho k \\ \check{\mu}C_{\xi^2}\frac{\partial v_z}{\partial \xi}+\frac{\check{\mu}}{3}C_{\xi\cdot v_\xi}\xi_z-\frac{2}{3}\xi_z\,Re\,\rho k \\ C_{\xi\cdot v}\left(\frac{\check{\mu}}{3}C_{\xi\cdot v_\xi}-\frac{2}{3}\,Re\,\rho k\right)+\cdots \\ \cdots+C_{\xi^2}\left[\frac{\kappa_\infty T_\infty}{\mu_\infty v_\infty^2}\kappa\left(1+\frac{Pr}{Pr_t}\frac{\mu_t}{\mu}\right)\frac{\partial T}{\partial \xi}+\cdots\right. \\ \cdots+\frac{1}{Sc}\sum_i \rho h_i\left(D_i+\frac{Sc}{Sc_t}\frac{\mu_t}{\rho}\right)\frac{\partial \omega_i}{\partial \xi}+\cdots \\ \left.\cdots+\check{\mu}C_{v\cdot v_\xi}+\mu^*\frac{\partial k}{\partial \xi}\right] \end{pmatrix},$$

with

$$\begin{aligned} C_{\xi^2}&=\xi_x^2+\xi_y^2+\xi_z^2,\\ C_{\xi\cdot v_\xi}&=\xi_x\frac{\partial v_x}{\partial \xi}+\xi_y\frac{\partial v_y}{\partial \xi}+\xi_z\frac{\partial v_z}{\partial \xi},\\ C_{\xi\cdot v}&=\xi_x v_x+\xi_y v_y+\xi_z v_z,\\ C_{v\cdot v_\xi}&=v_x\frac{\partial v_x}{\partial \xi}+v_y\frac{\partial v_y}{\partial \xi}+v_z\frac{\partial v_z}{\partial \xi}. \end{aligned}$$

10.6 Typical Thin Layer Navier-Stokes Solutions

A typical example with respect to a thin layer Navier-Stokes solution is given by the hypersonic flow past a blunt delta-wing, Fig. 10.3, [4]. The viscous part of this flow field is dominated by the velocity gradients normal to the surface of the configuration (boundary layer like effects), which means that no strong separation effects with flow bubbles, recirculation regions, etc. exist. The flight conditions are: $M_\infty = 8.7$, $\alpha =$

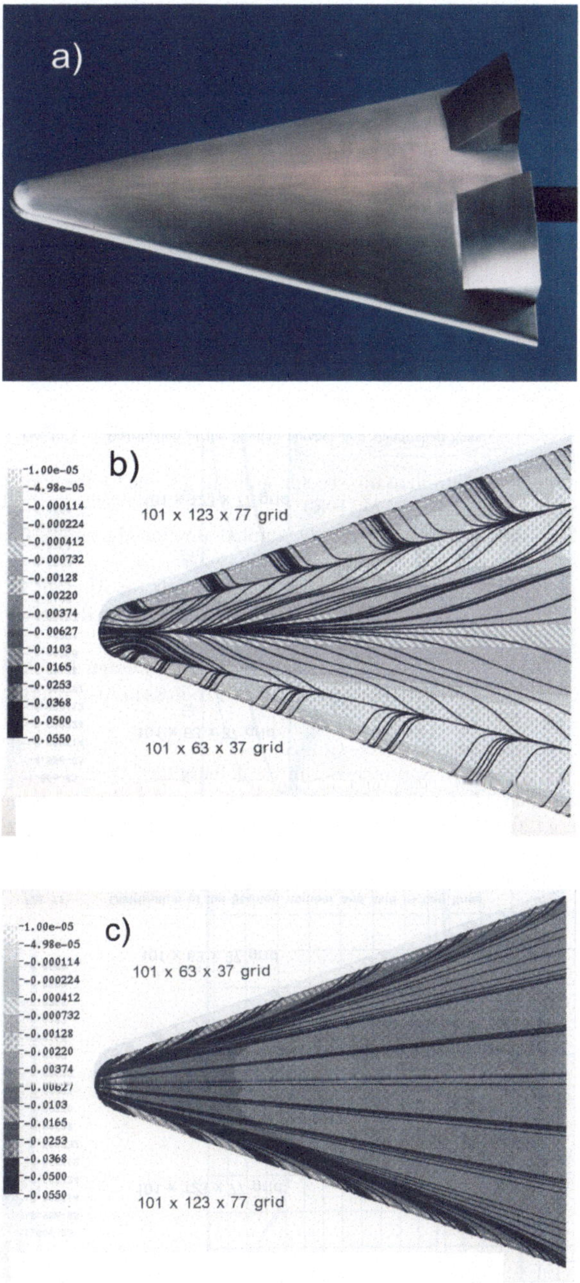

Fig. 10.3 Typical Thin Layer Navier-Stokes solution. Blunt delta-wing configuration, [4]. $M_\infty = 8.7, \alpha = 30^0, T_\infty = 55K, T_{wall} = 300K, Re/m = 2.25 \cdot 10^6$; Windtunnel model **a** skin-friction lines and Stanton number distribution on **b** leeward side and on **c** windward side

30^0, $T_\infty = 55\ K$, $T_{wall} = 300\ K$, $Re/m = 2.25 \cdot 10^6$. The upper photograph (figure a) in Fig. 10.3 shows a windtunnel model of the blunt delta wing with integrated body flaps (windward side view). The two figures (b) and (c) exhibit the Stanton number distributions and the skin friction lines (wall streamlines). The leeward side flow shows as a footprint on the surface the separation lines in the surroundings of the wing tip and a reattachment line in the symmetry plane (figure b)). On the windward side an attachment line can be observed near the wing tip (figure c)). More details to flow topologies can be found in [5–7].

References

1. Pulliam, T.H., Steger, J.L.: On implicit finite-difference simulations of three-dimensional flow. AIAA Paper 78-10 (1978)
2. Hänel, D., Schwane, R.: An implicit flux-vector splitting scheme for the computation of viscous hypersonic flow. AIAA paper 89-0274 (1989)
3. Molvik, G.A., Merkle, C.L.: A set of strongly coupled, upwind algorithms for computing flows in chemical nonequilibrium. AIAA Paper 89-0199 (1989)
4. Riedelbauch, S.: Hypersonic delta-wing flow, Case VII. 4. In: Abgrall, R., Desideri, J.-A., Glowinski, R., Mallet, M., Periaux, J. (eds.) Hypersonic Flows for Reentry Problems, vol. III. Springer Verlag, Berlin/Heidelberg (1992)
5. Hirschel, E.H., Weiland, C.: Selected Aerothermodynamic Design Problems of Hypersonic Flight Vehicles. Springer, Berlin/Heidelberg, and **229**, Progress in Astronautics and Aeronautics, AIAA, Reston, Va (2009)
6. Hirschel, E.H., Rizzi, A., Breitsamter, Ch., Staudacher, W.: Separated and Vortical Flow in Aircraft Wing Aerodynamics. Springer, Berlin/Heidelberg (2021)
7. Hirschel, E.H., Cousteix, J., Kordulla, W.: Three-Dimensional Attached Viscous Flow. Springer, Berlin/Heidelberg (2014)

Chapter 11
The Parabolized Navier-Stokes Equations (PNS)

Abstract In the second half of the 1970s and the whole 1980s there was a great interest in flow field solutions for space vehicles and military missiles flying at supersonic and hypersonic Mach numbers. With the computer power (CPU efficiency and storage capacity) at that time numerical solutions of the time dependent Navier-Stokes equations for complete three-dimensional configurations could not be achieved. Therefore fluid dynamic scientists did search for more efficient procedures to integrate the governing equations. One option was to employ so-called space marching methods, which are applicable, if the component of the velocity vector in marching direction is supersonic (valid for inviscid flow and viscous flow beyond the subsonic sublayer) and if in viscous flow no flow reversal exists [1–5]. For the Euler equations the condition for a space marching procedure is just that the velocity component with respect to the marching direction is supersonic. Then, if the time derivatives are dropped, the Euler equations are hyperbolic in space. For the Navier-Stokes equations also the time derivatives are cancelled and the viscous terms in marching direction are neglected, which leads to a mixed hyperbolic-parabolic system.

The Variables and Quantities Used in this Chapter

$\mathbf{E}, \mathbf{F}, \mathbf{G}$	vector: inviscid part of PNS equations
$\mathbf{E}_{visc}, \mathbf{F}_{visc}$	vector: viscous part of PNS equations
$\hat{\mathbf{E}}, \hat{\mathbf{F}}, \hat{\mathbf{G}}$	vector: inviscid part of Reynolds-averaged PNS equations
$\hat{\mathbf{E}}_{visc}, \hat{\mathbf{F}}_{visc}$	vector: viscous part of Reynolds-averaged PNS equations
$\hat{\mathbf{E}}_T, \hat{\mathbf{F}}_T, \hat{\mathbf{G}}_T$	vector: "inviscid part" of $k - \omega$ turbulence model for PNS equations
$\hat{\mathbf{E}}_{T,visc}, \hat{\mathbf{F}}_{T,visc}$	vector: "viscous part" of $k - \omega$ turbulence model for PNS equations
$\hat{\mathbf{S}}_T$	vector: source terms of $k - \omega$ turbulence model for PNS equations
$\hat{\tau}_{xx}, \hat{\tau}_{yy}, \hat{\tau}_{zz}$	elements of total viscous stress tensor in Cartesian
$\hat{\tau}_{xy}, \hat{\tau}_{xz}, \hat{\tau}_{yz}$	coordinates (laminar and turbulent)
$\hat{\tau}_{ij}^{turb}$	components of Reynolds stress tensor in tensor notation
$\hat{v}_x, \hat{v}_y, \hat{v}_z$	Favre-averaged velocity components in Cartesian coordinates

C. Weiland, *The Hierarchy of Fluid Dynamic Equations*,
https://doi.org/10.1007/978-3-031-71198-5_11

v''_x, v''_y, v''_z	fluctuating part of Favre-averaged velocity components
V_x, V_y, V_z	components of contravariant velocity vector
ξ, η, ζ	arbitrary coordinates for PNS equations
J	Jacobian of arbitrary coordinates
k	turbulent kinetic energy
ω	specific dissipation rate
$\hat{\beta}_x, \hat{\beta}_y, \hat{\beta}_z$	abbreviations (turbulent flow)
$\hat{e}$	$= \hat{e}_i + \frac{1}{2}\|\hat{\mathbf{v}}\|^2 + k$ total energy in turbulent flow
$\hat{e}_i$	internal energy in turbulent flow
$\bar{p}$	$= (\gamma - 1)\bar{\rho}(\hat{e} - \frac{1}{2}\|\hat{\mathbf{v}}\|^2 - k)$ pressure in turbulent flow, perfect gas
$\bar{\rho}$	density in turbulent flow

11.1 The PNS Equations for Laminar Perfect Gas Flow in Cartesian and Cylindrical Coordinates

The basis of our consideration is Eq. (5.1)[1] where we have cancelled the time derivative term $\partial Q/\partial t$ and the viscous vector $\mathbf{G}_{visc}$ in streamwise direction (z-direction).[2] This results in the PNS equations

$$\frac{\partial \mathbf{E}}{\partial x} + \frac{\partial \mathbf{F}}{\partial y} + \frac{\partial \mathbf{G}}{\partial z} = \frac{\partial \mathbf{E}_{visc}}{\partial x} + \frac{\partial \mathbf{F}_{visc}}{\partial y}, \tag{11.1}$$

where also all derivatives $\partial/\partial z$ in the remaining viscous vectors $\mathbf{E}_{visc}$ and $\mathbf{F}_{visc}$ are neglected. A very simple problem like that shown in Fig. 11.1, where the supersonic/hypersonic flow along a thick flat plate blunted with a half-cylinder is considered, can be calculated by Eq. (11.1) in Cartesian coordinates, if the computational region is bounded by $F(x, y) = \text{const}$ and $G(x, y) = \text{const}$. The same is valid for the axisymmetric flow around a hemisphere-cylinder configuration when using Eq. (11.1) in cylindrical coordinates,[3] if $F(r, \varphi) = \text{const}$ and $G(r, \varphi) = \text{const}$ holds. Since the flow in the blunted regime is partly subsonic usually a time dependent method is necessary to compute the flow in that area for generating the solution of the initial plane from which the marching procedure is started.

---

[1] $\mathbf{S}_{NS}$ is dropped.

[2] We change here the orientation of the Cartesian coordinates compared to the sections before from (x, y, z) to (z, x, y), where the z-coordinate denotes now the streamwise direction. This is done in order to be more compatible with the cylindrical coordinates (z, r, φ) which are often applied for flow cases, where the PNS approach comes into play (e.g.: missile flow).

[3] See also Sect. 4.3.

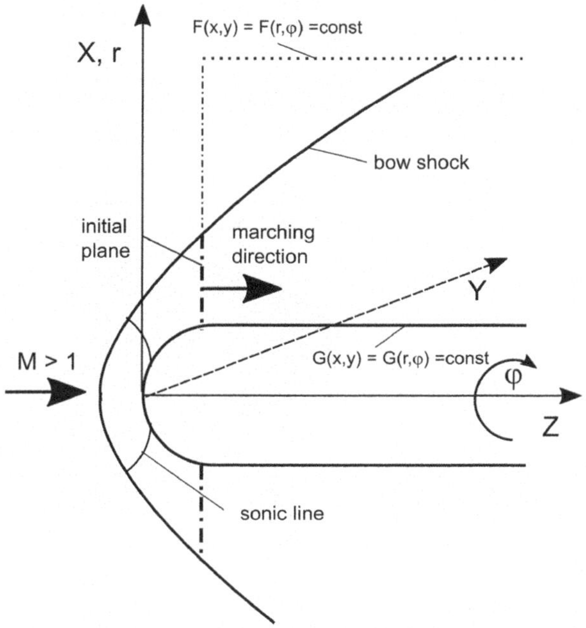

Fig. 11.1 Sketch showing an initial plane for starting the marching in z direction in a Cartesian (z, x, y) or a cylindrical (z, r, φ) frame

There exist also PNS solvers in hybrid form, which means that the full time dependent Navier-Stokes equations are considered, which are reduced to a pseudo-time iterative space marching procedure in flow regimes where the conditions for the application of such a space marching procedure are satisfied. These codes on the one hand are able to march in space where possible supersonic-hypersonic regimes are present and iterate the related marching plane in pseudo-time and can operate on the other hand in a fully time iteration version in subsonic flow field parts, for example subsonic pockets behind bow shocks, [6].

11.2 The PNS Equations for Laminar Perfect Gas Flow in Arbitrary Coordinates

In order to calculate flow solutions around real configurations, the introduction of arbitrary coordinates is mandatory. Therefore, the Eq. (6.3) are looked at and adjusted to the PNS approach.

$$\frac{\partial \mathbf{E}}{\partial \xi} + \frac{\partial \mathbf{F}}{\partial \eta} + \frac{\partial \mathbf{G}}{\partial \zeta} = \frac{\partial \mathbf{E}_{visc}}{\partial \xi} + \frac{\partial \mathbf{F}_{visc}}{\partial \eta}, \tag{11.2}$$

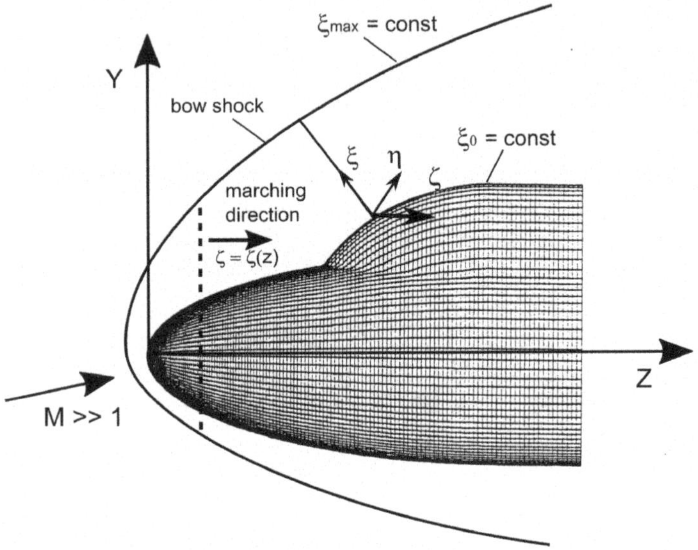

Fig. 11.2 Two-dimensional sketch showing a perpendicular initial plane for starting the marching in z direction in an arbitrary coordinate frame ξ, η, ζ

In a first step the coordinate transformation

$$\begin{aligned} \xi &= \xi(z, x, y), \\ \eta &= \eta(z, x, y), \\ \zeta &= \zeta(z), \end{aligned} \tag{11.3}$$

is applied, which means that the solution is achieved by marching with a plane perpendicular to the z-coordinate. Figures 11.2 and 11.3 illustrate the situation. Typically the flow is computed between the body contour (ξ_0 = const) and the bow shock (ξ_{max} = const). At the bow shock the boundary condition is often realized by a fitting technique on the basis of the Rankine-Hugoniot equations (see also Chap. 17).

The metric terms of the coordinate transformation Eq. (11.3) with $z_\xi = z_\eta = 0$ and $\zeta_y = \zeta_x = 0$ are given by (compare with Eqs. (6.1), (6.2))

$$\begin{aligned} &\xi_x = J(y_\eta z_\zeta), && \eta_x = J(-y_\xi z_\zeta), && \zeta_x = 0, \\ &\xi_y = J(-x_\eta z_\zeta), && \eta_y = J(x_\xi z_\zeta), && \zeta_y = 0, \\ &\xi_z = J(x_\eta y_\zeta - y_\eta x_\zeta), && \eta_z = J(y_\xi x_\zeta - x_\xi y_\zeta), && \zeta_z = J(x_\xi y_\eta - y_\xi x_\eta), \\ & && J^{-1} = x_\xi y_\eta z_\zeta - x_\eta y_\xi z_\zeta . \end{aligned} \tag{11.4}$$

The vectors $\mathbf{E}$, $\mathbf{F}$, $\mathbf{G}$, $\mathbf{E}_{visc}$, $\mathbf{F}_{visc}$ read

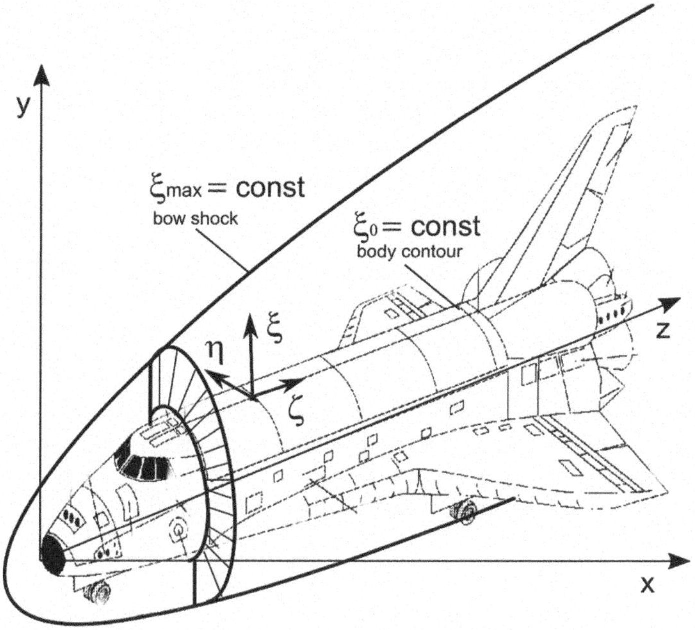

Fig. 11.3 Three-dimensional sketch showing a perpendicular initial plane for starting the marching in z direction in an arbitrary coordinate frame ξ, η, ζ

$$\mathbf{E} = J^{-1} \begin{pmatrix} \rho V_x \\ \rho v_x V_x + \xi_x p \\ \rho v_y V_x + \xi_y p \\ \rho v_z V_x + \xi_z p \\ (\rho e + p) V_x \end{pmatrix},$$

$$\mathbf{F} = J^{-1} \begin{pmatrix} \rho V_y \\ \rho v_x V_y + \eta_x p \\ \rho v_y V_y + \eta_y p \\ \rho v_z V_y + \eta_z p \\ (\rho e + p) V_y \end{pmatrix}, \mathbf{G} = J^{-1} \begin{pmatrix} \rho V_z \\ \rho v_x V_z \\ \rho v_y V_z \\ \rho v_z V_z + \zeta_z p \\ (\rho e + p) V_z \end{pmatrix},$$

$$\mathbf{E}_{visc} = J^{-1} \begin{pmatrix} 0 \\ \xi_x \tau_{xx} + \xi_y \tau_{xy} + \xi_z \tau_{xz} \\ \xi_x \tau_{yx} + \xi_y \tau_{yy} + \xi_z \tau_{yz} \\ \xi_x \tau_{zx} + \xi_y \tau_{zy} + \xi_z \tau_{zz} \\ \xi_x \beta_x + \xi_y \beta_y + \xi_z \beta_z \end{pmatrix},$$

$$
\mathbf{F}_{visc} = J^{-1} \begin{pmatrix} 0 \\ \eta_x \tau_{xx} + \eta_y \tau_{xy} + \eta_z \tau_{xz} \\ \eta_x \tau_{yx} + \eta_y \tau_{yy} + \eta_z \tau_{yz} \\ \eta_x \tau_{zx} + \eta_y \tau_{zy} + \eta_z \tau_{zz} \\ \eta_x \beta_x + \eta_y \beta_y + \eta_z \beta_z \end{pmatrix},
$$

with

$$
\begin{aligned}
V_x &= \xi_x v_x + \xi_y v_y + \xi_z v_z, \\
V_y &= \eta_x v_x + \eta_y v_y + \eta_z v_z, \\
V_z &= \zeta_z v_z.
\end{aligned} \tag{11.5}
$$

When evaluating the elements of the stress tensor τ_{ij} and the quantities β_i with respect to the coordinates ξ, η, ζ applying the chain rule Eq. (6.6) all derivatives $\partial/\partial\zeta$ must vanish.

In a second step the coordinate transformation is extended to

$$
\begin{aligned}
\xi &= \xi(z, x, y), \\
\eta &= \eta(z, x, y), \\
\zeta &= \zeta(z, x, y),
\end{aligned} \tag{11.6}
$$

which means that the marching plane must not be perpendicular to the z-coordinate, Figs. 11.4 and 11.5. The marching condition is now such that the velocity component in $\zeta(z, x, y)$ direction must be supersonic. In that case formally Eq. (11.2) holds with the metrics Eq. (6.1), while the definition of the vectors $\mathbf{E}$, $\mathbf{F}$, $\mathbf{E}_{visc}$, $\mathbf{F}_{visc}$ is the same as above and the vector $\mathbf{G}$ has the form

$$
\mathbf{G} = J^{-1} \begin{pmatrix} \rho V_z \\ \rho v_x V_z + \zeta_x p \\ \rho v_y V_z + \zeta_y p \\ \rho v_z V_z + \zeta_z p \\ (\rho e + p) V_z \end{pmatrix}.
$$

Further the set of contravariant velocities reads

$$
\begin{aligned}
V_x &= \xi_x v_x + \xi_y v_y + \xi_z v_z, \\
V_y &= \eta_x v_x + \eta_y v_y + \eta_z v_z, \\
V_z &= \zeta_x v_x + \zeta_y v_y + \zeta_z v_z,
\end{aligned}
$$

and as before all derivatives $\partial/\partial\zeta$ which arise, when the components of the stress tensor and the quantities β_i are evaluated, must vanish.

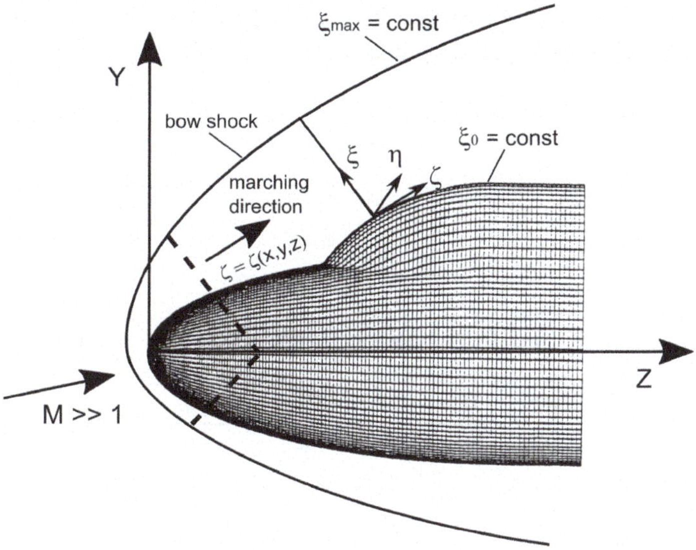

Fig. 11.4 Two-dimensional sketch showing a non-perpendicular initial plane for starting the marching in ζ direction in an arbitrary coordinate frame ξ, η, ζ

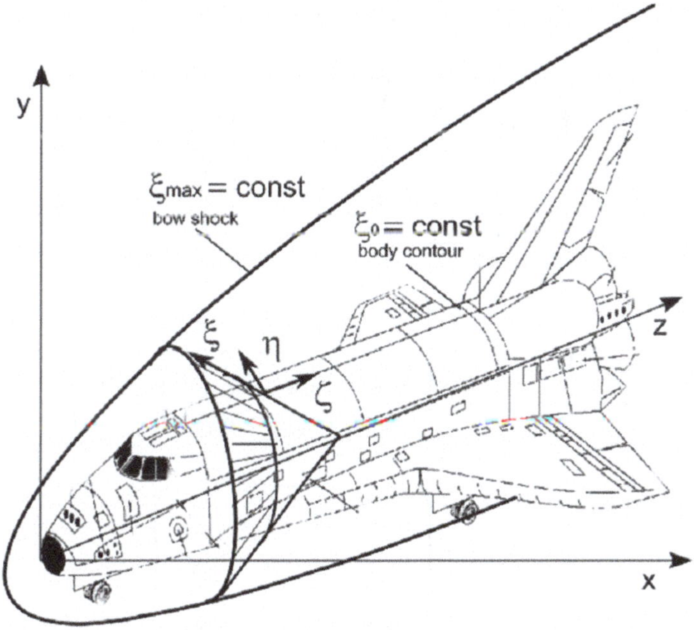

Fig. 11.5 Three-dimensional sketch showing a non-perpendicular initial plane for starting the marching in ζ direction in an arbitrary coordinate frame ξ, η, ζ

11.3 The PNS Equations for Turbulent Perfect Gas Flow in Arbitrary Coordinates

We consider only the case with the coordinate transformation

$$\begin{aligned}\xi &= \xi(z, x, y),\\ \eta &= \eta(z, x, y),\\ \zeta &= \zeta(z, x, y).\end{aligned}$$

The basic equation is Eq. (6.9), which is reduced to the PNS form and one obtains

$$\frac{\partial \hat{\mathbf{E}}}{\partial \xi} + \frac{\partial \hat{\mathbf{F}}}{\partial \eta} + \frac{\partial \hat{\mathbf{G}}}{\partial \zeta} = \frac{\partial \hat{\mathbf{E}}_{visc}}{\partial \xi} + \frac{\partial \hat{\mathbf{F}}_{visc}}{\partial \eta}, \tag{11.7}$$

with

$$\hat{\mathbf{E}} = J^{-1} \begin{pmatrix} \bar{\rho}\hat{V}_x \\ \bar{\rho}\hat{v}_x\hat{V}_x + \xi_x\bar{p} \\ \bar{\rho}\hat{v}_y\hat{V}_x + \xi_y\bar{p} \\ \bar{\rho}\hat{v}_z\hat{V}_x + \xi_z\bar{p} \\ (\bar{\rho}\hat{e} + \bar{p})\hat{V}_x \end{pmatrix},$$

$$\hat{\mathbf{F}} = J^{-1} \begin{pmatrix} \bar{\rho}\hat{V}_y \\ \bar{\rho}\hat{v}_x\hat{V}_y + \eta_x\bar{p} \\ \bar{\rho}\hat{v}_y\hat{V}_y + \eta_y\bar{p} \\ \bar{\rho}\hat{v}_z\hat{V}_y + \eta_z\bar{p} \\ (\bar{\rho}\hat{e} + \bar{p})\hat{V}_y \end{pmatrix}, \quad \hat{\mathbf{G}} = J^{-1} \begin{pmatrix} \bar{\rho}\hat{V}_z \\ \bar{\rho}\hat{v}_x\hat{V}_z + \zeta_x\bar{p} \\ \bar{\rho}\hat{v}_y\hat{V}_z + \zeta_y\bar{p} \\ \bar{\rho}\hat{v}_z\hat{V}_z + \zeta_z\bar{p} \\ (\bar{\rho}\hat{e} + \bar{p})\hat{V}_z \end{pmatrix},$$

$$\hat{\mathbf{E}}_{visc} = J^{-1} \begin{pmatrix} 0 \\ \xi_x\hat{\tau}_{xx} + \xi_y\hat{\tau}_{xy} + \xi_z\hat{\tau}_{xz} \\ \xi_x\hat{\tau}_{yx} + \xi_y\hat{\tau}_{yy} + \xi_z\hat{\tau}_{yz} \\ \xi_x\hat{\tau}_{zx} + \xi_y\hat{\tau}_{zy} + \xi_z\hat{\tau}_{zz} \\ \xi_x\hat{\beta}_x + \xi_y\hat{\beta}_y + \xi_z\hat{\beta}_z \end{pmatrix},$$

$$\hat{\mathbf{F}}_{visc} = J^{-1} \begin{pmatrix} 0 \\ \eta_x\hat{\tau}_{xx} + \eta_y\hat{\tau}_{xy} + \eta_z\hat{\tau}_{xz} \\ \eta_x\hat{\tau}_{yx} + \eta_y\hat{\tau}_{yy} + \eta_z\hat{\tau}_{yz} \\ \eta_x\hat{\tau}_{zx} + \eta_y\hat{\tau}_{zy} + \eta_z\hat{\tau}_{zz} \\ \eta_x\hat{\beta}_x + \eta_y\hat{\beta}_y + \eta_z\hat{\beta}_z \end{pmatrix},$$

$$\hat{\tau}_{xx} = 2\breve{\mu}\left(\frac{\partial \hat{v}_x}{\partial x} - \frac{1}{3}\left(\frac{\partial \hat{v}_x}{\partial x} + \frac{\partial \hat{v}_y}{\partial y} + \frac{\partial \hat{v}_z}{\partial z}\right)\right) - \frac{2}{3}\bar{\rho}k\,,$$

$$\hat{\tau}_{yy} = 2\breve{\mu}\left(\frac{\partial \hat{v}_y}{\partial y} - \frac{1}{3}\left(\frac{\partial \hat{v}_x}{\partial x} + \frac{\partial \hat{v}_y}{\partial y} + \frac{\partial \hat{v}_z}{\partial z}\right)\right) - \frac{2}{3}\bar{\rho}k\,,$$

$$\hat{\tau}_{zz} = 2\breve{\mu}\left(\frac{\partial \hat{v}_z}{\partial z} - \frac{1}{3}\left(\frac{\partial \hat{v}_x}{\partial x} + \frac{\partial \hat{v}_y}{\partial y} + \frac{\partial \hat{v}_z}{\partial z}\right)\right) - \frac{2}{3}\bar{\rho}k,$$

$$\hat{\tau}_{xy} = \hat{\tau}_{yx} = \breve{\mu}\left(\frac{\partial \hat{v}_x}{\partial y} + \frac{\partial \hat{v}_y}{\partial x}\right),$$

$$\hat{\tau}_{xz} = \hat{\tau}_{zx} = \breve{\mu}\left(\frac{\partial \hat{v}_x}{\partial z} + \frac{\partial \hat{v}_z}{\partial x}\right),$$

$$\hat{\tau}_{yz} = \hat{\tau}_{zy} = \breve{\mu}\left(\frac{\partial \hat{v}_y}{\partial z} + \frac{\partial \hat{v}_z}{\partial y}\right),$$

$$\hat{\beta}_x = \breve{\kappa}\frac{\partial \hat{T}}{\partial x} + \hat{v}_x\hat{\tau}_{xx} + \hat{v}_y\hat{\tau}_{xy} + \hat{v}_z\hat{\tau}_{xz} + \mu^*\frac{\partial k}{\partial x},$$

$$\hat{\beta}_y = \breve{\kappa}\frac{\partial \hat{T}}{\partial y} + \hat{v}_x\hat{\tau}_{yx} + \hat{v}_y\hat{\tau}_{yy} + \hat{v}_z\hat{\tau}_{yz} + \mu^*\frac{\partial k}{\partial y},$$

$$\hat{\beta}_z = \breve{\kappa}\frac{\partial \hat{T}}{\partial z} + \hat{v}_x\hat{\tau}_{zx} + \hat{v}_y\hat{\tau}_{zy} + \hat{v}_z\hat{\tau}_{zz} + \mu^*\frac{\partial k}{\partial z}.$$

The total energy $\hat{e}$, the pressure $\bar{p}$ and the turbulent kinetic energy k are

$$\hat{e} = \hat{e}_i + \frac{1}{2}|\hat{\mathbf{v}}|^2 + k\,,$$

$$\bar{p} = (\gamma - 1)\bar{\rho}(\hat{e} - \frac{1}{2}|\hat{\mathbf{v}}|^2 - k),$$

$$k = \frac{1}{2}((\widehat{v_x''})^2 + (\widehat{v_y''})^2 + (\widehat{v_z''})^2).$$

Again all derivatives $\partial/\partial\zeta$, which arise, when the components of the stress tensor and the quantities β_i are evaluated, are neglected and further the inner derivatives with respect to the z, x, y coordinates must be transformed by the chain rule regarding the ξ, η, ζ coordinates.

11.4 A Two-Equation Turbulence Model for the PNS Equations in Arbitrary Coordinates

There are no restrictions or assumptions necessary for the installation of algebraic turbulence models, like the Baldwin-Lomax model, Eqs. (8.1, 8.2), in the PNS equations. On the other hand one- and two-equation models require adjustments. For example the $k - \omega$ model in arbitrary coordinates, Eq. (8.6) reduces to, [7, 8]

$$\frac{\partial \hat{\mathbf{E}}_T}{\partial \xi} + \frac{\partial \hat{\mathbf{F}}_T}{\partial \eta} + \frac{\partial \hat{\mathbf{G}}_T}{\partial \zeta} = \frac{\partial \hat{\mathbf{E}}_{T,visc}}{\partial \xi} + \frac{\partial \hat{\mathbf{F}}_{T,visc}}{\partial \eta} + \hat{S}_T \,, \tag{11.8}$$

$$\hat{\mathbf{E}}_T = J^{-1}\begin{pmatrix} \bar{\rho}\hat{V}_x k \\ \bar{\rho}\hat{V}_x \omega \end{pmatrix}, \quad \hat{\mathbf{F}}_T = J^{-1}\begin{pmatrix} \bar{\rho}\hat{V}_y k \\ \bar{\rho}\hat{V}_y \omega \end{pmatrix}, \quad \hat{\mathbf{G}}_T = J^{-1}\begin{pmatrix} \bar{\rho}\hat{V}_z k \\ \bar{\rho}\hat{V}_z \omega \end{pmatrix},$$

$$\hat{\mathbf{E}}_{T,visc} = J^{-1}\mu^* \begin{pmatrix} \xi_x \dfrac{\partial k}{\partial x} + \xi_y \dfrac{\partial k}{\partial y} + \xi_z \dfrac{\partial k}{\partial z} \\ \xi_x \dfrac{\partial \omega}{\partial x} + \xi_y \dfrac{\partial \omega}{\partial y} + \xi_z \dfrac{\partial \omega}{\partial z} \end{pmatrix},$$

$$\hat{\mathbf{F}}_{T,visc} = J^{-1}\mu^* \begin{pmatrix} \eta_x \dfrac{\partial k}{\partial x} + \eta_y \dfrac{\partial k}{\partial y} + \eta_z \dfrac{\partial k}{\partial z} \\ \eta_x \dfrac{\partial \omega}{\partial x} + \eta_y \dfrac{\partial \omega}{\partial y} + \eta_z \dfrac{\partial \omega}{\partial z} \end{pmatrix},$$

$$\hat{\mathbf{S}}_T = \begin{pmatrix} \hat{\tau}_{ij}^{turb} \dfrac{\partial \hat{v}_i}{\partial x_j} - \beta^* \bar{\rho} \omega k \\ \alpha \dfrac{\omega}{k} \hat{\tau}_{ij}^{turb} \dfrac{\partial \hat{v}_i}{\partial x_j} - \beta \bar{\rho} \omega^2 \end{pmatrix}.$$

The term $\hat{\tau}_{ij}^{turb} \dfrac{\partial \hat{v}_i}{\partial x_j}$ in $\hat{\mathbf{S}}_T$ is expanded and reads

$$\begin{aligned} \hat{\tau}_{ij}^{turb} \frac{\partial \hat{v}_i}{\partial x_j} = \; & \hat{\tau}_{xx} \frac{\partial \hat{v}_x}{\partial x} + \hat{\tau}_{xy} \frac{\partial \hat{v}_x}{\partial y} + \hat{\tau}_{xz} \frac{\partial \hat{v}_x}{\partial z} + \\ & \hat{\tau}_{yx} \frac{\partial \hat{v}_y}{\partial x} + \hat{\tau}_{yy} \frac{\partial \hat{v}_y}{\partial y} + \hat{\tau}_{yz} \frac{\partial \hat{v}_y}{\partial z} + \\ & \hat{\tau}_{zx} \frac{\partial \hat{v}_x}{\partial z} + \hat{\tau}_{zy} \frac{\partial \hat{v}_z}{\partial y} + \hat{\tau}_{zz} \frac{\partial \hat{v}_z}{\partial z}, \end{aligned}$$

Fig. 11.6 Sketch of the Space Shuttle Orbiter forebody configuration, illustrating the PNS strategy. Figure taken from [9]

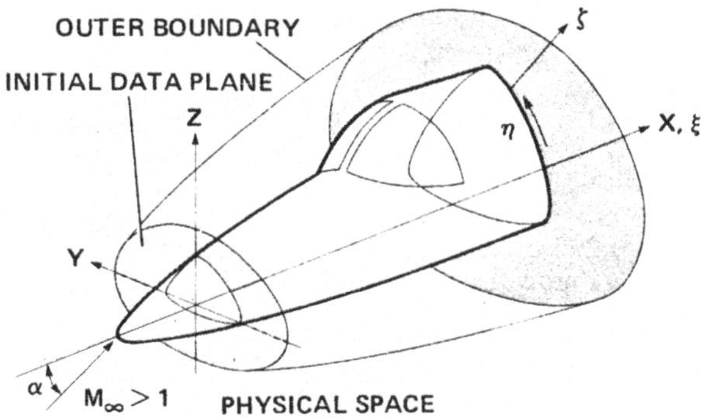

where all the inner derivatives with respect to the Cartesian coordinates (of $\hat{\tau}_{ij}^{turb}$ and $\frac{\partial \hat{v}_i}{\partial x_j}$) must be transformed by the chain rule, see Eq. (6.10).

11.5 Typical Parabolized Navier-Stokes Solutions

One of the motivations for the development of the PNS equations based numerical methods was the Space Shuttle Orbiter. At the end of the 1970s and during the 1980s several papers were concerned with the computation of the flow past the forebody of the Space Shuttle Orbiter, [3, 4, 9, 10]. In the regime of the stagnation point, where a subsonic flow region exists, the PNS method can not be applied, instead a time-dependent method has to be used, which provides then the initial conditions (initial data plane) for the PNS computation, for illustration see Fig. 11.6.

Other examples of application of the PNS approach are given by the flows through ducts and inlets of propulsions systems, [11, 12]. As an example a two-dimensional inlet flow is presented. The sketch, Fig. 11.7, exhibits the geometry, which consists of two horizontal parts (ducts) connected with a 15^0 ramp part. The ramp compresses the flow and leads to a complicated flow field with interactions between shocks, shear layers and boundary layers. The flow field is laminar and no separation occurs neither in the compression corner nor elsewhere. The freestream conditions are: $M_\infty = 15.0$, $T_\infty = 10\,K$, $Re = 8 \cdot 10^4$, $T_{wall} = 1000K$, $\gamma = 1.4$, $Pr = 0.72$.

Figure 11.8 presents the presssure distribution in the duct2 part of the configuration obtained by the numerical solution of the PNS equations. The flow pattern with the production and reflections of the shocks as well as the interactions between boundary layers and shocks can clearly be identified.[4]

[4] Such flowfields contain usually also weak discontinuities like shear layers. But these shear layers can not be detected by pressure distributions plots, because the pressure is continuous across these interfaces. Instead density or entropy sensors would be able to detect such weak discontinuities, see Chap. 16.

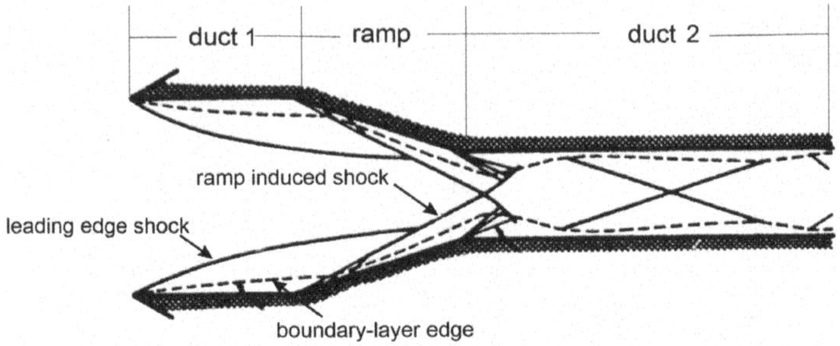

Fig. 11.7 Sketch of the inlet flow generated by a combined duct/ramp configuration. Schematic view of the flow structure. Figure taken from [11]

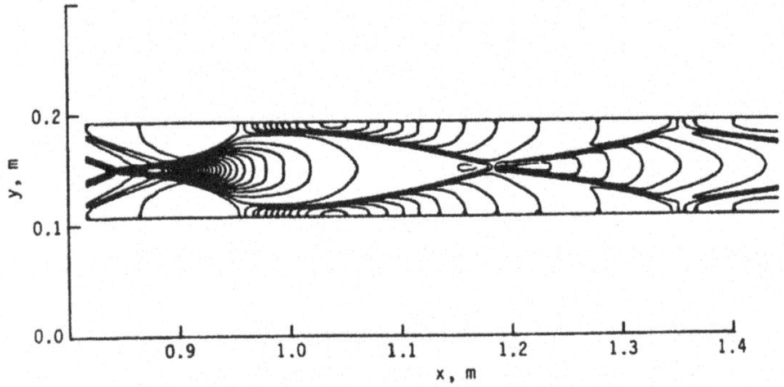

Fig. 11.8 Inlet flow: pressure contours in duct2 of the inlet configuration. $M_\infty = 15.0$, $T_\infty = 10K$, $Re = 8 \cdot 10^4$, $T_{wall} = 1000K$, $\gamma = 1.4$, $Pr = 0.72$. Figure taken from [11]

References

1. Hirsch, C.: Numerical Computation of Internal and External Flows, vol. I. Wiley, New York, Reprint (1997)
2. Tannehill, J.C., Venkatapathy, E., Rakich, J.V.: Numerical solution of supersonic viscous flow over blunt delta wings. AIAA J. **20**, 203–210 (1982)
3. Rakich, J.V., Venkatapathy, E., Tannehill, J.C., Prabhu, D.: Numerical solution of space shuttle orbiter flowfield. AIAA J. Spacecraft **21**, 9–15 (1984)
4. Chaussee, D.S.: High speed viscous flow calculation about complex configurations. AGARD 58th Fluid Dynamics Panel Symposium, Aix-en-Provence, Paper No. 29 (1986)
5. Tannehill, J.C., Buelow, P.E., Ievalts, J.O., Lawrence, S.L.: A three-dimensional upwind parabolized Navier-Stokes code for real gas flows. AIAA Paper 89-1651 (1989)
6. White, J.A., Morrison, J.H.: A pseudo-temporal multi-grid relaxation scheme for solving the parabolized Navier-Stokes equations. AIAA Paper 99-3360 (1999)
7. Sinha, N., Dash, S.M.: Parabolized Navier-stokes analysis of ducted turbulent mixing problems with finite-rate chemistry. AIAA Paper 86-0004 (1986)
8. Dash, S.M., Wolf, D.E., Seiner, J.M.: Analysis of turbulent under expanded jets, Part I: Parabolized Navier-Stokes Model. SCIPVIS. AIAA J. **23**, 505–514 (1985)
9. Schiff, L.B., Steger, J.L.: Numerical simulation of steady supersonic viscous flow. AIAA J. **18**, 1421–1430 (1980)

10. Szema, K.Y., Griffith, B.J., Maus, J.R., Best, J.T.: Laminar viscous flow field prediction of shuttle-like vehicle aerodynamics. AIAA Paper 83-0211 (1983)
11. Lawrence, S.L., Tannehill, J.C., Chaussee, D.S.: Upwind algorithm for the parabolized Navier-Stokes equations. AIAA J. **27**, 1175–1183 (1989)
12. Prince, S.A., Williams, M.J., Edwards, J.A.: Application of a parabolized Navier-Stokes solver to some problems in hypersonic propulsion aerodynamics. AIAA Paper 2001-4066 (2001)

Chapter 12
Incompressible Forms of the Fluid Dynamic Equations

Abstract Incompressible flow primarily means that the density ρ is a constant. In general, when the flow is isothermal, no dependency between the velocity field and the temperature field exists, which requires also that the dynamic viscosity is a constant. Of course, if the flow field permits temperature variations, for example by heat conduction or by other external heat sources as well as by variations of the viscosity with temperature, a coupling between the fields of velocity and temperature is necessary.

The Variables and Quantities Used in this Chapter

v_x, v_y, v_z	components of velocity vector $\mathbf{v}$ in Cartesian coordinates
x, y, z	Cartesian coordinates
p	pressure
ρ	density
ν	kinematic viscosity
Re	Reynolds number
$\mathbf{f}_e$	external forces
$\bar{\bar{\tau}}$	viscous stress tensor
$\boldsymbol{\omega}$	vorticity vector
U	potential of external force in conservative field

12.1 The Incompressible Navier-Stokes Equations

In the following we state the Navier-Stokes equations for flow which are incompressible and isothermal. The continuity Eq. (4.12) reduces to

$$\frac{\partial v_x}{\partial x} + \frac{\partial v_y}{\partial y} + \frac{\partial v_z}{\partial z} = 0, \tag{12.1}$$

C. Weiland, *The Hierarchy of Fluid Dynamic Equations*,
https://doi.org/10.1007/978-3-031-71198-5_12

which means that an incompressible flow field is solenoidal.[1]

For the momentum Eqs. (4.14) together with the relations for the elements of the stress tensor Eqs. (4.22) we find

$$\begin{aligned}\frac{\partial v_x}{\partial t}+v_x\frac{\partial v_x}{\partial x}+v_y\frac{\partial v_x}{\partial y}+v_z\frac{\partial v_x}{\partial z}&=-\frac{1}{\rho}\frac{\partial p}{\partial x}+\nu\left(\frac{\partial^2 v_x}{\partial x^2}+\frac{\partial^2 v_x}{\partial y^2}+\frac{\partial^2 v_x}{\partial z^2}\right)\\ \frac{\partial v_y}{\partial t}+v_x\frac{\partial v_y}{\partial x}+v_y\frac{\partial v_y}{\partial y}+v_z\frac{\partial v_y}{\partial z}&=-\frac{1}{\rho}\frac{\partial p}{\partial y}+\nu\left(\frac{\partial^2 v_y}{\partial x^2}+\frac{\partial^2 v_y}{\partial y^2}+\frac{\partial^2 v_y}{\partial z^2}\right)\\ \frac{\partial v_z}{\partial t}+v_x\frac{\partial v_z}{\partial x}+v_y\frac{\partial v_z}{\partial y}+v_z\frac{\partial v_z}{\partial z}&=-\frac{1}{\rho}\frac{\partial p}{\partial z}+\nu\left(\frac{\partial^2 v_z}{\partial x^2}+\frac{\partial^2 v_z}{\partial y^2}+\frac{\partial^2 v_z}{\partial z^2}\right)\end{aligned} \tag{12.2}$$

In formal vector notation the momentum equation for an incompressible flow field reads

$$\frac{\partial \mathbf{v}}{\partial t}+(\mathbf{v}\cdot\mathrm{grad})\mathbf{v}=-\frac{1}{\rho}\mathrm{grad}\,p+\nu\Delta\mathbf{v}. \tag{12.3}$$

With Eqs. (12.1), (12.2) we have four equations for the four unknowns v_x, v_y, v_z, p (ρ = const). But Eq. (12.1) has no time evolutionary character and can not be used for the computation of the pressure. A possible way out is to apply the divergence operation **div** and the solenoidality of the velocity field on Eq. (12.3), which results in the following relation for the pressure

$$\underbrace{\mathrm{div}\left(\frac{\partial \mathbf{v}}{\partial t}\right)}_{=0}+\mathrm{div}(\mathbf{v}\cdot\mathrm{grad})\mathbf{v}=-\frac{1}{\rho}\mathrm{div\ grad}\,p+\underbrace{\mathrm{div}(\nu\Delta\mathbf{v})}_{=0},$$

$$\Longrightarrow\quad \frac{1}{\rho}\Delta\,p=-\mathrm{div}(\mathbf{v}\cdot\mathrm{grad})\mathbf{v}, \tag{12.4}$$

where $\Delta \equiv$ (div grad) is the Laplace operator with $\Delta p=\frac{\partial^2 p}{\partial x^2}+\frac{\partial^2 p}{\partial y^2}+\frac{\partial^2 p}{\partial z^2}$. Equation (12.4) constitutes a Poisson equation for which specific numerical solution procedures exist.

12.2 A Typical Solution of the Incompressible Navier-Stokes Equations

A square cylinder was placed between parallel walls, where the distance H between the walls and the width B of the square cylinder ($B = A$) determines the blockage ratio of the flow, Fig. 12.1. The upstream velocity profile was measured in an

[1] Solenoidal means without sources, in this case without mass sources.

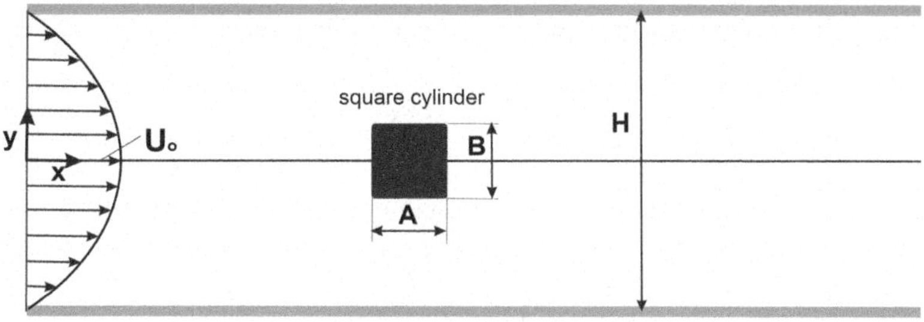

Fig. 12.1 Sketch of the configuration regarding the flow around a square cylinder

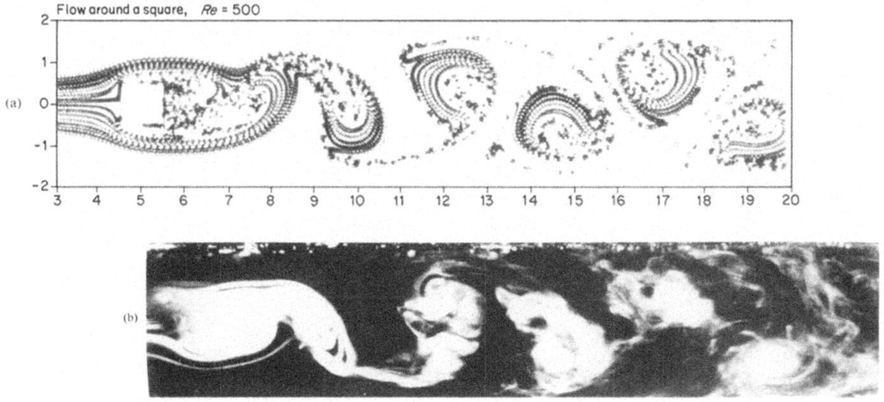

Fig. 12.2 Flow around a square cylinder. Comparison of a numerical result of the incompressible Navier-Stokes equations (above) with experimental data (below). Figures taken from [1]

experiment by a hot-wire anemometry probe and then applied as freestream condition in the computation. Figure 12.2 show for the blockage ratio $H^{-1} = 1/4$ and the Reynolds numbers $Re = 500$ (computation) respectively $Re = 550$ (experiment), the streakline plot of the computation (upper figure) and the corresponding experimental smoke-wire photograph (lower figure). The flow structure behind the square cylinder indicates the well-known vortex street, which is typical at such Reynolds numbers. The patterns in both cases, experimentally and theoretically, are qualitatively in a very nice accordance [1].

12.3 The Vorticity Equation

We consider the momentum equation in the version (3.19)

$$\frac{\partial \mathbf{v}}{\partial t} + \text{grad}\frac{\mathbf{v}^2}{2} - (\mathbf{v} \times \text{rot}\mathbf{v}) = \mathbf{f}_e - \frac{1}{\rho}\text{grad}\, p + \frac{1}{\rho}\text{div}\,\bar{\bar{\tau}}$$

where the friction term is formulated for $\mu = \text{const}$ and reads

$$\frac{1}{\rho}\text{div}\bar{\bar{\tau}} = \nu \Delta \mathbf{v}.$$

For the derivation of the vorticity equation the following relations and definitions are utilized

$$\text{rot}\left(\frac{1}{\rho}\text{grad}\, p\right) = -\text{grad}\, p \times \text{grad}\,\frac{1}{\rho},$$
$$\boldsymbol{\omega} = \text{rot}\,\mathbf{v} : \text{vorticity vector},$$

$$\text{rot grad}\, a = 0, \quad a : \text{scalar}, \quad \text{div rot}\,\mathbf{b} = 0, \quad \mathbf{b} : \text{vector},$$
$$\text{rot}(\mathbf{v} \times \boldsymbol{\omega}) = (\boldsymbol{\omega} \cdot \text{grad})\,\mathbf{v} - (\mathbf{v} \cdot \text{grad})\,\boldsymbol{\omega} + \mathbf{v}\,(\text{div}\,\boldsymbol{\omega}) - \boldsymbol{\omega}\,(\text{div}\,\mathbf{v}), \quad (12.5)$$

Taking[2] the *curl* of the momentum equation by the vector operation "rot" given above and assume that $\mathbf{f}_e$ is an external force in a conservative field, which holds a potential U with $\mathbf{f}_e = \text{grad}\, U$, one obtains

$$\frac{\partial \boldsymbol{\omega}}{\partial t} + \text{rot grad}\,\frac{\mathbf{v}^2}{2} - \text{rot}(\mathbf{v} \times \boldsymbol{\omega}) = \text{rot}\,\mathbf{f}_e - \text{rot}\left(\frac{1}{\rho}\text{grad}\, p\right) + \nu\, \Delta \text{rot}\mathbf{v}. \quad (12.6)$$

Applying the relations Eq. (12.5) to Eq. (12.6) one finds

$$\frac{D\boldsymbol{\omega}}{Dt} = \frac{\partial \boldsymbol{\omega}}{\partial t} + (\mathbf{v} \cdot \text{grad})\boldsymbol{\omega} = (\boldsymbol{\omega} \cdot \text{grad})\,\mathbf{v} - \boldsymbol{\omega}(\text{div}\,\mathbf{v}) + \text{grad}\, p \times \text{grad}\frac{1}{\rho} + \nu\, \Delta \boldsymbol{\omega}. \quad (12.7)$$

This is the general vorticity transport equation of von Helmholtz for viscous fluids and conservatively external forces. Incompressible viscous fluids reduce the equation to

$$\frac{D\boldsymbol{\omega}}{Dt} = (\boldsymbol{\omega} \cdot \text{grad})\,\mathbf{v} + \nu\, \Delta \boldsymbol{\omega}, \quad (12.8)$$

and incompressible inviscid fluids to

$$\frac{D\boldsymbol{\omega}}{Dt} = (\boldsymbol{\omega} \cdot \text{grad})\,\mathbf{v}. \quad (12.9)$$

The last equation is called the vorticity differential equation of von Helmholtz.

Remark In the case of a plane flow the vectors $\mathbf{v}$ and $\boldsymbol{\omega}$ are perpendicular to each other ($\mathbf{v} \perp \boldsymbol{\omega}$), where the vorticity vector has only the z-component $\boldsymbol{\omega} = (0, 0, \omega_z)^T$. The consequence is that the term $(\boldsymbol{\omega} \cdot \text{grad})\mathbf{v}$ in Eq. (12.8) vanishes and it remains

---

[2] $\boldsymbol{\omega}$ is also denoted as angular velocity.

$$\frac{D\omega_z}{Dt} = \nu \Delta \omega_z.$$

This equation describes the diffusion of the vorticity.

Reference

1. Davis, R.W., Moore, E.F., Purtell, L.P.: A numerical-experimental study of confined flow around rectangular cylinders. Phys. Fluids **27**, 46–59 (1984)

Chapter 13
Boundary Layer Equations

Abstract The boundary layer concept of Ludwig Prandtl was a breakthrough in the understanding of flow behavior. The boundary layer equations can be derived from the Navier-Stokes equations by physical estimations of the development of the boundary layer as well as the magnitude of the corresponding fluid dynamic variables.

The Variables and Quantities Used in this Chapter

v_x, v_y, v_z	components of velocity vector $\mathbf{v}$ in Cartesian coordinates
x, y, z	Cartesian coordinates
p	pressure
ρ	density
T	temperature
l	reference length
δ	boundary layer thickness
μ	dynamic viscosity
κ	thermal conductivity
c_p	specific heat at constant pressure
Re	Reynolds number
Pr	Prandtl number
Ec	Eckert number

13.1 The Derivation of the Boundary Layer Equations

We consider the continuity Eq. (4.12), the momentum Eqs. (4.14) and the energy equation in the form Eq. (4.21) for steady flow. L. Prandtl did observe, [1], that the flow along a flat plate develops due to the no slip condition at the wall and for small values of the viscosity towards a very small (friction) layer of thickness δ at the wall,

C. Weiland, *The Hierarchy of Fluid Dynamic Equations*,
https://doi.org/10.1007/978-3-031-71198-5_13

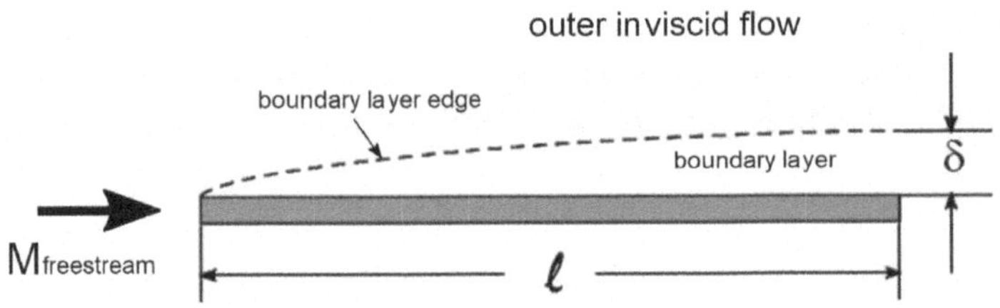

Fig. 13.1 Illustration of boundary layer development on a flat plate

Fig. 13.1. The thickness of this layer is small compared, for example, to the length of the flat plate $\delta \ll l$.

With the dimensionsless variables

$$v'_x = \frac{v_x}{\bar{v}},\ v'_y = \frac{v_y}{\bar{v}},\ v'_z = \frac{v_z}{\bar{v}},\ p' = \frac{p}{\bar{\rho}\bar{v}^2},\ \rho' = \frac{\rho}{\bar{\rho}},\ x' = \frac{x}{l},\ y' = \frac{y}{\delta},$$
$$z' = \frac{z}{l},\ \kappa' = \frac{\kappa}{\bar{\kappa}},\ \mu' = \frac{\mu}{\bar{\mu}},\ T' = \frac{T}{\bar{T}},\ c'_p = \frac{c_p}{\bar{c}_p},\ t' = \frac{t}{\bar{t}},$$

the above mentioned equations are made dimensionless whereby the dimensionless quantities Reynolds number Re, Prandtl number Pr and Eckert Ec number are introduced, which appear in the boundary layer equations derived below

$$Re = \frac{\bar{\rho}\bar{v}l}{\bar{\mu}},\quad Pr = \frac{\bar{\mu}\bar{c}_p}{\bar{\kappa}},\quad Ec = \frac{\bar{v}^2}{\bar{c}_p\bar{T}} = (\gamma - 1)M_\infty^2\,.$$

Further, the orders of magnitude are supposed by

$$v'_x \sim 1,\quad v'_y \sim 1,\quad v'_z \sim 1,\quad p' \sim 1,\quad \rho' \sim 1,\quad x' \sim 1,$$
$$y' \sim 1,\quad z' \sim 1,\quad \kappa' \sim 1,\quad \mu' \sim 1,\quad T' \sim 1\ \ c'_p \sim 1.$$

The three-dimensional continuity equation in dimensionless form then reads

$$\frac{\partial \rho' v'_x}{\partial x'} + \frac{\partial \rho' v'_y}{\partial y'}\frac{l}{\delta} + \frac{\partial \rho' v'_z}{\partial z'} = 0.$$

The two-dimensional continuity equation gives a better insight into the concept of the estimation and it follows

$$\frac{\partial \rho' v'_x}{\partial x'} + \frac{\partial \rho' v'_y}{\partial y'}\frac{l}{\delta} = 0\,.$$

It is simple to see that the second term of the left hand side is only of order $O(1)$ if $v''_y = v'_y \frac{l}{\delta} \sim O(1)$ is true. Further we assume that the differential operators $\frac{\partial}{\partial x'}, \frac{\partial}{\partial y'}, \frac{\partial}{\partial z'}$ do not change the order of magnitude.

The continuity equation then has the form

$$\frac{\partial \rho' v'_x}{\partial x'} + \frac{\partial \rho' v''_y}{\partial y'} + \frac{\partial \rho' v'_z}{\partial z'} \approx 0\,. \tag{13.1}$$

The x-momentum equation (without gravitation term) is regarded in the form of Eq. (4.14)

$$\rho v_x \frac{\partial v_x}{\partial x} + \rho v_y \frac{\partial v_x}{\partial y} + \rho v_z \frac{\partial v_x}{\partial z} = -\frac{\partial p}{\partial x} + \frac{\partial \tau_{xx}}{\partial x} + \frac{\partial \tau_{xy}}{\partial y} + \frac{\partial \tau_{xz}}{\partial z}.$$

Inserting the above estimation in the dimensionless x-momentum equation yields

$$\begin{aligned}
& v'_x \frac{\partial v'_x}{\partial x'} + v''_y \frac{\partial v'_x}{\partial y'} + v'_z \frac{\partial v'_x}{\partial z'} = \\
& -\frac{1}{\rho'}\frac{\partial p'}{\partial x'} + \underbrace{\frac{1}{Re}\frac{l^2}{\delta^2}}_{\sim 1} 2 \underbrace{\frac{1}{\rho'}\frac{\delta^2}{l^2}\frac{\partial}{\partial x'}\mu'\left(v\frac{\partial v'_x}{\partial x'} - \frac{1}{3}\left(\frac{\partial v'_x}{\partial x'} + \frac{\partial v''_y}{\partial y'} + \frac{\partial v'_z}{\partial z'}\right)\right)}_{\ll 1} + \\
& \underbrace{\frac{1}{Re}\frac{l^2}{\delta^2}}_{\sim 1}\underbrace{\frac{1}{\rho'}\frac{\partial}{\partial y'}\mu'\frac{\partial v'_x}{\partial y'}}_{\sim 1} + \underbrace{\frac{1}{Re}\frac{l^2}{\delta^2}}_{\sim 1}\underbrace{\frac{1}{\rho'}\frac{\partial}{\partial y'}\mu'\frac{\delta^2}{l^2}\frac{\partial v''_y}{\partial x'}}_{\ll 1} + \underbrace{\frac{1}{Re}\frac{l^2}{\delta^2}}_{\sim 1}\underbrace{\frac{1}{\rho'}\frac{\delta^2}{l^2}\frac{\partial}{\partial z'}\mu'\left(\frac{\partial v'_z}{\partial x'} + \frac{\partial v'_x}{\partial z'}\right)}_{\ll 1} \\
& \approx -\frac{1}{\rho'}\frac{\partial p'}{\partial x'} + \frac{1}{Re}\frac{l^2}{\delta^2}\frac{1}{\rho'}\frac{\partial}{\partial y'}\mu'\frac{\partial v'_x}{\partial y'},
\end{aligned} \tag{13.2}$$

and it follows, assumed that the viscous term is active and that the convective terms and the pressure term are proportional to unity (~ 1) for $Re \gg 1$

$$\frac{\delta}{l} \sim \frac{1}{\sqrt{Re}}. \tag{13.3}$$

With Eq. (13.3) the dimensionless quantities y' and v''_y are rearranged by

$$y' = \frac{y}{l}\sqrt{Re} \text{ and } v''_y = \frac{v_y}{\bar{v}}\sqrt{Re}. \tag{13.4}$$

The y-momentum equation

$$\rho v_x \frac{\partial v_y}{\partial x} + \rho v_y \frac{\partial v_y}{\partial y} + \rho v_z \frac{\partial v_y}{\partial z} = -\frac{\partial p}{\partial y} + \frac{\partial \tau_{yx}}{\partial x} + \frac{\partial \tau_{yy}}{\partial y} + \frac{\partial \tau_{yz}}{\partial z}$$

is made dimensionless and one obtains

$$\begin{aligned}
&\frac{1}{Re}\left(v'_x \frac{\partial v''_y}{\partial x'} + v''_y \frac{\partial v''_y}{\partial y'} + v'_z \frac{\partial v''_y}{\partial z'}\right) = -\frac{1}{\rho'}\frac{\partial p'}{\partial y'} \\
&+\frac{1}{Re}\frac{1}{\rho'}\frac{\partial}{\partial x'}\mu'\frac{\partial v'_x}{\partial y'} + \frac{1}{Re^2}\frac{1}{\rho'}\frac{\partial}{\partial x'}\mu'\frac{\partial v''_y}{\partial x'} \\
&+\frac{1}{Re}\frac{1}{\rho'}\frac{\partial}{\partial y'}2\mu'\left(\frac{\partial v''_y}{\partial y'} - \frac{1}{3}\left(\frac{\partial v'_x}{\partial x'} + \frac{\partial v''_y}{\partial y'} + \frac{\partial v'_z}{\partial z'}\right)\right) + \\
&\frac{1}{Re^2}\frac{1}{\rho'}\frac{\partial}{\partial z'}\mu'\frac{\partial v''_y}{\partial z'} + \frac{1}{Re}\frac{1}{\rho'}\frac{\partial}{\partial z'}\mu'\frac{\partial v'_z}{\partial y'}.
\end{aligned} \tag{13.5}$$

For $Re \gg 1$ this results in

$$\frac{\partial p'}{\partial y'} \approx 0. \tag{13.6}$$

The dimensionless z-momentum equation with the estimations Eqs. (13.4) then is

$$v'_x \frac{\partial v'_z}{\partial x'} + v''_y \frac{\partial v'_z}{\partial y'} + v'_z \frac{\partial v'_z}{\partial z'} \approx -\frac{1}{\rho'}\frac{\partial p'}{\partial z'} + \underbrace{\frac{1}{Re}\frac{l^2}{\delta^2}}_{\sim 1}\frac{1}{\rho'}\frac{\partial}{\partial y'}\mu'\frac{\partial v'_z}{\partial y'}, \tag{13.7}$$

Finally, the energy Eq. (4.21)

$$\begin{aligned}
&\rho c_p\left(v_x \frac{\partial T}{\partial x} + v_y \frac{\partial T}{\partial y} + v_z \frac{\partial T}{\partial z}\right) = v_x \frac{\partial p}{\partial x} + v_y \frac{\partial p}{\partial y} + v_z \frac{\partial p}{\partial z} \\
&+ \frac{\partial}{\partial x}\left(\kappa \frac{\partial T}{\partial x}\right) + \frac{\partial}{\partial y}\left(\kappa \frac{\partial T}{\partial y}\right) + \frac{\partial}{\partial z}\left(\kappa \frac{\partial T}{\partial z}\right) \\
&+ \tau_{xx}\frac{\partial v_x}{\partial x} + \tau_{xy}\frac{\partial v_x}{\partial y} + \tau_{xz}\frac{\partial v_x}{\partial z} + \tau_{yx}\frac{\partial v_y}{\partial x} + \tau_{yy}\frac{\partial v_y}{\partial y} + \tau_{yz}\frac{\partial v_y}{\partial z} \\
&+ \tau_{zx}\frac{\partial v_z}{\partial x} + \tau_{zy}\frac{\partial v_z}{\partial y} + \tau_{zz}\frac{\partial v_z}{\partial z}
\end{aligned}$$

is made dimensionless applying again the estimations Eqs. (13.4).

$$
\begin{aligned}
&\rho' c'_p \left(v'_x \frac{\partial T'}{\partial x'} + v''_y \frac{\partial T'}{\partial y'} + v'_z \frac{\partial T'}{\partial z'} \right) = Ec \left(v'_x \frac{\partial p'}{\partial x'} + v''_y \frac{\partial p'}{\partial y'} + v'_z \frac{\partial p'}{\partial z'} \right) \\
&+ \frac{1}{Re\, Pr} \left(\underbrace{\frac{\partial}{\partial x'} \kappa' \frac{\partial T'}{\partial x'}}_{a} + Re \frac{\partial}{\partial y'} \kappa' \frac{\partial T'}{\partial y'} + \underbrace{\frac{\partial}{\partial z'} \kappa' \frac{\partial T'}{\partial z'}}_{b} \right) \\
&+ \frac{Ec}{Re} 2\mu' \left\{ \left(\frac{\partial v'_x}{\partial x'} - \frac{1}{3} \left(\frac{\partial v'_x}{\partial x'} + \frac{\partial v''_y}{\partial y'} + \frac{\partial v'_z}{\partial z'} \right) \right) \frac{\partial v'_x}{\partial x'} \right. \\
&+ \left(\frac{\partial v''_y}{\partial y'} - \frac{1}{3} \left(\frac{\partial v'_x}{\partial x'} + \frac{\partial v''_y}{\partial y'} + \frac{\partial v'_z}{\partial z'} \right) \right) \frac{\partial v''_y}{\partial y'} \\
&+ \left. \left(\frac{\partial v'_z}{\partial z'} - \frac{1}{3} \left(\frac{\partial v'_x}{\partial x'} + \frac{\partial v''_y}{\partial y'} + \frac{\partial v'_z}{\partial z'} \right) \right) \frac{\partial v'_z}{\partial z'} \right\} \\
&+ \frac{Ec}{Re} \mu' \left\{ Re \left(\frac{\partial v'_x}{\partial y'} \right)^2 + \frac{\partial v''_y}{\partial x'} \frac{\partial v'_x}{\partial y'} + \frac{\partial v'_z}{\partial x'} \frac{\partial v'_x}{\partial z'} + \left(\frac{\partial v'_x}{\partial z'} \right)^2 + \right. \\
&+ \frac{1}{Re} \left(\frac{\partial v''_y}{\partial x'} \right)^2 + \frac{\partial v'_x}{\partial y'} \frac{\partial v''_y}{\partial x'} + \frac{\partial v'_z}{\partial y'} \frac{\partial v''_y}{\partial z'} + \frac{1}{Re} \left(\frac{\partial v''_y}{\partial z'} \right)^2 \\
&+ \left. \left(\frac{\partial v'_z}{\partial x'} \right)^2 + \frac{\partial v'_x}{\partial z'} \frac{\partial v'_z}{\partial x'} + \frac{\partial v''_y}{\partial z'} \frac{\partial v'_z}{\partial y'} + Re \left(\frac{\partial v'_z}{\partial y'} \right)^2 \right\} .
\end{aligned}
\tag{13.8}
$$

For high Reynolds numbers ($Re \gg 1$) it remains

$$
\begin{aligned}
&\rho' c'_p \left(v'_x \frac{\partial T'}{\partial x'} + v''_y \frac{\partial T'}{\partial y'} + v'_z \frac{\partial T'}{\partial z'} \right) \approx Ec \left(v'_x \frac{\partial p'}{\partial x'} + v''_y \frac{\partial p'}{\partial y'} + v'_z \frac{\partial p'}{\partial z'} \right) \\
&+ \frac{1}{Pr} \left(\frac{\partial}{\partial y'} \kappa' \frac{\partial T'}{\partial y'} \right) + Ec\, \mu' \left(\left(\frac{\partial v'_x}{\partial y'} \right)^2 + \left(\frac{\partial v'_z}{\partial y'} \right)^2 \right) .
\end{aligned}
\tag{13.9}
$$

Equations (13.1), (13.2), (13.6), (13.7), (13.9) are the boundary layer equations for three-dimensional steady flows.[1] The terms *a* and *b* in Eq. (13.8) are sometimes retained in the equations. The reason for that is discussed in detail in [4].

---

[1] In the literature the reader can find different estimation forms regarding the derivation of the boundary layer equations from the Navier-Stokes equations, e.g.: [2–7].

13.2 A Three-Dimensional Boundary Layer Solution for an Industrial Project

The flow past a helicopter model in the frame of an industrial project was investigated experimentally as well as numerically by applying for the inviscid part of the flow field a panel method and for the viscous part an integral method for the solution of the three-dimensional boundary layer equations, [8]. The configuration of the fuselage of the helicopter model is presented in Fig. 13.2. The freestream and boundary conditions were: $M_\infty = 0.184$, $\alpha = -5^0$, $Re = 6.558 \cdot 10^6$, $T_\infty = 300\,K$, $T_{wall} = 300\,K$. The

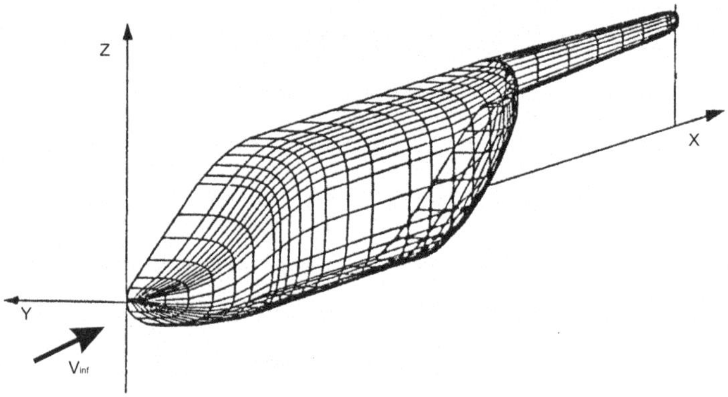

Fig. 13.2 Geometrical presentation of the fuselage of a helicopter model. Figure taken from [9]

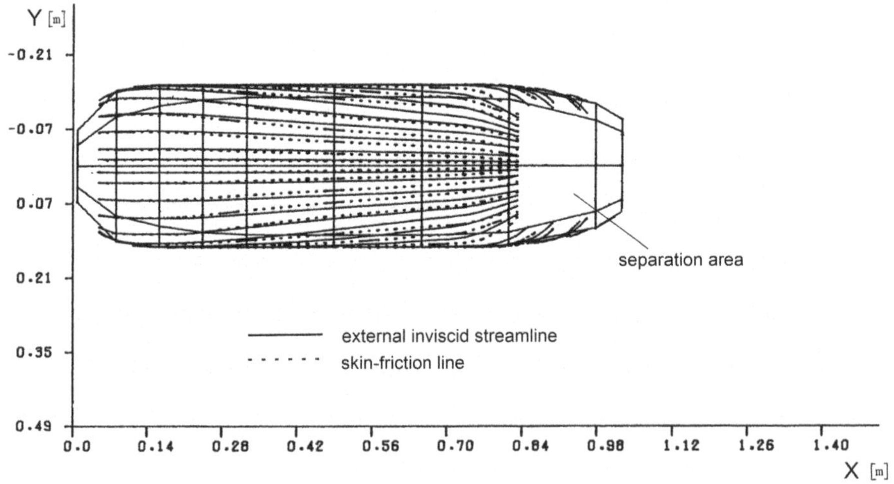

Fig. 13.3 Three-dimensional boundary layer solution on helicopter fuselage. Skin-friction lines and external inviscid streamlines. Lower side view. Figure taken from [9]

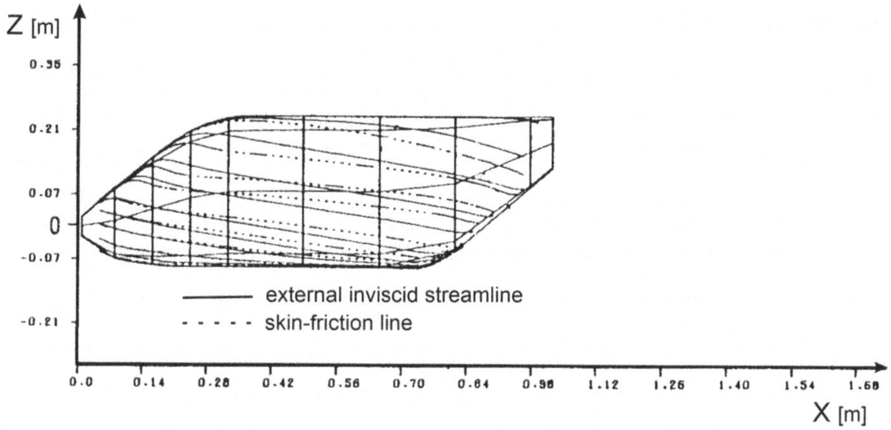

Fig. 13.4 Three-dimensional boundary layer solution on helicopter fuselage. Skin-friction lines and external inviscid streamlines. Left side view. Figure taken from [9]

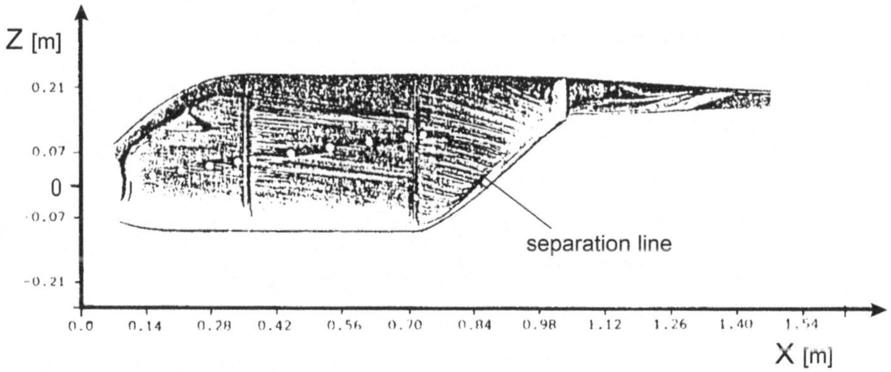

Fig. 13.5 Oil flow picture on helicopter fuselage. Left side view. Figure taken from [10]

boundary layer computation was performed under the assumption of fully turbulent flow.

External inviscid streamlines and skin-friction lines are plotted in Figs. 13.3 and 13.4, where the first figure shows the view at the lower side.[2] The second figure presents the external inviscid streamlines and skin-friction lines in a side view, Fig. 13.4. The comparison of this pattern with an experimental oil flow picture exhibits reasonable agreement, Fig. 13.5, [9, 10].

[2] The concept of the boundary layer theory does not allow the computation of regimes where the flow separates from the body surface.

References

1. Prandtl, L.: Über Flüssigkeitsbewegung bei sehr kleiner Reibung. Verhand. III. Int. Math. Kongress, Heidelberg, B.G. Teubner Verlag, Leipzig, pp. 485–491 (1904)
2. Krause, E.: Strömungslehre. Gasdynamik und aerodynamisches Laboratorium. Teubner Verlag, Stuttgart (2003)
3. Weiland, C.: Mechanics of Flow Similarities. Springer, Berlin (2020)
4. Hirschel, E.H.: Basics of Aerothermodynamics, 2nd edn. Springer, Berlin (2015)
5. Anderson, J.D.: Fundamentals of Aerothermodynamics, 5th edn. McGraw-Hill, New York (2011)
6. Schlichting, H., Gersten, K.: Boundary Layer Theorie. 8th revised and enlarged Edition, Springer, Berlin (2000)
7. Zierep, J.: Ähnlichkeitsgesetze und Modellregeln der Strömungslehre. G. Braun Verlag, Karlsruhe (1972)
8. Hirschel, E.H., Cousteix, J., Kordulla, W.: Three-Dimensional Attached Viscous Flow. Springer, Berlin/Heidelberg (2014)
9. Hirschel, E.H.: Computation of Three-Dimensional Boundary Layers on Fuselages. J. Aircraft **21**, 23–29 (1984)
10. Polz, G., Quentin, J., Amtsberg, J., Kühn, A.: Windkanaluntersuchungen zur optimalen Gestaltung von Transporthubschrauberzellen hinsichtlich Leistungsbedarf und Stabilitätsverhalten.. MBB-UD-299/80, Ottobrunn, Germany (1981)

Chapter 14
Inviscid Forms of the Fluid Dynamic Equations

Abstract The aerospace industry until mid of the 1980s has used mainly inviscid analytical, semi-empirical and numerical methods for the prediction of flows past elements of airplanes, missiles, helicopters and space planes. At that time computer power (computing capacity and storage space) for the calculation of complete viscous flow fields past complex configurations was not sufficient. On the other hand the available numerical approximation methods (stability and robustness) had not the quality of the discretisation schemes we know today. The same was true regarding the grid generation procedure. Therefore, inviscid solutions for airfoils, isolated wings and fuselages, etc., applying linear and non-linear potential equations for disturbed potentials (panel methods) and later Euler solutions were the options available. Today, these methods are in use as design tools for preliminary studies during the pre-design phase of future projects. Other applications are given in the frame of the establishment of aerodynamic data sets, where the inviscid approach often is sufficient. In the following several versions of the Euler equations are presented for flows with gases in the various thermodynamic states: perfect, equilibrium and chemical non-equilibrium. Quasi-linear forms of the Euler equations are derived with diagonalisation of the Jacobians to allow for the application of splitting procedures regarding numerical discretization methods. Finally, some versions of the potential equation are given.

The Variables and Quantities Used in this Chapter

$\mathbf{E}, \mathbf{F}, \mathbf{G}$	vector: conservatively formulated Euler equations
$\mathbf{Q}$	vector: conservative variables
$\mathbb{A}, \mathbb{B}, \mathbb{C}$	matrices of non-conservatively formulated Euler equations
$\mathbf{X}$	vector: non-conservative variables
$\mathbb{J}(\mathbf{Q}), \mathbb{K}(\mathbf{Q}), \mathbb{L}(\mathbf{Q})$	Jacobians of conservatively formulated Euler equations in Cartesian coordinates
$\mathsf{K}^\xi, \mathsf{K}^\eta, \mathsf{K}^\zeta$	Jacobians of conservatively formulated Euler equations in arbitrary coordinates
$\mathsf{A}^\xi, \mathsf{A}^\eta, \mathsf{A}^\zeta$	Jacobians of non-conservatively formulated Euler equations in arbitrary coordinates

C. Weiland, *The Hierarchy of Fluid Dynamic Equations*,
https://doi.org/10.1007/978-3-031-71198-5_14

$\mathbb{M}$	$= \frac{\partial \mathbf{Q}}{\partial \mathbf{X}}$ jacobian conservative to non-conservative variables
$\mathbb{I}$	Identity matrix
$\mathbb{T}_k$	matrix of eigenvectors, conservative variables
$\mathbb{P}_k$	matrix of eigenvectors, non-conservative variables
x, y, z	Cartesian coordinates
ξ, η, ζ	arbitrary coordinates
J	Jacobian of coordinate transformation $x, y, z \Longleftrightarrow \xi, \eta, \zeta$
v_x, v_y, v_z	components of velocity vector in Cartesian coordinates
V_x, V_y, V_z	components of contravariant velocity vector
Θ_k	$= k_x v_x + k_y v_y + k_z v_z$, $k = \xi, \eta, \zeta$ abbreviation
ψ_R	$= \bar{p}_{e_i}(\lvert\mathbf{v}\rvert^2 - \frac{\bar{e}}{\rho}) + p_\rho$ abbreviation (real gas), $\mathbb{J}, \mathbb{K}, \mathbb{L}$ matrices
ψ	$= \frac{\gamma-1}{2}\lvert\mathbf{v}\rvert^2$ abbreviation (perfect gas), $\mathbb{J}, \mathbb{K}, \mathbb{L}$ matrices
$\bar{p}_{e_i}$	$= \frac{1}{\rho}\frac{\partial p}{\partial e_i}\Big\rvert_\rho$ abbreviation
Ω_1	$= \frac{1}{\rho}\frac{\partial p}{\partial \rho} + \frac{\partial e_i}{\partial \rho}\bar{p}_{e_i}$ abbreviation (real gas), $\mathbb{A}, \mathbb{B}, \mathbb{C}$ matrices
Ω_2	$= (p - \frac{\partial e_i}{\partial \rho}\rho^2)\bar{p}_{e_i}$ abbreviation (real gas), $\mathbb{A}, \mathbb{B}, \mathbb{C}$ matrices
Σ_R	$= e + \rho\frac{\partial e_i}{\partial \rho}$ abbreviation (real gas), $\mathbb{M}, \mathbb{M}^{-1}$ matrices
Σ	$= \frac{\gamma-1}{2}\lvert\mathbf{v}\rvert^2$ abbreviation (perfect gas), $\mathbb{M}, \mathbb{M}^{-1}$ matrices
c_R	$c_R^2 = \frac{p}{\rho^2}p_{e_i} + p_\rho$ speed of sound (real gas)
c	$c^2 = \gamma\frac{p}{\rho}$ speed of sound (perfect gas)
c_0	speed of sound at rest (perfect gas)
U_k	$= k_x v_x + k_y v_y + k_z v_z + k_t$ contravariant velocity vector, $k = \xi, \eta, \zeta$
$\lambda_k^{1,\ldots,5}$	eigenvalues, $k = \xi, \eta, \zeta$
p	pressure
ρ	density
T	temperature
$\bar{e}$	$= \rho e$ total energy per unit volume
e	total energy per unit mass
e_i	internal energy per unit mass
h_0	total enthalpy per unit mass
s	entropy per unit mass
R	specific gas constant
$\kappa_1, \kappa_2, \kappa_3, \alpha_1$	abbreviations
Ω	$= \frac{\bar{e}+p}{\rho} - \frac{c_R^2}{\bar{p}_{e_i}}$ abbreviation (real gas), $\mathbb{T}, \mathbb{T}^{-1}$ matrices
$\mathbf{E}_S, \mathbf{F}_S, \mathbf{G}_S$	vector: species conservation
$\mathbf{Q}_S$	vector: species densities
$\mathbf{S}_S$	vector: source term of species conservation
α_i	source term of conservation of species i
Φ	potential of irrotational flow field
γ	ratio of specific heats
M_∞	free-stream Mach number
φ	perturbation velocity potential
$\bar{y}, \bar{z}$	constants (small perturbation theory)

14.1 The Euler Equations

14.1.1 General Considerations

Leonhard Euler was the man who in 1755 did formulate the set of partial differential equations which describes the momentum of ideal[1] fluid in three dimensions,[2] which is uniquely amazing since the mathematical disciplines vector algebra and vector analysis were known only in a rudimentary state at that time. This was one of the remarkable steps in the evolution of the discipline fluid mechanics. Considering Eq. (4.14) and neglecting the viscous terms one obtains

$$\begin{aligned}
\rho\frac{\partial v_x}{\partial t} + \rho v_x\frac{\partial v_x}{\partial x} + \rho v_y\frac{\partial v_x}{\partial y} + \rho v_z\frac{\partial v_x}{\partial z} &= \rho g_x - \frac{\partial p}{\partial x}\\
\rho\frac{\partial v_y}{\partial t} + \rho v_x\frac{\partial v_y}{\partial x} + \rho v_y\frac{\partial v_y}{\partial y} + \rho v_z\frac{\partial v_y}{\partial z} &= \rho g_y - \frac{\partial p}{\partial y}\\
\rho\frac{\partial v_z}{\partial t} + \rho v_x\frac{\partial v_z}{\partial x} + \rho v_y\frac{\partial v_z}{\partial y} + \rho v_z\frac{\partial v_z}{\partial z} &= \rho g_z - \frac{\partial p}{\partial z}.
\end{aligned} \tag{14.1}$$

This set of equations is valid for incompressible but also for compressible flow. In the case of incompressible flow the system of the four unknowns, the components of the velocity vector v_x, v_y, v_z and the pressure p is closed by the incompressibility condition (continuity equation) Eq. (12.1)

$$\frac{\partial v_x}{\partial x} + \frac{\partial v_y}{\partial y} + \frac{\partial v_z}{\partial z} = 0.$$

An interaction between the velocity field and the temperature field is not possible in inviscid, incompressible flow in contrast to viscous, incompressible flow, see Chap. 12. For compressible flow the energy equation comes into play. Considering the energy equation in the form of Eq. (4.19) and neglecting both the power of the friction force and the heat conduction,[3] it follows

$$\begin{aligned}
&\frac{\partial \rho e}{\partial t} + \frac{\partial(\rho e + p)v_x}{\partial x} + \frac{\partial(\rho e + p)v_y}{\partial y} + \frac{\partial(\rho e + p)v_z}{\partial z} =\\
&\rho\,(g_x v_x + g_y v_y + g_z v_z).
\end{aligned} \tag{14.2}$$

Further, the continuity equation in the form of Eq. (4.12) is in force to determine the density

[1] Ideal means inviscid and incompressible.

[2] The Euler equations can also be derived by applying Hamilton's principle with the pressure acting as a Lagrange multiplier. For details see [1].

[3] Of course in inviscid flow fields the temperature field can also be influenced by heat conduction and external heat sources.

$$\frac{\partial \rho}{\partial t} + \frac{\partial \rho v_x}{\partial x} + \frac{\partial \rho v_y}{\partial y} + \frac{\partial \rho v_z}{\partial z} = 0.$$

Compressible flow fields are described by six unknowns, the components of the velocity vector v_x, v_y, v_z, the density ρ, the temperature T and the pressure p. With the continuity equation, the three momentum equations and the energy equation five relations are available to determine ρ, v_x, v_y, v_z, T, while the pressure p is calculated by a thermodynamic equation of state, for example for perfect gas by $p = \rho R T$. Often the energy equation is used in the form of Eq. (3.29)

$$\begin{aligned}&\frac{\partial h_0}{\partial t} + v_x \frac{\partial h_0}{\partial x} + v_y \frac{h_0}{\partial y} + v_z \frac{\partial h_0}{\partial z} = \\ &\frac{1}{\rho}\frac{\partial p}{\partial t} + (g_x v_x + g_y v_y + g_z v_z),\end{aligned} \tag{14.3}$$

with $h_0 = (h + |\mathbf{v}|^2/2) = (e + p/\rho)$. Another form of the inviscid energy equation (without the power of the inertial force) is given by Eq. (3.37) (and Eq. (3.38))

$$\frac{Ds}{Dt} = 0,$$

which says that an inviscid, isoenergetic flow behaves isentropic. Fluid elements travelling along streamlines (and path lines) do not change the entropy.

Today, for incompressible flows the set of Eqs. (14.1) and (12.1) and for compressible flows the set of Eqs. (14.1), (14.2) and (4.12) bear the name Euler equations.

14.1.2 Conservative Formulation of the Euler Equations

We consider Eq. (4.12) for the continuity, Eq. (14.1) for the components of the momentum and Eq. (14.2) for the energy, where we have dropped the terms regarding the inertial force. These equations are written in formal vector notation (see Eq. (5.1)), where conservative variables $\mathbf{Q}^T = (\rho, \rho v_x, \rho v_y, \rho v_z, \rho e)$ are used. The equations below therefore are called the conservatively formulated Euler equations in Cartesian coordinates.

$$\frac{\partial \mathbf{Q}}{\partial t} + \frac{\partial \mathbf{E}}{\partial x} + \frac{\partial \mathbf{F}}{\partial y} + \frac{\partial \mathbf{G}}{\partial z} = 0, \tag{14.4}$$

and

$$\mathbf{Q} = \begin{pmatrix} \rho \\ \rho v_x \\ \rho v_y \\ \rho v_z \\ \rho e \end{pmatrix},$$

$$\mathbf{E} = \begin{pmatrix} \rho v_x \\ \rho v_x^2 + p \\ \rho v_x v_y \\ \rho v_x v_z \\ (\rho e + p) v_x \end{pmatrix}, \quad \mathbf{F} = \begin{pmatrix} \rho v_y \\ \rho v_y v_x \\ \rho v_y^2 + p \\ \rho v_y v_z \\ (\rho e + p) v_y \end{pmatrix}, \quad \mathbf{G} = \begin{pmatrix} \rho v_z \\ \rho v_z v_x \\ \rho v_z v_y \\ \rho v_z^2 + p \\ (\rho e + p) v_z \end{pmatrix}.$$

For arbitrary coordinates (ξ, η, ζ, τ) by using the relations Eqs. (6.1), (6.2) the Euler equations (14.4) are transformed to

$$J^{-1} \frac{\partial \mathbf{Q}}{\partial \tau} + \frac{\partial \mathbf{E}}{\partial \xi} + \frac{\partial \mathbf{F}}{\partial \eta} + \frac{\partial \mathbf{G}}{\partial \zeta} = 0, \tag{14.5}$$

$$\mathbf{Q} = \begin{pmatrix} \rho \\ \rho v_x \\ \rho v_y \\ \rho v_z \\ \rho e \end{pmatrix}, \quad \mathbf{E} = J^{-1} \begin{pmatrix} \rho V_x \\ \rho v_x V_x + \xi_x p \\ \rho v_y V_x + \xi_y p \\ \rho v_z V_x + \xi_z p \\ (\rho e + p) V_x - \xi_t p \end{pmatrix},$$

$$\mathbf{F} = J^{-1} \begin{pmatrix} \rho V_y \\ \rho v_x V_y + \eta_x p \\ \rho v_y V_y + \eta_y p \\ \rho v_z V_y + \eta_z p \\ (\rho e + p) V_y - \eta_t p \end{pmatrix}, \quad \mathbf{G} = J^{-1} \begin{pmatrix} \rho V_z \\ \rho v_x V_z + \zeta_x p \\ \rho v_y V_z + \zeta_y p \\ \rho v_z V_z + \zeta_z p \\ (\rho e + p) V_z - \zeta_t p \end{pmatrix},$$

with

$$\begin{aligned} V_x &= \xi_t + \xi_x v_x + \xi_y v_y + \xi_z v_z, \\ V_y &= \eta_t + \eta_x v_x + \eta_y v_y + \eta_z v_z, \\ V_z &= \zeta_t + \zeta_x v_x + \zeta_y v_y + \zeta_z v_z. \end{aligned}$$

14.1.3 Quasi-Conservative Formulation of the Euler Equations

The application of successful numerical methods on finite-difference basis requires that the governing equations are formulated in a vector-matrix form like Eq. (14.10) with the primitive variables $\mathbf{X}^T = (\rho, v_x, v_y, v_z, p)$. Emanating from the conservative form of the Euler equations (14.4) and by using the conservative variables $\mathbf{Q}^T = (\rho, \rho v_x, \rho v_y, \rho v_z, \rho e)$, a so-called quasi-conservative form of the Euler equations in vector-matrix form can be derived (see [2–4]). One obtains by using

$$\frac{\partial \mathbf{Q}}{\partial t} + \frac{\partial \mathbf{E}}{\partial x} + \frac{\partial \mathbf{F}}{\partial y} + \frac{\partial \mathbf{G}}{\partial z} = 0,$$

with the Jacobians $\mathbb{J}(\mathbf{Q}) = \partial \mathbf{E}/\partial \mathbf{Q}$, $\mathbb{K}(\mathbf{Q}) = \partial \mathbf{F}/\partial \mathbf{Q}$, $\mathbb{L}(\mathbf{Q}) = \partial \mathbf{G}/\partial \mathbf{Q}$, which are formulated for a general real gas, the vector-matrix form

$$\frac{\partial \mathbf{Q}}{\partial t} + \mathbb{J}(\mathbf{Q})\frac{\partial \mathbf{Q}}{\partial x} + \mathbb{K}(\mathbf{Q})\frac{\partial \mathbf{Q}}{\partial y} + \mathbb{L}(\mathbf{Q})\frac{\partial \mathbf{Q}}{\partial z} = 0, \tag{14.6}$$

with

$$\mathbb{J}(\mathbf{Q}) = \begin{pmatrix} 0 & 1 & 0 & 0 & 0 \\ \psi_R - v_x^2 & v_x(2 - \bar{p}_{e_i}) & -v_y \bar{p}_{e_i} & -v_z \bar{p}_{e_i} & \bar{p}_{e_i} \\ -v_x v_y & v_y & v_x & 0 & 0 \\ -v_x v_z & v_z & 0 & v_x & 0 \\ v_x(\psi_R - \frac{\bar{e}+p}{\rho}) & (\frac{\bar{e}+p}{\rho} - v_x^2 \bar{p}_{e_i}) & -v_x v_y \bar{p}_{e_i} & -v_x v_z \bar{p}_{e_i} & v_x(\bar{p}_{e_i} + 1) \end{pmatrix},$$

$$\mathbb{K}(\mathbf{Q}) = \begin{pmatrix} 0 & 0 & 1 & 0 & 0 \\ -v_x v_y & v_y & v_x & 0 & 0 \\ \psi_R - v_y^2 & -v_x \bar{p}_{e_i} & v_y(2 - \bar{p}_{e_i}) & -v_z \bar{p}_{e_i} & \bar{p}_{e_i} \\ -v_y v_z & 0 & v_z & v_y & 0 \\ v_y(\psi_R - \frac{\bar{e}+p}{\rho}) & -v_x v_y \bar{p}_{e_i} & (\frac{\bar{e}+p}{\rho} - v_y^2 \bar{p}_{e_i}) & -v_y v_z \bar{p}_{e_i} & v_y(\bar{p}_{e_i} + 1) \end{pmatrix},$$

$$\mathbb{L}(\mathbf{Q}) = \begin{pmatrix} 0 & 0 & 0 & 1 & 0 \\ -v_x v_z & v_z & 0 & v_x & 0 \\ -v_y v_z & 0 & v_z & v_y & 0 \\ \psi_R - v_z^2 & -v_x \bar{p}_{e_i} & -v_y \bar{p}_{e_i} & v_z(2 - \bar{p}_{e_i}) & \bar{p}_{e_i} \\ v_z(\psi_R - \frac{\bar{e}+p}{\rho}) & -v_x v_z \bar{p}_{e_i} & -v_y v_z \bar{p}_{e_i} & (\frac{\bar{e}+p}{\rho} - v_z^2 \bar{p}_{e_i}) & v_z(\bar{p}_{e_i} + 1) \end{pmatrix}.$$

Equation (14.6) is equivalent to the conservative formulation [5]

$$\frac{\partial \mathbf{Q}}{\partial t} + \frac{\partial \mathbb{J}(\mathbf{Q})\mathbf{Q}}{\partial x} + \frac{\partial \mathbb{K}(\mathbf{Q})\mathbf{Q}}{\partial y} + \frac{\partial \mathbb{L}(\mathbf{Q})\mathbf{Q}}{\partial z} = 0,$$

due to the homogeneity relation according to L. Euler [6].

The total energy in a volume is $\bar{e} = \rho e = \rho(e_i + 1/2|\mathbf{v}|^2)$ with e being the total energy per unit mass and e_i the internal energy per unit mass. For a gas in thermodynamic equilibrium the pressure p and the speed of sound c_R are defined by (see Sect. 9.4)

$$p = p(\rho, e_i), \qquad c_R^2(\rho, e_i) = \frac{p}{\rho^2}\left(\frac{\partial p}{\partial e_i}\right)_\rho + \left(\frac{\partial p}{\partial \rho}\right)_{e_i}. \tag{14.7}$$

The establishment of the Jacobians requires the derivatives of the pressure p and the internal energy e_i with respect to the five conservative variables ρ, $\rho v_x = m_1$, $\rho v_y = m_2$, $\rho v_z = m_3$, $\rho e = \bar{e}$ (more details can be found in Chap. 15). These are incorporated in the matrices above and lead to the abbreviations

$$\begin{aligned}\psi_R &= \bar{p}_{e_i}\left(|\mathbf{v}|^2 - \frac{\bar{e}}{\rho}\right) + p_\rho \\ &= \bar{p}_{e_i}\left(|\mathbf{v}|^2 - \frac{\bar{e}+p}{\rho}\right) + c_R^2, \\ c_R^2 &= \frac{p}{\rho}\bar{p}_{e_i} + p_\rho, \\ \bar{p}_{e_i} &= \frac{1}{\rho}p_{e_i}.\end{aligned} \tag{14.8}$$

For perfect gas these abbreviations reduce with $p = (\gamma - 1)\rho e_i$ to

$$\begin{aligned}\psi &= \frac{\gamma - 1}{2}|\mathbf{v}|^2, \\ c^2 &= \gamma\frac{p}{\rho}, \\ \bar{p}_{e_i} &= \gamma - 1.\end{aligned} \tag{14.9}$$

14.1.4 Non-conservative Formulation of the Euler Equations

A non-conservative formulation of the Euler equations is obtained by using the continuity Eq. (3.14), the momentum Eq. (14.1) (without the inertial terms) and the energy Eq. (3.38) or (4.23). These equations are compiled in a vector-matrix form, too and read for perfect gas

$$\frac{\partial \mathbf{X}}{\partial t} + \mathbb{A}\frac{\partial \mathbf{X}}{\partial x} + \mathbb{B}\frac{\partial \mathbf{X}}{\partial y} + \mathbb{C}\frac{\partial \mathbf{X}}{\partial z} = 0, \tag{14.10}$$

$$\mathbf{X} = \begin{pmatrix} \rho \\ v_x \\ v_y \\ v_z \\ p \end{pmatrix}, \quad \mathbb{A} = \begin{pmatrix} v_x & \rho & 0 & 0 & 0 \\ 0 & v_x & 0 & 0 & 1/\rho \\ 0 & 0 & v_x & 0 & 0 \\ 0 & 0 & 0 & v_x & 0 \\ 0 & \rho c^2 & 0 & 0 & v_x \end{pmatrix},$$

$$\mathbb{B} = \begin{pmatrix} v_y & 0 & \rho & 0 & 0 \\ 0 & v_y & 0 & 0 & 0 \\ 0 & 0 & v_y & 0 & 1/\rho \\ 0 & 0 & 0 & v_y & 0 \\ 0 & 0 & \rho c^2 & 0 & v_y \end{pmatrix}, \quad \mathbb{C} = \begin{pmatrix} v_z & 0 & 0 & \rho & 0 \\ 0 & v_z & 0 & 0 & 0 \\ 0 & 0 & v_z & 0 & 0 \\ 0 & 0 & 0 & v_z & 1/\rho \\ 0 & 0 & 0 & \rho c^2 & v_z \end{pmatrix}. \tag{14.11}$$

For a real gas approach the matrices $\mathbb{A}$, $\mathbb{B}$, $\mathbb{C}$ have to be extended to the form

$$\mathbb{A} = \begin{pmatrix} v_x & \rho & 0 & 0 & 0 \\ \Omega_1 & v_x & 0 & 0 & 1/\rho \\ 0 & 0 & v_x & 0 & 0 \\ 0 & 0 & 0 & v_x & 0 \\ 0 & \Omega_2 & 0 & 0 & v_x \end{pmatrix},$$

$$\mathbb{B} = \begin{pmatrix} v_y & 0 & \rho & 0 & 0 \\ 0 & v_y & 0 & 0 & 0 \\ \Omega_1 & 0 & v_y & 0 & 1/\rho \\ 0 & 0 & 0 & v_y & 0 \\ 0 & 0 & \Omega_2 & 0 & v_y \end{pmatrix}, \quad \mathbb{C} = \begin{pmatrix} v_z & 0 & 0 & \rho & 0 \\ 0 & v_z & 0 & 0 & 0 \\ 0 & 0 & v_z & 0 & 0 \\ \Omega_1 & 0 & 0 & v_z & 1/\rho \\ 0 & 0 & 0 & \Omega_2 & v_z \end{pmatrix}. \tag{14.12}$$

with

$$\Omega_1 = \frac{1}{\rho}\frac{\partial p}{\partial \rho} + \frac{\partial e_i}{\partial \rho}\bar{p}_{e_i}, \qquad \Omega_2 = (p - \frac{\partial e_i}{\partial \rho}\rho^2)\bar{p}_{e_i}$$
$$= \rho\left(\frac{p}{\rho}\bar{p}_{e_i} + \frac{\partial p}{\partial \rho}\right) = \rho c_R^2$$

and for perfect gas the quantities Ω_1, Ω_2 are $\Omega_1 = 0, \Omega_2 = \rho c^2$.

14.1.5 Diagonalization of Quasi-Linear Forms of the Euler Equations

The modern numerical approximation methods regarding the convective part of the Navier-Stokes equations, respectively the Euler equations, are based on the method of characteristics (as long as the partial differential equations are hyperbolic). This means that the discretisations are performed such that the directions of the information fluxes, given by the eigenvalues of the Jacobians of the systems, are respected. This has led to the "Split-Coefficient Matrix Methods", [2, 7], the "Flux Vector Splitting Methods", [8, 9], the "Riemann Solvers", [10, 11] and generally the "Upwind Methods", [12].

In the following subsections we demonstrate this procedure on the Euler equations, where quasi-linear forms are generated by determining the corresponding Jacobians. Eigenvalues and eigenvectors of the Jacobians have to be calculated in order to allow the diagonalization and the subsequent splitting of the system of equations.

A first quasi-linear form

The quasi-conservative form of the Euler equations (14.6) in arbitrary coordinates $k = (\xi, \eta, \zeta, \tau)$ is given by

$$\frac{\partial \mathbf{Q}}{\partial \tau} + \mathsf{K}^{\xi}(\mathbf{Q})\frac{\partial \mathbf{Q}}{\partial \xi} + \mathsf{K}^{\eta}(\mathbf{Q})\frac{\partial \mathbf{Q}}{\partial \eta} + \mathsf{K}^{\zeta}(\mathbf{Q})\frac{\partial \mathbf{Q}}{\partial \zeta} = 0, \tag{14.13}$$

where for perfect gas the Jacobians read[4]

$$\mathsf{K}^k = \mathbb{J}(\mathbf{Q})k_x + \mathbb{K}(\mathbf{Q})k_y + \mathbb{L}(\mathbf{Q})k_z + k_t\mathbb{I},$$

with k_x, k_y, k_z, k_t; $\quad k = (\xi, \eta, \zeta)$ given by Eq. (6.1) and

$$\psi = \frac{\gamma - 1}{2}|\mathbf{v}|^2, \quad \Theta_k = k_x v_x + k_y v_y + k_z v_z$$

$$\mathsf{K}^k = \begin{pmatrix} 0 & k_x & k_y & k_z & 0 \\ k_x\psi - v_x\Theta_k & k_x v_x(2-\gamma) + \Theta_k & k_y v_x - k_x(\gamma-1)v_y & k_z v_x - k_x(\gamma-1)v_z & k_x(\gamma-1) \\ k_y\psi - v_y\Theta_k & k_x v_y - k_y(\gamma-1)v_x & k_y v_y(2-\gamma) + \Theta_k & k_z v_y - k_y(\gamma-1)v_z & k_y(\gamma-1) \\ k_z\psi - v_z\Theta_k & k_x v_z - k_z(\gamma-1)v_x & k_y v_z - k_z(\gamma-1)v_y & k_z v_z(2-\gamma) + \Theta_k & k_z(\gamma-1) \\ (\psi - \frac{\bar{e}+p}{\rho})\Theta_k & k_x(\frac{\bar{e}+p}{\rho}) - (\gamma-1)v_x\Theta_k & k_y(\frac{\bar{e}+p}{\rho}) - (\gamma-1)v_y\Theta_k & k_z(\frac{\bar{e}+p}{\rho}) - (\gamma-1)v_z\Theta_k & \gamma\Theta_k \end{pmatrix} + k_t\mathbb{I}.$$

Applying a general equation of state for real gas flow one obtains for the Jacobians

$$\mathsf{K}^k = \begin{pmatrix} 0 & k_x & k_y & k_z & 0 \\ k_x\psi_R - v_x\Theta_k & k_x v_x(1-\bar{p}_{e_i}) + \Theta_k & k_y v_x - k_x v_y\bar{p}_{e_i} & k_z v_x - k_x v_z\bar{p}_{e_i} & k_x\bar{p}_{e_i} \\ k_y\psi_R - v_y\Theta_k & k_x v_y - k_y v_x\bar{p}_{e_i} & k_y v_y(1-\bar{p}_{e_i}) + \Theta_k & k_z v_y - k_y v_z\bar{p}_{e_i} & k_y\bar{p}_{e_i} \\ k_z\psi_R - v_z\Theta_k & k_x v_z - k_z v_x\bar{p}_{e_i} & k_y v_z - k_z v_y\bar{p}_{e_i} & k_z v_z(1-\bar{p}_{e_i}) + \Theta_k & k_z\bar{p}_{e_i} \\ (\psi_R - \frac{\bar{e}+p}{\rho})\Theta_k & k_x(\frac{\bar{e}+p}{\rho}) - \bar{p}_{e_i}v_x\Theta_k & k_y(\frac{\bar{e}+p}{\rho}) - \bar{p}_{e_i}v_y\Theta_k & k_z(\frac{\bar{e}+p}{\rho}) - \bar{p}_{e_i}v_z\Theta_k & (\bar{p}_{e_i}+1)\Theta_k \end{pmatrix} + k_t\mathbb{I}, \tag{14.14}$$

with

$$\bar{p}_{e_i} = \frac{1}{\rho}\left(\frac{\partial p}{\partial e_i}\right)_\rho, \quad \psi_R = \bar{p}_{e_i}\left(|\mathbf{v}|^2 - \frac{\bar{e}+p}{\rho}\right) + c_R^2, \quad c_R^2 = \frac{p}{\rho}\bar{p}_{e_i} + \left(\frac{\partial p}{\partial \rho}\right)_{e_i}.$$

[4] For time dependent grids the Jacobian has to be expanded by $k_t\mathbb{I}$ with $\mathbb{I}$ the identity matrix, [2].

A second quasi-linear form

Another quasi-linear form of the Euler equations is given by the non-conservative variant Eq. (14.10), where the vector $\mathbf{X}$ and the matrices $\mathbb{A}$, $\mathbb{B}$, $\mathbb{C}$ for perfect gas and Cartesian coordinates are defined by Eq. (14.11). The diagonalization procedure is demonstrated at the matrix $\mathbb{A}$. The eigenvalues of $\mathbb{A}$ are

$$\lambda^{1,2,3} = v_x, \quad \lambda^4 = v_x + c, \quad \lambda^5 = v_x - c. \tag{14.15}$$

With the computation of the right eigenvectors which are arranged in columns of a matrix $\mathbb{P}_x$ and the left eigenvectors arranged in rows of a matrix $\mathbb{P}_x^{-1}$ the following relation is valid

$$\mathbb{A} = \mathbb{P}_x \Lambda \mathbb{P}_x^{-1} \tag{14.16}$$

with

$$\mathbb{P}_x = \begin{pmatrix} 1 & 0 & 0 & \frac{1}{c^2} & \frac{1}{c^2} \\ 0 & 0 & 0 & \frac{1}{\rho c} & -\frac{1}{\rho c} \\ 0 & 1 & 0 & 0 & 0 \\ 0 & 0 & 1 & 0 & 0 \\ 0 & 0 & 0 & 1 & 1 \end{pmatrix}, \quad \mathbb{P}_x^{-1} = \begin{pmatrix} 1 & 0 & 0 & 0 & -\frac{1}{c^2} \\ 0 & 0 & 1 & 0 & 0 \\ 0 & 0 & 0 & 1 & 0 \\ 0 & \frac{1}{2}\rho c & 0 & 0 & \frac{1}{2} \\ 0 & -\frac{1}{2}\rho c & 0 & 0 & \frac{1}{2} \end{pmatrix},$$

and Λ the diagonal matrix of the eigenvalues.[5]

Turning to the arbitrary coordinates $k = (\xi, \eta, \zeta, \tau)$ we obtain for perfect gas flow

$$\frac{\partial \mathbf{X}}{\partial \tau} + \mathsf{A}^{\xi}(\mathbf{X})\frac{\partial \mathbf{X}}{\partial \xi} + \mathsf{A}^{\eta}(\mathbf{X})\frac{\partial \mathbf{X}}{\partial \eta} + \mathsf{A}^{\zeta}(\mathbf{X})\frac{\partial \mathbf{X}}{\partial \zeta} = 0, \tag{14.17}$$

with

$$\mathsf{A}^k = \mathbb{A}(\mathbf{X})k_x + \mathbb{B}(\mathbf{X})k_y + \mathbb{C}(\mathbf{X})k_z + k_t\mathbb{I},$$

$$\mathsf{A}^k = \begin{pmatrix} \Theta_k & \rho k_x & \rho k_y & \rho k_z & 0 \\ 0 & \Theta_k & 0 & 0 & 1/\rho k_x \\ 0 & 0 & \Theta_k & 0 & 1/\rho k_y \\ 0 & 0 & 0 & \Theta_k & 1/\rho k_z \\ 0 & \rho c^2 k_x & \rho c^2 k_y & \rho c^2 k_z & \Theta_k \end{pmatrix} + k_t\mathbb{I},$$

and

$$\Theta_k = k_x v_x + k_y v_y + k_z v_z, \quad c^2 = \gamma\frac{p}{\rho} \text{ the speed of sound.}$$

[5] Further downstream the diagonalization procedure is presented for non-conservative and quasi-conservative formulations each with arbitrary coordinates and real gas flow.

Further Eq. (14.13) can be written as

$$\frac{\partial \mathbf{Q}}{\partial \mathbf{X}}\frac{\partial \mathbf{X}}{\partial \tau}+\mathsf{K}^{\xi}(\mathbf{Q})\frac{\partial \mathbf{Q}}{\partial \mathbf{X}}\frac{\partial \mathbf{X}}{\partial \xi}+\mathsf{K}^{\eta}(\mathbf{Q})\frac{\partial \mathbf{Q}}{\partial \mathbf{X}}\frac{\partial \mathbf{X}}{\partial \eta}+\mathsf{K}^{\zeta}(\mathbf{Q})\frac{\partial \mathbf{Q}}{\partial \mathbf{X}}\frac{\partial \mathbf{X}}{\partial \zeta}=0, \tag{14.18}$$

where the matrix $\mathbb{M}=\dfrac{\partial \mathbf{Q}}{\partial \mathbf{X}}$ and its inverse $\mathbb{M}^{-1}$ for perfect gas are

$$\mathbb{M}=\begin{pmatrix} 1 & 0 & 0 & 0 & 0\\ v_x & \rho & 0 & 0 & 0\\ v_y & 0 & \rho & 0 & 0\\ v_z & 0 & 0 & \rho & 0\\ \frac{\Sigma}{\gamma-1} & \rho v_x & \rho v_y & \rho v_z & \frac{1}{\gamma-1}\end{pmatrix},$$

$$\mathbb{M}^{-1}=\begin{pmatrix} 1 & 0 & 0 & 0 & 0\\ -v_x/\rho & 1/\rho & 0 & 0 & 0\\ -v_y/\rho & 0 & 1/\rho & 0 & 0\\ -v_z/\rho & 0 & 0 & 1/\rho & 0\\ \Sigma & -v_x(\gamma-1) & -v_y(\gamma-1) & -v_z(\gamma-1) & \gamma-1\end{pmatrix}, \tag{14.19}$$

with

$$\Sigma=\frac{\gamma-1}{2}|\mathbf{v}|^2\equiv\psi.$$

Multiplying Eq. (14.18) on the left with $\mathbb{M}^{-1}$ results in

$$\mathbb{I}\frac{\partial \mathbf{X}}{\partial \tau}+\mathbb{M}^{-1}\mathsf{K}^{\xi}\mathbb{M}\frac{\partial \mathbf{X}}{\partial \xi}+\mathbb{M}^{-1}\mathsf{K}^{\eta}\mathbb{M}\frac{\partial \mathbf{X}}{\partial \eta}+\mathbb{M}^{-1}\mathsf{K}^{\zeta}\mathbb{M}\frac{\partial \mathbf{X}}{\partial \zeta}=0, \tag{14.20}$$

with $\mathbb{I}$ the identity matrix. Comparing Eq. (14.17) with Eq. (14.20) leads to

$$\mathsf{A}^k=\mathbb{M}^{-1}\mathsf{K}^k\mathbb{M}, \tag{14.21}$$

which is another way to derive the matrix A^k. This means that the matrices A^k and K^k are similar and therefore have the same eigenvalues.

For completeness, we list the matrices $\mathbb{M}$ and $\mathbb{M}^{-1}$ also for real gas flow.

$$\mathbb{M}=\begin{pmatrix} 1 & 0 & 0 & 0 & 0\\ v_x & \rho & 0 & 0 & 0\\ v_y & 0 & \rho & 0 & 0\\ v_z & 0 & 0 & \rho & 0\\ \Sigma_R & \rho v_x & \rho v_y & \rho v_z & \frac{1}{\bar{p}_{e_i}}\end{pmatrix},$$

$$\mathbb{M}^{-1} = \begin{pmatrix} 1 & 0 & 0 & 0 & 0 \\ -v_x/\rho & 1/\rho & 0 & 0 & 0 \\ -v_y/\rho & 0 & 1/\rho & 0 & 0 \\ -v_z/\rho & 0 & 0 & 1/\rho & 0 \\ (|\mathbf{v}|^2 - \Sigma_R)\bar{p}_{e_i} & -v_x\bar{p}_{e_i} & -v_y\bar{p}_{e_i} & -v_z\bar{p}_{e_i} & \bar{p}_{e_i} \end{pmatrix},$$

with

$$\Sigma_R = e + \rho\frac{\partial e_i}{\partial \rho}.$$

It is a simple task to determine the eigenvalues with respect to the matrix A^k, which are also the ones of the matrix K^k. They read for perfect gas

$$\begin{aligned} \lambda_k^{1,2,3} &= U_k, & k &= (\xi, \eta, \zeta), \\ \lambda_k^4 &= U_k + c\,|\nabla k|, & |\nabla k| &\equiv |\mathrm{grad}\, k|, \\ \lambda_k^5 &= U_k - c\,|\nabla k|, & |\nabla k| &= (k_x^2 + k_y^2 + k_z^2)^{1/2}, \\ & & U_k &= k_x v_x + k_y v_y + k_z v_z + k_t. \end{aligned} \tag{14.22}$$

For real gas the speed of sound c in λ_k^4, λ_k^5 has to be replaced by the real gas definition c_R.

We state that the matrices A^k and K^k are quadratic and have n rows. Further, we assume that these matrices possess n independent eigenvectors and belong therefore to the class of "diagonal similar" matrices, [13]. Then, the relation which diagonalize for example the matrix K^k

$$\mathbb{T}_k^{-1}\mathsf{K}^k\mathbb{T}_k = \Lambda_k \tag{14.23}$$

holds, where the matrix $\mathbb{T}_k$ is built by the right eigenvectors, which are arranged in columns, whereas the matrix $\mathbb{T}_k^{-1}$ contains the left eigenvectors arranged in rows. Λ_k is the diagonal matrix consisting of the eigenvalues. Multiplying Eq. (14.23) from the left with $\mathbb{T}_k$ and from the right with $\mathbb{T}_k^{-1}$ leads to

$$\mathsf{K}^k = \mathbb{T}_k\Lambda_k\mathbb{T}_k^{-1}. \tag{14.24}$$

This matrix is split due to the sign of the eigenvalues into a positive and a negative part in order to care for the proper influence of the information fluxes in a discretized version of the convective part of the governing equations, as mentioned above. The direct determination of the eigenvectors of K^k is a laborious work because the matrix has only few zeros and its elements are lengthy. Much better is the situation for the matrix A^k. We therefore construct first a set of eigenvectors for A^k with

$$\mathsf{A}^k = \mathbb{P}_k\Lambda_k\mathbb{P}_k^{-1}, \tag{14.25}$$

where the columns of $\mathbb{P}_k$ represent the right eigenvectors and the rows of $\mathbb{P}_k^{-1}$ the left eigenvectors and obtain

$$\mathbb{P}_k = \begin{pmatrix} \hat{k}_x & \hat{k}_y & \hat{k}_z & \kappa_1 & \kappa_1 \\ 0 & -\hat{k}_z & \hat{k}_y & \frac{\hat{k}_x}{\sqrt{2}} & -\frac{\hat{k}_x}{\sqrt{2}} \\ \hat{k}_z & 0 & -\hat{k}_x & \frac{\hat{k}_y}{\sqrt{2}} & -\frac{\hat{k}_y}{\sqrt{2}} \\ -\hat{k}_y & \hat{k}_x & 0 & \frac{\hat{k}_z}{\sqrt{2}} & -\frac{\hat{k}_z}{\sqrt{2}} \\ 0 & 0 & 0 & \kappa_1 c^2 & \kappa_1 c^2 \end{pmatrix}$$

$$\mathbb{P}_k^{-1} = \begin{pmatrix} \hat{k}_x & 0 & \hat{k}_z & -\hat{k}_y & -\frac{\hat{k}_x}{c^2} \\ \hat{k}_y & -\hat{k}_z & 0 & \hat{k}_x & -\frac{\hat{k}_y}{c^2} \\ \hat{k}_z & \hat{k}_z & -\hat{k}_x & 0 & -\frac{\hat{k}_z}{c^2} \\ 0 & \frac{\hat{k}_x}{\sqrt{2}} & \frac{\hat{k}_y}{\sqrt{2}} & \frac{\hat{k}_z}{\sqrt{2}} & \kappa_2 \\ 0 & -\frac{\hat{k}_x}{\sqrt{2}} & -\frac{\hat{k}_y}{\sqrt{2}} & -\frac{\hat{k}_z}{\sqrt{2}} & \kappa_2 \end{pmatrix} \tag{14.26}$$

with

$$\kappa_1 = \frac{\rho}{\sqrt{2}c}, \quad \kappa_2 = \frac{1}{\sqrt{2}\rho c}, \quad \hat{\Theta}_k = \frac{\Theta_k}{|\nabla k|}, \quad \hat{k}_{x_i} = \frac{k_{x_i}}{|\nabla k|}, \quad x_i = x, y, z.$$

Due to the triple eigenvalue $\lambda_k^{1,2,3}$ the number of independent equations (in this case: five) for the computation of the eigenvectors is reduced. Therefore, the determination of the associate set of eigenvectors is not unique. To overcome this problem some parts of the eigenvectors are defined expediently–by making a guess– while subsequently the completion of the set of eigenvectors is performed by solving the remaining independent characteristic equations using the assumption of that guess. Below another set of eigenvectors, compared to Eq. (14.26), is presented which of course meets also the relation $\mathsf{A}^k = \mathbb{P}_k \Lambda_k \mathbb{P}_k^{-1}$.

$$\mathbb{P}_k = \begin{pmatrix} \dfrac{1}{c^2} & \dfrac{1}{c^2} & 1 & 0 & 0 \\ k_x\kappa_3 & -k_x\kappa_3 & 0 & -\dfrac{k_y}{k_x} & -\dfrac{k_z}{k_x} \\ k_y\kappa_3 & -k_y\kappa_3 & 0 & 1 & 0 \\ k_z\kappa_3 & -k_z\kappa_3 & 0 & 0 & 1 \\ 1 & 1 & 0 & 0 & 0 \end{pmatrix},$$

$$\mathbb{P}_k^{-1} = \begin{pmatrix} 0 & k_x\kappa_4 & k_y\kappa_4 & k_z\kappa_4 & \dfrac{1}{2} \\ 0 & -k_x\kappa_4 & -k_y\kappa_4 & -k_z\kappa_4 & \dfrac{1}{2} \\ 1 & 0 & 0 & 0 & -\dfrac{1}{c} \\ 0 & -k_xk_y\kappa_5 & (k_x^2+k_z^2)\kappa_5 & -k_yk_z\kappa_5 & 0 \\ 0 & -k_xk_z\kappa_5 & -k_yk_z\kappa_5 & (k_x^2+k_y^2)\kappa_5 & 0 \\ 1 & 1 & 0 & 0 & 0 \end{pmatrix},$$

with

$$\kappa_3 = \frac{1}{\rho c|\nabla k|}, \quad \kappa_4 = \frac{c\rho}{2|\nabla k|}, \quad \kappa_5 = \frac{1}{|\nabla k|^2}.$$

The matrices of the right and left eigenvectors of A^k for real gas, based on the Eq. (14.26), have the form

$$\mathbb{P}_k = \begin{pmatrix} \hat{k}_x & \hat{k}_y & \hat{k}_z & 1 & 1 \\ 0 & -\hat{k}_z & \hat{k}_y & \hat{k}_x\dfrac{c_R}{\rho} & -\hat{k}_x\dfrac{c_R}{\rho} \\ \hat{k}_z & 0 & -\hat{k}_x & \hat{k}_y\dfrac{c_R}{\rho} & -\hat{k}_y\dfrac{c_R}{\rho} \\ -\hat{k}_y & \hat{k}_x & 0 & \hat{k}_z\dfrac{c_R}{\rho} & -\hat{k}_z\dfrac{c_R}{\rho} \\ \hat{k}_x\alpha_1 & \hat{k}_y\alpha_1 & \hat{k}_z\alpha_1 & \alpha_1+c_r^2 & \alpha_1+c_r^2 \end{pmatrix},$$

$$\mathbb{P}_k^{-1} = \begin{pmatrix} \hat{k}_x + \hat{k}_x \dfrac{\alpha_1}{c_R^2} & 0 & \hat{k}_z & -\hat{k}_y & -\dfrac{\hat{k}_x}{c_r^2} \\ \hat{k}_y + \hat{k}_y \dfrac{\alpha_1}{c_R^2} & -\hat{k}_z & 0 & \hat{k}_x & -\dfrac{\hat{k}_y}{c_R^2} \\ \hat{k}_z + \hat{k}_z \dfrac{\alpha_1}{c_R^2} & \hat{k}_z & -\hat{k}_x & 0 & -\dfrac{\hat{k}_z}{c_R^2} \\ \dfrac{\alpha_1}{2c_R^2} & \hat{k}_x \dfrac{\rho}{2c_R} & \hat{k}_y \dfrac{\rho}{2c_R} & \hat{k}_z \dfrac{\rho}{2c_R} & \dfrac{1}{2c_R^2} \\ \dfrac{\alpha_1}{2c_R^2} & -\hat{k}_x \dfrac{\rho}{2c_R} & -\hat{k}_y \dfrac{\rho}{2c_R} & -\hat{k}_z \dfrac{\rho}{2c_R} & \dfrac{1}{2c_R^2} \end{pmatrix}, \tag{14.27}$$

with $\alpha_1 = \bar{p}_{e_i}(|\mathbf{v}|^2 - 2\psi_R)$.

From the relations (14.21), (14.24) and (14.25) one can derive

$$\mathsf{A}^k = \mathbb{M}^{-1}\mathsf{K}^k\mathbb{M} = \mathbb{M}^{-1}\mathbb{T}^k\Lambda_k\mathbb{T}_K^{-1}\mathbb{M} = \mathbb{P}_k\Lambda_k\mathbb{P}_k^{-1}, \tag{14.28}$$

where the matrices of the eigenvectors $\mathbb{T}_k$ and $\mathbb{T}_k^{-1}$ of K^k can now be evaluated by

$$\mathbb{T}_k = \mathbb{M}\mathbb{P}_k, \quad \mathbb{T}_k^{-1} = \mathbb{P}_k^{-1}\mathbb{M}^{-1},$$

and read for perfect gas[6]

$$\mathbb{T}_k = \left(\begin{array}{c|c|} \hat{k}_x & \hat{k}_y \\ v_x\hat{k}_x & v_x\hat{k}_y - \rho\hat{k}_z \\ v_y\hat{k}_x + \rho\hat{k}_z & v_y\hat{k}_y \\ v_z\hat{k}_x - \rho\hat{k}_y & v_z\hat{k}_y + \rho\hat{k}_x \\ \frac{\psi}{\gamma-1}\hat{k}_x + \rho(v_y\hat{k}_z - v_z\hat{k}_y) & \frac{\psi}{\gamma-1}\hat{k}_y + \rho(v_z\hat{k}_x - v_x\hat{k}_z) \end{array} \right.$$

$$\left. \begin{array}{|c|c|c} \hat{k}_z & \kappa_1 & \kappa_1 \\ v_x\hat{k}_z + \rho\hat{k}_y & \kappa_1(v_x + c\hat{k}_x) & \kappa_1(v_x - c\hat{k}_x) \\ v_y\hat{k}_z - \rho\hat{k}_x & \kappa_1(v_y + c\hat{k}_y) & \kappa_1(v_y - c\hat{k}_y) \\ v_z\hat{k}_z & \kappa_1(v_z + c\hat{k}_z) & \kappa_1(v_z - c\hat{k}_z) \\ \frac{\psi}{\gamma-1}\hat{k}_z + \rho(v_x\hat{k}_y - v_y\hat{k}_x) & \kappa_1(\frac{\psi+c^2}{\gamma-1} + c\,\hat{\Theta}_k) & \kappa_1(\frac{\psi+c^2}{\gamma-1} - c\,\hat{\Theta}_k) \end{array} \right),$$

[6] The matrices of the eigenvectors $\mathbb{T}_k$ and $\mathbb{T}_k^{-1}$ are the same for time dependent and time independent coordinate systems.

$$
\mathbb{T}_k^{-1} = \begin{pmatrix}
\hat{k}_x(1-\frac{\psi}{c^2})+\frac{1}{\rho}(v_z\hat{k}_y - v_y\hat{k}_z) & \hat{k}_x\frac{v_x(\gamma-1)}{c^2} & \hat{k}_z\frac{1}{\rho}+\hat{k}_x\frac{v_y(\gamma-1)}{c^2} & -\hat{k}_y\frac{1}{\rho}+\hat{k}_x\frac{v_z(\gamma-1)}{c^2} & -\hat{k}_x\frac{\gamma-1}{c^2} \\
\hat{k}_y(1-\frac{\psi}{c^2})+\frac{1}{\rho}(v_x\hat{k}_z - v_z\hat{k}_x) & -\hat{k}_z\frac{1}{\rho}+\hat{k}_y\frac{v_x(\gamma-1)}{c^2} & \hat{k}_y\frac{v_y(\gamma-1)}{c^2} & \hat{k}_x\frac{1}{\rho}+\hat{k}_y\frac{v_z(\gamma-1)}{c^2} & -\hat{k}_y\frac{\gamma-1}{c^2} \\
\hat{k}_z(1-\frac{\psi}{c^2})+\frac{1}{\rho}(v_y\hat{k}_x - v_x\hat{k}_y) & \hat{k}_y\frac{1}{\rho}+\hat{k}_z\frac{v_x(\gamma-1)}{c^2} & -\hat{k}_x\frac{1}{\rho}+\hat{k}_z\frac{v_y(\gamma-1)}{c^2} & \hat{k}_z\frac{v_z(\gamma-1)}{c^2} & -\hat{k}_z\frac{\gamma-1}{c^2} \\
\kappa_2(\psi - c\,\hat{\Theta}_k) & \kappa_2(c\hat{k}_x - v_x(\gamma-1)) & \kappa_2(c\hat{k}_y - v_y(\gamma-1)) & \kappa_2(c\hat{k}_z - v_z(\gamma-1)) & \kappa_2(\gamma-1) \\
\kappa_2(\psi + c\,\hat{\Theta}_k) & -\kappa_2(c\hat{k}_x + v_x(\gamma-1)) & -\kappa_2(c\hat{k}_y + v_y(\gamma-1)) & -\kappa_2(c\hat{k}_z + v_z(\gamma-1)) & \kappa_2(\gamma-1)
\end{pmatrix}. \quad (14.29)
$$

Based on the Jacobian K^k for real gas (Eq. (14.14)) the matrices $\mathbb{T}_k$ and $\mathbb{T}_k^{-1}$ have the form[7]

$$
\mathbb{T}_k = \begin{pmatrix}
\hat{k}_x & \hat{k}_y & \hat{k}_z & 1 & 1 \\
v_x\hat{k}_x & v_x\hat{k}_y - \rho\hat{k}_z & v_x\hat{k}_z + \rho\hat{k}_y & v_x + c_R\hat{k}_x & v_x - c_R\hat{k}_x \\
v_y\hat{k}_x + \rho\hat{k}_z & v_y\hat{k}_y & v_y\hat{k}_z - \rho\hat{k}_x & v_y + c_R\hat{k}_y & v_y - c_R\hat{k}_y \\
v_z\hat{k}_x - \rho\hat{k}_y & v_z\hat{k}_y + \rho\hat{k}_x & v_z\hat{k}_z & v_z + c_R\hat{k}_z & v_z - c_R\hat{k}_z \\
\Omega\hat{k}_x + \rho(v_y\hat{k}_z - v_z\hat{k}_y) & \Omega\hat{k}_y + \rho(v_z\hat{k}_x - v_x\hat{k}_z) & \Omega\hat{k}_z + \rho(v_x\hat{k}_y - v_y\hat{k}_x) & (\frac{\bar{e}+p}{\rho} + c_R\,\hat{\Theta}_k) & (\frac{\bar{e}+p}{\rho} - c_R\,\hat{\Theta}_k)
\end{pmatrix},
$$

$$
\mathbb{T}_k^{-1} = \begin{pmatrix}
\hat{k}_x(1-\frac{\psi_R}{c_R^2})+\frac{1}{\rho}(v_z\hat{k}_y - v_y\hat{k}_z) & \hat{k}_x\frac{v_x\bar{p}_{e_i}}{c_R^2} & \hat{k}_z\frac{1}{\rho}+\hat{k}_x\frac{v_y\bar{p}_{e_i}}{c_R^2} & -\hat{k}_y\frac{1}{\rho}+\hat{k}_x\frac{v_z\bar{p}_{e_i}}{c_R^2} & -\hat{k}_x\frac{\bar{p}_{e_i}}{c_R^2} \\
\hat{k}_y(1-\frac{\psi_R}{c_R^2})+\frac{1}{\rho}(v_x\hat{k}_z - v_z\hat{k}_x) & -\hat{k}_z\frac{1}{\rho}+\hat{k}_y\frac{v_x\bar{p}_{e_i}}{c_R^2} & \hat{k}_y\frac{v_y\bar{p}_{e_i}}{c_R^2} & \hat{k}_x\frac{1}{\rho}+\hat{k}_y\frac{v_z\bar{p}_{e_i}}{c_R^2} & -\hat{k}_y\frac{\bar{p}_{e_i}}{c_R^2} \\
\hat{k}_z(1-\frac{\psi_R}{c_R^2})+\frac{1}{\rho}(v_y\hat{k}_x - v_x\hat{k}_y) & \hat{k}_y\frac{1}{\rho}+\hat{k}_z\frac{v_x\bar{p}_{e_i}}{c_R^2} & -\hat{k}_x\frac{1}{\rho}+\hat{k}_z\frac{v_y\bar{p}_{e_i}}{c_R^2} & \hat{k}_z\frac{v_z\bar{p}_{e_i}}{c_R^2} & -\hat{k}_z\frac{\bar{p}_{e_i}}{c_R^2} \\
\kappa_3(\psi_R - c_R\,\hat{\Theta}_k) & \kappa_3(c_R\hat{k}_x - v_x\bar{p}_{e_i}) & \kappa_3(c_R\hat{k}_y - v_y\bar{p}_{e_i}) & \kappa_3(c_R\hat{k}_z - v_z\bar{p}_{e_i}) & \kappa_3\bar{p}_{e_i} \\
\kappa_3(\psi_R + c_R\,\hat{\Theta}_k) & -\kappa_3(c_R\hat{k}_x + v_x\bar{p}_{e_i}) & -\kappa_3(c_R\hat{k}_y + v_y\bar{p}_{e_i}) & -\kappa_3(c_R\hat{k}_z + v_z\bar{p}_{e_i}) & \kappa_3\bar{p}_{e_i}
\end{pmatrix}, \quad (14.30)
$$

[7] Note: The structure of the matrices $\mathbb{T}_k$ and $\mathbb{T}_k^{-1}$ for real gas differs sligtly from the ones for perfect gas.

where

$$\kappa_1 = \frac{\rho}{\sqrt{2}c}, \quad \kappa_2 = \frac{1}{\sqrt{2}\rho c}, \quad \kappa_3 = \frac{1}{2c_R^2}, \quad \psi = \frac{1}{\gamma - 1}|\mathbf{v}|^2, \quad \Omega = \frac{\bar{e} + p}{\rho} - \frac{c_R^2}{\bar{p}_{e_i}},$$

$$\psi_R = \bar{p}_{e_i}\left(|\mathbf{v}|^2 - \frac{\bar{e} + p}{\rho}\right) + c_R^2, \quad \hat{\Theta}_k = (k_x v_x + k_y v_y + k_z v_z)/|\nabla k|, \quad \hat{k}_{x_i} = \frac{k_{x_i}}{|\nabla k|}.$$

Remark Above we have presented the matrices of the eigenvectors (right and left) for the matrices $\mathbb{J}(\mathbf{Q})$ and $\mathbb{A}(\mathbf{X})$ for arbitrary coordinates and for both the perfect gas and the real gas approach. Missing are the right and left eigenvector matrices of $\mathbb{J}(\mathbf{Q})$ (Eq. (14.6) with Eq. (14.9)) for Cartesian coordinates and perfect gas.[8] With $\mathbb{M}$ of Eq. (14.19) and $\mathbb{P}_x$ of Eq. (14.16) (see also Eq. (14.28)) one obtains

$$\mathbb{T}_x = \mathbb{M} \cdot \mathbb{P}_x,$$

$$\mathbb{T}_x = \begin{pmatrix} 1 & 0 & 0 & \frac{1}{c^2} & \frac{1}{c^2} \\ v_x & 0 & 0 & \frac{v_x}{c^2} + \frac{1}{c} & \frac{v_x}{c^2} + \frac{1}{c} \\ v_y & \rho & 0 & \frac{v_y}{c^2} & \frac{v_y}{c^2} \\ v_z & 0 & \rho & \frac{v_z}{c^2} & \frac{v_z}{c^2} \\ \frac{|\mathbf{v}|^2}{2} & \rho v_y & \rho v_z & \left(\frac{1}{\gamma-1} + \frac{|\mathbf{v}|^2}{2c^2} + \frac{v_x}{c}\right) & \left(\frac{1}{\gamma-1} + \frac{|\mathbf{v}|^2}{2c^2} - \frac{v_x}{c}\right) \end{pmatrix}, \quad (14.31)$$

$$\mathbb{T}_x^{-1} = \mathbb{P}_x^{-1} \cdot \mathbb{M}^{-1},$$

$$\mathbb{T}_x^{-1} = \begin{pmatrix} 1 - (\gamma - 1)\frac{|\mathbf{v}|^2}{2c^2} & (\gamma - 1)\frac{v_x}{c^2} & (\gamma - 1)\frac{v_y}{c^2} & (\gamma - 1)\frac{v_z}{c^2} & -(\gamma - 1)\frac{1}{c^2} \\ -\frac{v_y}{\rho} & 0 & \frac{1}{\rho} & 0 & 0 \\ -\frac{v_z}{\rho} & 0 & 0 & \frac{1}{\rho} & 0 \\ (\gamma - 1)\frac{|\mathbf{v}|^2}{4} - \frac{c v_x}{2} & \frac{c-(\gamma-1)v_x}{2} & -\frac{(\gamma-1)v_y}{2} & -\frac{(\gamma-1)v_z}{2} & -\frac{(\gamma-1)}{2} \\ (\gamma - 1)\frac{|\mathbf{v}|^2}{4} + \frac{c v_x}{2} & \frac{-c-(\gamma-1)v_x}{2} & -\frac{(\gamma-1)v_y}{2} & -\frac{(\gamma-1)v_z}{2} & -\frac{(\gamma-1)}{2} \end{pmatrix}. \quad (14.32)$$

---

[8] The right and left eigenvector matrices $\mathbb{T}_y$, $\mathbb{T}_y^{-1}$, $\mathbb{T}_z$, $\mathbb{T}_z^{-1}$ for the Jacobians $\mathbb{K}(\mathbf{Q})$ and $\mathbb{L}(\mathbf{Q})$ (Eq. (14.6)) can be calculated in the same sense as $\mathbb{T}_x$, $\mathbb{T}_x^{-1}$.

14.1.6 Euler Equations for Gases in Thermodynamic Equilibrium

In the second half of the 1980s and the 1990s there was a great word-wide interest in designing winged space vehicles for the reason of space exploration in connection with the International Space Station ISS. Such space vehicles should ensure on the one hand the autonomous access to space for the engaged countries and on the other hand were intended to be the successor of the Space Shuttle Orbiter. In Europe the Hermes project was launched, in Japan the Hope and later the Hope X programs. The activities in the United States are related to the X-33, X-34 and X-37 vehicles. Later, in a cooperation between Europe and the US the crew return vehicle X-38 on the basis of NASA's X-24A vehicle was thus far developed that a demonstration flight with a fully integrated vehicle was close to be realized. But unfortunately the US government with president G.W. Bush decided to cancel this project due to political reasons just before the scheduled test flight, see [14, 15]. Germany developed the flight demonstrator PHOENIX, which completed successfully a first drop test in May 2004 on the ESRANGE in Vidsel (near Kiruna) in Sweden. Since then nothing has happened with respect to the development of a winged space vehicle on the basis of an industrial project.[9]

The establishment of the aerodynamic data base of such vehicles requires due to the ambitious flight trajectories (during re-entry from space) with high temperatures and low densities the observation of real gas effects. For a part of the trajectory the gas can be treated as being in a thermodynamic equilibrium state.

In Sect. 9.4 we have presented the relations with which for the general viscous case the thermodynamic equilibrium state of a flow can be described. For an inviscid flow, where the energy equation is formulated by Eq. (14.2), it is convenient to define the pressure p and the speed of sound c as functions of ρ and e_i as shown in Eq. (14.7).

The pressure $p(\rho, e_i)$ can be determined by a computer program [16], or a table look-up [17, 18]. The speed of sound is then calculated by numerical differentiation of Eq. (14.7).

In the frame of the industrial projects Hermes, X-38 and PHOENIX Euler calculations were performed in order to contribute to the establishment of the aerodynamic data bases, [14]. Comparison with windtunnel tests indicated reasonable agreement with the numerical results of the Euler equations. The reason for that is that flow fields past space vehicles mentioned above are mainly pressure dominated, compared for example to hypersonic cruise and acceleration flight vehicles, which are more viscous effects dominated, [14, 19].

[9] The development of the US Dream Chaser, which started in the early 2000s, was not completed until 2023.

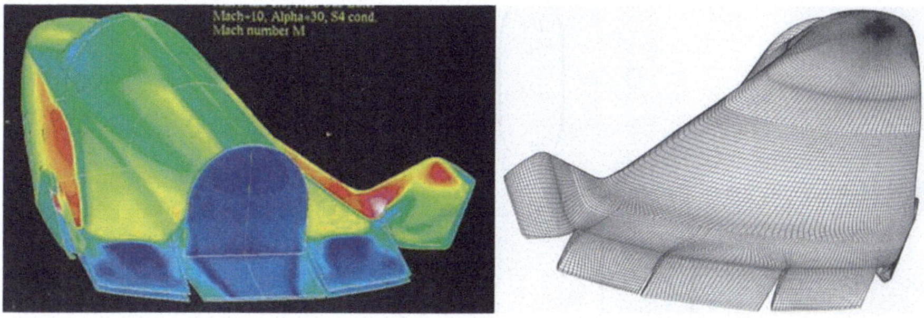

Fig. 14.1 Hermes configuration: Numerical grid on surface with deflected body and wing flaps as well as winglet rudders (right), Mach number contours of an Euler solution with equilibrium real gas approach for $M_\infty = 10, \alpha = 30^0$ (left), [20]

An Euler solution with real gas effects in thermodynamic equilibrium past the ambitious configuration of Hermes with deflected body and wing flaps as well as winglet rudders is presented in Fig. 14.1. The left part of the figure shows the Mach number distribution on the surface of the Hermes vehicle, whereas the right part gives an impression of the complex grid generation requirements, [20].

14.1.7 Euler Equations for Gases in Chemical Non-equilibrium

The basis of our consideration is the conservative formulated Euler Eq. (14.4)

$$\frac{\partial \mathbf{Q}}{\partial t} + \frac{\partial \mathbf{E}}{\partial x} + \frac{\partial \mathbf{F}}{\partial y} + \frac{\partial \mathbf{G}}{\partial z} = 0.$$

For chemical non-equilibrium the equations of species conservation have to be added and read, if we neglect the transport phenomena (diffusion of the species, see Eq. (9.1))

$$\rho \frac{\partial \omega_i}{\partial t} + \rho \mathbf{v} \operatorname{grad} \omega_i = \frac{\partial \rho_i}{\partial t} + \operatorname{div} \rho_i \mathbf{v} = \alpha_i . \tag{14.33}$$

The definitions of the mass fraction ω_i and the source term α_i as well as the rules how to determine the total pressure p and the total density ρ can be found in Chap. 9. For consistency reasons Eq. (14.33) is written also in a formal vector notation (see Eq. (9.3))

$$\frac{\partial \mathbf{Q}_S}{\partial t} + \frac{\partial \mathbf{E}_S}{\partial x} + \frac{\partial \mathbf{F}_S}{\partial y} + \frac{\partial \mathbf{G}_S}{\partial z} = \mathbf{S}_\mathbf{S} \tag{14.34}$$

$$\mathbf{Q}_S = \begin{pmatrix} \rho_1 \\ \rho_2 \\ \rho_3 \\ \vdots \end{pmatrix}, \; \mathbf{S}_S = \begin{pmatrix} \alpha_1 \\ \alpha_2 \\ \alpha_3 \\ \vdots \end{pmatrix},$$

$$\mathbf{E}_S = \begin{pmatrix} \rho_1 v_x \\ \rho_2 v_x \\ \rho_3 v_x \\ \vdots \end{pmatrix}, \; \mathbf{F}_S = \begin{pmatrix} \rho_1 v_y \\ \rho_2 v_y \\ \rho_3 v_y \\ \vdots \end{pmatrix}, \; \mathbf{G}_S = \begin{pmatrix} \rho_1 v_z \\ \rho_2 v_z \\ \rho_3 v_z \\ \vdots \end{pmatrix}.$$

Winged space vehicles performing a re-entry through the Earth's atmosphere move along a trajectory, which is designed with respect to mechanically and thermally restrictive condition, see e.g., [19]. Figure 14.2 (below) shows two of such trajectories, one for the operational Space Shuttle Orbiter and one for the X-38 experimental vehicle. A space flight vehicle during re-entry from space encounters specific physical phenomena, which are presented in Fig. 14.2 (above). Due to the high Mach numbers and relatively low densities in the altitude regime $55\,km \lesssim H \lesssim 80\,km$ and the Mach number regime $12 \lesssim M_\infty \lesssim 25$ non-equilibrium real gas effects must be taken into account, which govern the flow field and therefore also the aerodynamic data. In this regard Fig. 14.3 exhibits an Euler solution with chemical non-equilibrium real gas effects for the flight conditions $M_\infty = 20$, $\alpha = 30^0$, $p_\infty = 2.106\,Pa$, $\rho_\infty = 3.738 \cdot 10^{-5}\,kg/m^3$, $H = 76\,km$. Mach number contours at the surface are given in the left figure, whereas isobars in a cross-sectional plane are presented in the right figure. Both figures show very clearly the complexity of the embedded shock structures, the footprints on the surface as well as their behavior inside the flow domain, [21]. Note: The bow shock is fitted by the numerical code, which is probably the only one for complex 3-D applications, [22].

14.2 The Potential Equation

The Eqs. (4.12), (14.1), (3.37) describe inviscid flow fields, which can contain vorticity (rotational flow). An equivalent system is given, if one replaces Eq. (14.1) by Eq. (3.23). For steady flows one obtains

$$\begin{aligned} \operatorname{div}(\rho \mathbf{v}) &= 0 \qquad \Longrightarrow \qquad \left(\frac{\partial \rho v_x}{\partial x} + \frac{\partial \rho v_y}{\partial y} + \frac{\partial \rho v_z}{\partial z} = 0 \right), \\ -(\mathbf{v} \times \operatorname{rot} \mathbf{v}) &= T \operatorname{grad} s, \qquad (14.35) \\ \mathbf{v} \cdot \operatorname{grad} s &= 0. \end{aligned}$$

For irrotational flow fields it holds $\operatorname{rot} \mathbf{v} = 0$ and with Eq. (14.35) it follows grad s $= 0$, which means that the complete flow field is isentropic (often called homentropic).

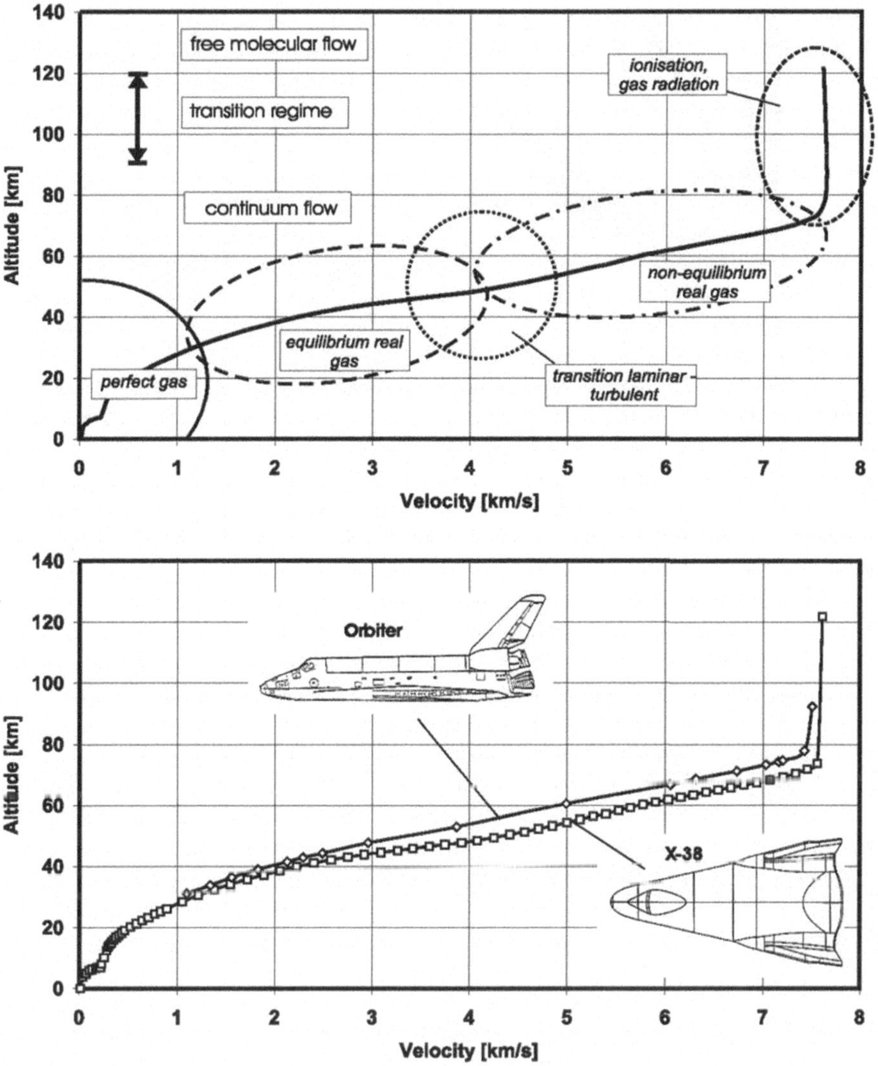

Fig. 14.2 Flow regimes and physical phenomena along a typical re-entry trajectory (X-38 vehicle, [23]), (above). Re-entry trajectories of X-38 and Space Shuttle Orbiter (STS2), (below) [24]

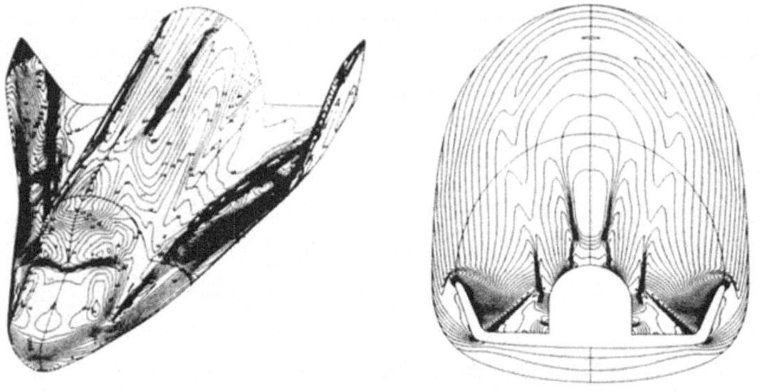

Fig. 14.3 Hermes configuration, chemical non-equilibrium Euler solution: Surface Mach number contours (left), isobars ($\log(p/p_\infty)$) in a cross-section of $x = 14\,m$ (right), $M_\infty = 20, \alpha = 30^0$, $p_\infty = 2.106\,Pa$, $\rho_\infty = 3.738 \cdot 10^{-5}\,kg/m^3$, [21]

Irrotational vector fields, for example for the velocity, allow for the definition of a potential $\Phi(x, y, z)$ with

$$\mathbf{v} = \text{grad}\ \Phi \implies \text{rot grad}\ \Phi = 0.$$

From the above set of equations for irrotational flow fields it only remains

$$\frac{\partial \rho v_x}{\partial x} + \frac{\partial \rho v_y}{\partial y} + \frac{\partial \rho v_z}{\partial z} = \frac{\partial \rho \Phi_x}{\partial x} + \frac{\partial \rho \Phi_y}{\partial y} + \frac{\partial \rho \Phi_z}{\partial z} = 0, \tag{14.36}$$

which is called the conservatively formulated full potential equation. The energy equation in the frame of the flow filament theory and for perfect gas reads

$$\frac{\gamma}{\gamma - 1} \frac{p_\infty}{\rho_\infty} + \frac{v_\infty^2}{2} = \frac{\gamma}{\gamma - 1} \frac{p}{\rho} + \frac{|\mathbf{v}|^2}{2}. \tag{14.37}$$

Further the isentropic relation is given by

$$\frac{p}{p_\infty} = \left(\frac{\rho}{\rho_\infty}\right)^\gamma. \tag{14.38}$$

Making the variables $v_x, v_y, v_z, v_\infty, \rho, \rho_\infty, p, p_\infty$ dimensionless by

$$v_x' = \frac{v_x}{v_\infty},\ v_y' = \frac{v_y}{v_\infty},\ v_z' = \frac{v_z}{v_\infty},\ v_\infty' = \frac{v_\infty}{v_\infty} = 1,$$
$$\rho' = \frac{\rho}{\rho_\infty},\ \rho_\infty' = \frac{\rho_\infty}{\rho_\infty} = 1,\ p' = \frac{p}{\rho_\infty v_\infty^2},\ p_\infty' = \frac{p_\infty}{\rho_\infty v_\infty^2},$$

one obtains for the density, the pressure and the speed of sound as function of the freestream Machnumber and the velocity

$$\rho' = \left[1 + \frac{\gamma - 1}{2} M_\infty^2 \left(1 - (v'^2_x + v'^2_y + v'^2_z)\right)\right]^{\frac{1}{\gamma-1}},$$
$$p' = \frac{\rho'^\gamma}{\gamma M_\infty^2},$$
$$c^2 = \frac{\rho'^{\gamma-1}}{M_\infty^2}. \tag{14.39}$$

The solution of Eq. (14.36) together with the relations (14.39) provide the user with the needed quantities of the flow field considered.

With the following operations

1. differentiation of the continuity equation $\frac{\partial \rho v_x}{\partial x} + \frac{\partial \rho v_y}{\partial y} + \frac{\partial \rho v_z}{\partial z} = 0$,
2. use of the derivatives of the density with respect to the x, y, z coordinates of Eq. (14.39)[10] $\rho = \left[1 + \frac{\gamma-1}{2} M_\infty^2 \left(1 - (v_x^2 + v_y^2 + v_z^2)\right)\right]^{\frac{1}{\gamma-1}}$,
3. application of the irrotational condition $\operatorname{rot} \mathbf{v} = 0$ on the equation received,

the relation is obtained

$$\left(1 - \frac{v_x^2}{c^2}\right)\Phi_{xx} + \left(1 - \frac{v_y^2}{c^2}\right)\Phi_{yy} + \left(1 - \frac{v_z^2}{c^2}\right)\Phi_{zz} -$$
$$-2\frac{v_x v_y}{c^2}\Phi_{xy} - 2\frac{v_x v_z}{c^2}\Phi_{xz} - 2\frac{v_y v_z}{c^2}\Phi_{yz} = 0, \tag{14.40}$$

and

$$c^2 = c_0^2 - \frac{\gamma - 1}{2}(\Phi_x^2 + \Phi_y^2 + \Phi_z^2), \tag{14.41}$$

which is called the non-conservatively formulated full potential equation.

Figure 14.4 shows the solution of the three-dimensional full potential equation around the Butler wing. The freestream quantities were $M_\infty = 2.5$, $\alpha = 5^0$, [25]. A comparison between the potential solution and the Euler solution is presented in the figure. Of course, there are deviations between the two solutions as can be observed in the pressure distribution on the body surface along the plane of symmetry and also in the isobar plots in the cross sections $x/L = 0.7$ (Note: The bow shock is captured in the potential solution and fitted in the Euler solution. Further the captured shock in the potential solution does not meet the Rankine-Hugoniot condition regarding the entropy jump.). But these differences are obviously small, as the presented cases reveal.

---

[10] For convenience we have dropped the prime.

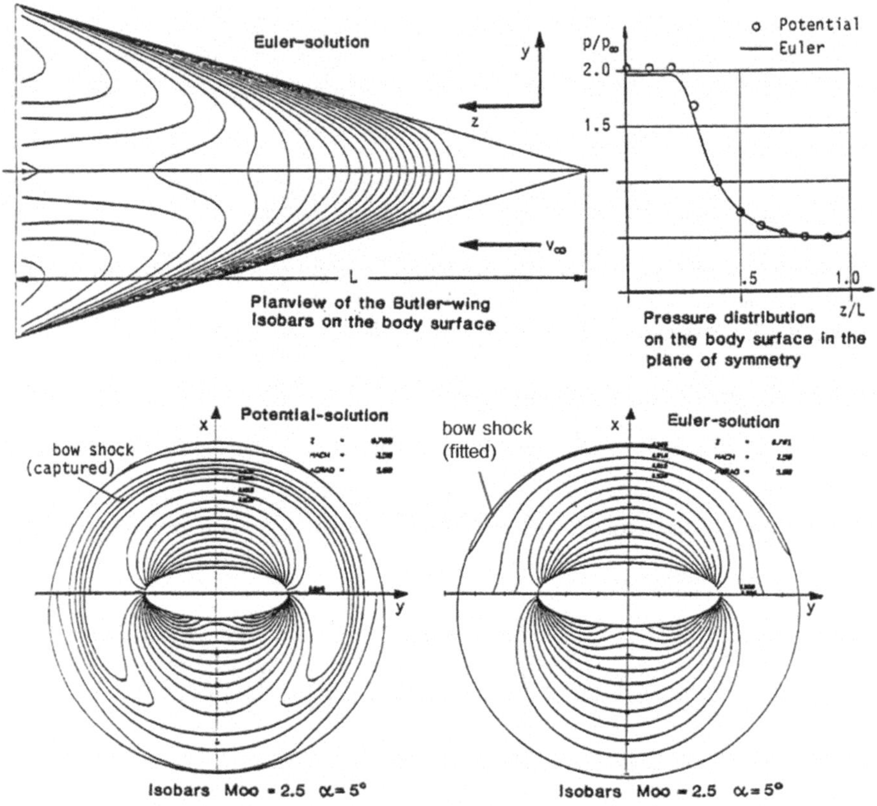

Fig. 14.4 Three-dimensional flow past the Butler wing. $M_\infty = 2.5, \alpha = 5^0$. Comparison of the solutions of the potential equations with the Euler equations, [25]

14.2.1 Potential Equations in Perturbation Form

Several forms are known of the potential equation when applied for disturbed potentials. The application of a perturbation potential on the conservative full potential equation Eq. (14.36) gives another result as on the non-conservative full potential equation Eq. (14.40). Additionally, restricting assumptions with respect to the asymptotic behavior of the geometry and the Mach number lead to specific forms of disturbed potential equations. In the following we present some of them.[11]

Introducing the perturbation velocity potential[12] $\Phi = x + \varphi$ the perturbation form of the non-conservative potential equation based on the Eqs. (14.40), (14.41) reads

[11] A comprehensive survey about the potential equation and their perturbation forms in particular for transonic flows can be found in [26].

[12] Normalized definitions of the potentials are used.

$$
\begin{aligned}
&(1 - M_\infty^2)\varphi_{xx} + \varphi_{yy} + \varphi_{zz} \\
&= \left((\gamma + 1)\varphi_x + \frac{\gamma + 1}{2}\varphi_x^2 + \frac{\gamma - 1}{2}(\varphi_y^2 + \varphi_z^2) \right) M_\infty^2 \varphi_{xx} \\
&+ \left((\gamma - 1)\varphi_x + \frac{\gamma + 1}{2}\varphi_y^2 + \frac{\gamma - 1}{2}(\varphi_x^2 + \varphi_z^2) \right) M_\infty^2 \varphi_{yy} \\
&+ \left((\gamma - 1)\varphi_x + \frac{\gamma + 1}{2}\varphi_z^2 + \frac{\gamma - 1}{2}(\varphi_x^2 + \varphi_y^2) \right) M_\infty^2 \varphi_{zz} \\
&+ 2M_\infty^2(1 + \varphi_x)\varphi_y\varphi_{xy} + 2M_\infty^2(1 + \varphi_x)\varphi_z\varphi_{xz} + 2M_\infty^2\varphi_y\varphi_z\varphi_{yz}, \quad (14.42)
\end{aligned}
$$

14.2.2 Linearized Small Perturbation Equation

It can easely be seen that Eq. (14.42) reduces to the well known first order accurate Prandtl-Glauert equation, when linearized, namely[13]

$$
(1 - M_\infty^2)\varphi_{xx} + \varphi_{yy} + \varphi_{zz} = 0. \quad (14.43)
$$

14.2.3 Transonic Small Perturbation Equation

Considering thin wings and airfoils at small inclinations with respect to the free-stream, the boundary condition for potential flow with the contour function $F(x, y, z) = 0$ can be approximated by satisfying $\mathbf{v} \cdot \mathbf{n} = 0$ (see Fig. 14.5) on the plane $z = 0$. Then it is assumed that

$$
z = f(x, y) = \delta g(x, y), \qquad \delta \ll 1 \quad (14.44)
$$

is valid. Stretching the y and z coordinates for example by

$$
\begin{aligned}
\bar{y} &= \delta^{1/3} y, \\
\bar{z} &= \delta^{1/3} z, \quad (14.45)
\end{aligned}
$$

and assuming that $(1 - M_\infty^2)$ is of $\mathcal{O}(\delta^{2/3})$ results in

$$
(1 - M_\infty^2)\varphi_{xx} - (3 - (2 - \gamma)M_\infty^2)M_\infty^2\varphi_x\varphi_{xx} + \varphi_{yy} + \varphi_{zz} = 0 + \mathcal{O}(\delta^2). \quad (14.46)
$$

[13] Other derivations of this equation can be found in [27, 28].

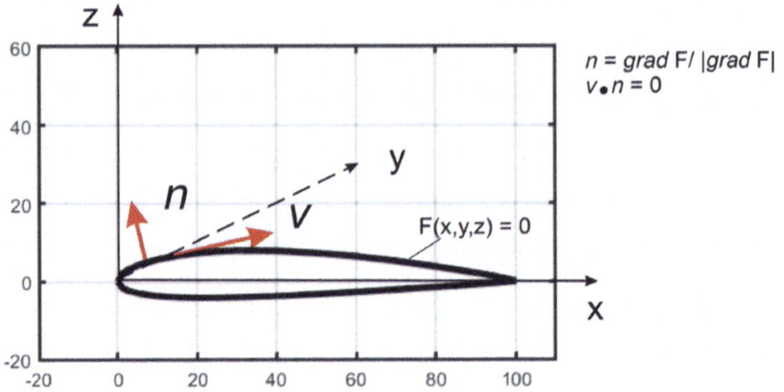

Fig. 14.5 Boundary condition for potential flows using the function of the surface contour $F(x, y, z) = 0$. **n** surface normal, **v** flow velocity at surface. The airfoil shown is the NACA 2412 profile

This is the first-order transonic small perturbation equation based on the conservative potential equation Eq. (14.36). Applying the same assumptions and approximations as above to the non-conservative full potential equation Eq. (14.40) (or Eq. (14.42)), one obtains the classical Guderley-v.Kármán transonic small perturbation equation which reads

$$(1 - M_\infty^2)\varphi_{xx} - (\gamma + 1)M_\infty^2 \varphi_x \varphi_{xx} + \varphi_{yy} + \varphi_{zz} = 0 + \mathcal{O}(\delta^2) \tag{14.47}$$

and is denoted as the first-order transonic small perturbation equation based on the non-conservative potential equation.

Abandoning the scaling Eq. (14.45) and the related estimation for $(1 - M_\infty^2)$ leads to the second order accurate transonic small perturbation equation, which corresponds to Eq. (14.46).

$$\begin{aligned}(1 - M_\infty^2)\varphi_{xx} - (3 - (2 - \gamma)M_\infty^2)M_\infty^2 \varphi_x \varphi_{xx} + \varphi_{yy} + \varphi_{zz} - \\ - M_\infty^2(2\varphi_y\varphi_{xy} + 2\varphi_z\varphi_{xz} + \varphi_x\varphi_{yy} + \varphi_x\varphi_{zz}) = 0 + \mathcal{O}(\delta^3).\end{aligned} \tag{14.48}$$

References

1. Sommerfeld, A.: Mechanik der Deformierbaren Medien. Akademische Verlagsgesellschaft Geest & Portig, Leipzig (1954)
2. Pfitzner, M., Weiland, C.: 3-D Euler Solutions for Hypersonic Freestream Mach Numbers. AGARD-CP-428, Paper 22, Bristol, UK (1987)
3. Weiland, C., Pfitzner, M.: 3-D and 2-D Solutions of the Quasi-Conservative Euler Equations. Lecture Notes in Physics 264, Springer, pp. 654–659 (1986)

4. Weiland, C.: A split-matrix Runge-Kutta type space marching procedure. J. Comp. Phys. **102**(2), 319–335 (1992)
5. Weiland, C.: Calculation of three-dimensional stationary supersonic flow fields by applying the “Progonka” process to a conservative formulation of the governing equations. J. Comp. Phys. **29**(2), 173–198 (1978)
6. Courant, R., Hilbert, D.: Methoden der Mathematischen Physik. Springer Verlag, Berlin/Heidelberg (1968)
7. Chakravarthy, S.R., Anderson, D.A., Salas, M.D.: The split-coefficient matrix method for hyperbolic systems of gas dynamic equations. AIAA Paper 80-0268 (1980)
8. Steger, J.L., Warming, R.F.: Flux vector splitting of the inviscid gasdynamic equations with application to finite-difference methods. J. Comp. Phys. **40**, 263–293 (1984)
9. Whitfield, D.L., Janus, J.M.: Three-dimensional unsteady Euler equations solution using flux vector splitting. AIAA Paper 84-1552 (1984)
10. Roe, P.L.: Approximate Riemann solvers, parameter vectors, and difference schemes. J. Comp. Phys. **43**, 357–372 (1981)
11. van Leer, B.: Towards the ultimate conservation difference scheme. V. A second-order sequel to Godunov’s method. J. Comp. Phys. **32**, 101–136 (1979)
12. Yee, H.C.: Upwind and symmetric shock-capturing schemes. NASA Technical Memorandum 89464 (1987)
13. Zurmühl, R.: Matrizen und ihre Technischen Anwendungen. Springer Verlag, Berlin/Heidelberg (1964)
14. Weiland, C.: Aerodynamic Data of Space Vehicles. Springer Verlag, Berlin/Heidelberg (2014)
15. Kuczera, H., Sacher, P.: Reusable Space Transportation Systems. Springer Verlag, Berlin/Heidelberg (2011)
16. Gordon, S., McBride, B.J.: Computer Program for Calculation of Complex Chemical Equilibrium Compositions, Rocket Performance, Incident and Reflected Shocks, and Chapman-Jouguet Detonations. NASA TN D-7056, Lewis Research Center (1973)
17. Srinivasan, S., Tannehill, J.C., Weilmuenster, K.J.: Simplified Curve Fits for the Thermodynamic Properties of Equilibrium Air. ISU-ERI-Ames-86401, Iowa State University (1986)
18. Mundt, Ch., Keraus, R., Fischer, J.: New, accurate vectorized approximations of the state surfaces for the thermodynamic and transport properties of equilibrium air. Zeitschrift fuer Flugwissenschaften und Weltraumforschung (ZfW) **15**(3), 179–184 (1991)
19. Hirschel, E.H., Weiland, C.: Selected Aerothermodynamic Design Problems of Hypersonic Flight Vehicles. Springer, Berlin/Heidelberg, and **229**, Progress in Astronautics and Aeronautics, AIAA, Reston, Va (2009)
20. Hartmann, G., Menne, S., Schröder, W.: Uncertainties Analysis/Critical Points. HERMES Report, H-NT-1-0329-DASA, Dasa, München/Ottobrunn, Germany (1994)
21. Menne, S., Weiland, C., Pfitzner, M.: Computation of 3-D hypersonic flows in chemical non-equilibrium including transport phenomena. AIAA Paper 92-2876, 27th Thermophysics Conference, Nashville USA (1992)
22. Pfitzner, M.: A 3-D Non-equilibrium shock-fitting algorithm using effective Rankine-Hugoniot relations. AIAA Paper 91-1467 (1991)
23. NN: X-38 Data Base. Industrial communication, Dassault Aviation-NASA-EADS (1999)
24. Weiland, C.: Computational Space Flight Mechanics. Springer Verlag, Berlin/Heidelberg (2010)
25. Weiland, C.: A Comparison of Potential and Euler Methods for the Calculation of 3-D Supersonic Flows Past Wings. Notes on Numerical Fluid Mechanics, **7**, Vieweg Verlag, pp.363-369 (1984)
26. Slooff, J.: Applied computational transonic aerodynamics. AGARDograph No. 266 (1982)
27. Zierep, J.: Theoretische Gasdynamik. G. Braun Verlag, Karlsruhe (1972)
28. Weiland, C.: Mechanics of Flow Similarities. Springer, Berlin (2020)

Chapter 15
Conservation Forms

Abstract Through the groundbreaking work of Ernst Mach and Peter Salcher, [1, 2], the experimental proof was provided that projectiles flying faster than the speed of sound create shock waves. The same is true for supersonic jet flows in ambient air [3]. Since these findings, a lot of investigations of supersonic and hypersonic flows around real configurations like aircraft and spacecraft as well as flow through propulsion systems, inlets and diffusors etc. have revealed that in such flow fields, very complex structures of discontinuities exist. These discontinuities are not only shock waves, but also vortex sheets and contact discontinuities and their interactions.

The Variables and Quantities Used in this Chapter

$\mathbf{E}$	vector: conservatively formulated Euler equations
$\mathbf{Q}$	vector: conservative variables
$\mathbb{A}$	matrix of non-conservatively formulated Euler equations
$\mathbf{X}$	vector: non-conservative variables
$\mathbb{J}(\mathbf{Q})$	Jacobian of conservatively formulated Euler equations
ρ	density
p	pressure
e	total energy per unit mass
v_x	one-dimensional velocity component
$\rho, m, \bar{e}$	dependent variables
c	speed of sound
γ	ratio of specific heats
n	homogeneity parameter

In the 1960s and 1970s, due to the intensive space activities of the United States of America and the Soviet Union, the solution of the blunt body problem was of eminent interest. In those days, simple configurations with a blunt nose like spheres,

C. Weiland, *The Hierarchy of Fluid Dynamic Equations*,
https://doi.org/10.1007/978-3-031-71198-5_15

cylinders and ellipsoids were looked at to study the possibilities for predicting accurate numerical solutions. Supersonic flow generates a bow shock, which was mainly computed with the so-called shock fitting technique at that time, [5–7]. This technique uses a coordinate transformation in the sense that the bow shock matches with a coordinate surface, where the Rankine-Hugoniot relations are met across it[1].

As mentioned above, the supersonic flow field past real configurations contains not only the bow shock but also a variety of embedded shocks, which in particular in three-dimensional flow cannot be treated with the shock fitting technique. Therefore, one looked for another procedure for predicting discontinuities in particular of embedded shocks. Such a procedure is given by the shock capturing technique, which is based on a specific kind of conservative formulation of the governing equations. The theoretical background of this method was given by the work of P.D. Lax [8, 9][2]. He did show that the conservatively formulated governing equations (divergence form, see Eqs. (3.6), (3.9), (3.12)) enclose the jump conditions across discontinuities as weak solutions. The divergence term is defined such that the volume integral does not change, if the argument function inside the volume is varied. Therefore, it is irrelevant for the magnitude of the volume integral, if the argument function has any discontinuities in the volume.

But the classical divergence formulation is not the only possibility to deal with discontinuities using the shock capturing technique. In the following another formulation is defined, which is called the quasi-conservative formulation.

To highlight some aspects of this subject, we restrict ourselves to one-dimensional unsteady inviscid flow. Three formulations of the related Euler equations are considered and their numeral approximations are investigated.

15.1 Formulations of the Euler Equations

Conservative form

In vector notation the conservative Euler equations for perfect gas are given by (see Eq. (14.4))

---

[1] This procedure delivers the most accurate results for bow shocks. But for other discontinuities like embedded shocks it is not of practical value.

[2] Of course, from the mathematical point of view the existence of discontinuities as part of the solutions of hyperbolic partial differential equations, like the Burgers equation and the onedimensional Euler equations, are known since the seminal work of B. Riemann (1826–1866). The Riemann problem describes the unsteady flow field in a pipe, where at $t = 0$ in a left and in a right part of the pipe different initial conditions are available, which produce moving discontinuities for $t > 0$ (shock tube problem), [3]. But at that time, the physical meaning of this solution was not completely recognized and B. Riemann had not realized that the flow was not isentropic across the discontinuity. This drawback was smoothed out firstly by the works ofW.J.M. Rankine (1820–1872) and P.H. Hugoniot (1851–1887).

$$\frac{\partial \mathbf{Q}}{\partial t} + \frac{\partial \mathbf{E}}{\partial x} = 0, \qquad \mathbf{Q} = \begin{pmatrix} \rho \\ \rho v_x \\ \rho e \end{pmatrix}, \qquad \mathbf{E} = \begin{pmatrix} \rho v_x \\ \rho v_x^2 + p \\ (\rho e + p) v_x \end{pmatrix},$$

$$\rho e = \frac{p}{\gamma - 1} + \frac{1}{2} \rho v_x^2 . \tag{15.1}$$

For the subsequent mathematical operations, it is convenient to introduce the dependent variables $m = \rho v_x$ and $\bar{e} = \rho e$ and one obtains

$$\mathbf{Q} = \begin{pmatrix} \rho \\ m \\ \bar{e} \end{pmatrix}, \qquad \mathbf{E} = \begin{pmatrix} m \\ m^2/\rho + p \\ (\bar{e} + p) m/\rho \end{pmatrix}. \tag{15.2}$$

Non-conservative form

In vector-matrix notation the non-conservative Euler equations for perfect gas are given by (see Eq. (14.10))

$$\frac{\partial \mathbf{X}}{\partial t} + \mathbb{A} \frac{\partial \mathbf{X}}{\partial x} = 0, \qquad \mathbf{X} = \begin{pmatrix} \rho \\ v_x \\ p \end{pmatrix}, \qquad \mathbb{A} = \begin{pmatrix} v_x & \rho & 0 \\ 0 & v_x & 1/\rho \\ 0 & \rho c^2 & v_x \end{pmatrix},$$

$$c^2 = \gamma \frac{p}{\rho}. \tag{15.3}$$

Quasi-conservative form

In vector-matrix notation, the quasi-conservative Euler equations for perfect gas with $\mathbb{J}(\mathbf{Q})$ the Jacobian are given by (see Eq. (14.6))

$$\frac{\partial \mathbf{Q}}{\partial t} + \mathbb{J}(\mathbf{Q}) \frac{\partial \mathbf{Q}}{\partial x} = 0, \qquad J_{lm} = \frac{\partial E_l}{\partial Q_m}. \tag{15.4}$$

Equation (15.4) is called quasi-conservative, if the vector $\mathbf{E}(\mathbf{Q})$ satisfies Euler's homogeneity relation $\mathbf{E}(\lambda \mathbf{Q}) = \lambda^n \, \mathbf{E}(\mathbf{Q})$ and it holds $\mathbf{E} = n \, \mathbb{J} \, \mathbf{Q}$.

Proof:

From Eqs. (15.1) and (15.4) it follows

$$\frac{\partial \mathbf{Q}}{\partial t} + \mathbb{J} \frac{\partial \mathbf{Q}}{\partial x} = \frac{\partial \mathbf{Q}}{\partial t} + n \, \frac{\partial (\mathbb{J} \mathbf{Q})}{\partial x} = 0 \tag{15.5}$$

and

$$(n-1) \, \mathbb{J} \frac{\partial \mathbf{Q}}{\partial x} + n \, \frac{\partial \mathbb{J}}{\partial x} \mathbf{Q} = 0 . \tag{15.6}$$

If $\mathbb{J}\dfrac{\partial \mathbf{Q}}{\partial x}$ and $\dfrac{\partial \mathbb{J}}{\partial x}\mathbf{Q}$ are independent of each other, Eq. (15.6) can only be satisfied for $\mathbb{J}\dfrac{\partial \mathbf{Q}}{\partial x} \neq 0$, if simultaneously $n = 1$ and $\dfrac{\partial \mathbb{J}}{\partial x}\mathbf{Q} \equiv 0$ holds. Indeed, $\mathbf{E}$ is homogeneous of degree one and by simple evaluation one verifies $\dfrac{\partial \mathbb{J}}{\partial x}\mathbf{Q} \equiv 0$. Further, it is also $\dfrac{\partial \mathbb{J}}{\partial t}\mathbf{Q} \equiv 0$. $\mathbb{J}(\mathbf{Q})$ has the form

$$\mathbb{J}(\mathbf{Q}) = \begin{pmatrix} 0 & 1 & 0 \\ \dfrac{(\gamma-3)}{2}\dfrac{m^2}{\rho^2} & (3-\gamma)\dfrac{m}{\rho} & (\gamma-1) \\ \left[-\gamma\dfrac{m\bar{e}}{\rho^2} + (\gamma-1)\dfrac{m^3}{\rho^3}\right] & \left[\dfrac{\gamma\bar{e}}{\rho} - \dfrac{3(\gamma-1)}{2}\dfrac{m^2}{\rho^2}\right] & \gamma\dfrac{m}{\rho} \end{pmatrix}. \tag{15.7}$$

15.2 Integral Versions of the Various Forms of the Euler Equations

The nature of the integral and differential forms of the fluid dynamic equations are described in Sects. 3.2 and 3.3. From there, we extract the statement that the Eqs. (15.1), (15.3), (15.4) are inappropriate for calculating discontinuities because the derivatives do not exist at the position of any discontinuity with respect to the spatial coordinate .

Integration of the Eqs. (15.1), (15.3), (15.4) in the intervals $t_0 \leq t \leq t_1$ and $x_1 \leq x \leq x_2$ results in:

Conservative form

$$\int_{x_1}^{x_2}\Big[\mathbf{Q}(t_1,x) - \mathbf{Q}(t_0,x)\Big]dx + \int_{t_o}^{t_1}\Big[\mathbf{E}(t,x_2) - \mathbf{E}(t,x_1)\Big]dt = 0\,. \tag{15.8}$$

Non-conservative form

$$\int_{x_1}^{x_2}\Big[\mathbf{X}(t_1,x) - \mathbf{X}(t_0,x)\Big]dx + \int_{t_o}^{t_1}\int_{x_1}^{x_2}\Big[\mathbb{A}\frac{\partial \mathbf{X}}{\partial x}\Big]dxdt =$$

$$\int_{x_1}^{x_2}\Big[\mathbf{X}(t_1,x) - \mathbf{X}(t_0,x)\Big]dx + \int_{t_0}^{t_1}\Big[\mathbb{A}\mathbf{X}|_{x_1}^{x_2} - \int_{x_1}^{x_2}\frac{\partial \mathbb{A}}{\partial x}\mathbf{X}dx\Big]dt = 0\,. \tag{15.9}$$

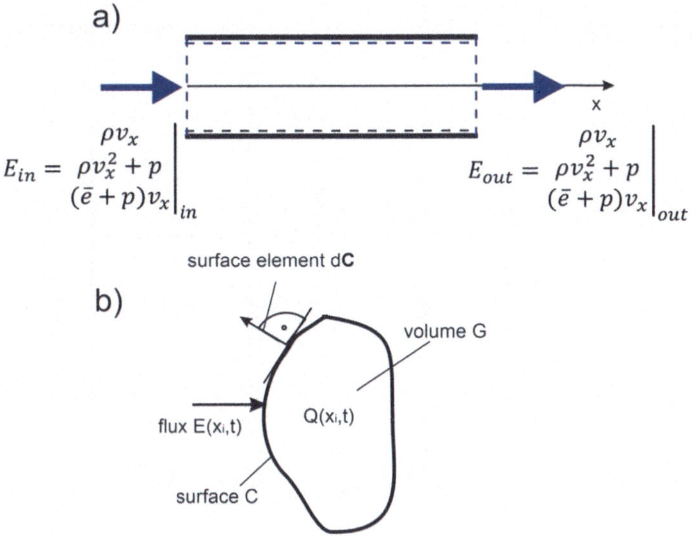

Fig. 15.1 Sketch of the control volume. **a** One-dimensional flow, **b** general

Quasi-conservative form

$$\int_{x_1}^{x_2}\Big[\mathbf{Q}(t_1,x)-\mathbf{Q}(t_0,x)\Big]dx+\int_{t_0}^{t_1}\Big[\mathbb{J}\mathbf{Q}|_{x_1}^{x_2}-\int_{x_1}^{x_2}\frac{\partial\mathbb{J}}{\partial x}\mathbf{Q}dx\Big]dt$$

$$=\int_{x_1}^{x_2}\Big[\mathbf{Q}(t_1,x)-\mathbf{Q}(t_0,x)\Big]dx$$

$$+\int_{t_0}^{t_1}\Big[\mathbb{J}(t,x_2)\mathbf{Q}(t,x_2)-\mathbb{J}(t,x_1)\mathbf{Q}(t,x_1)\Big]dt=0\,. \qquad (15.10)$$

When we examine Eq. (15.8), it is obvious that the temporal progress of the variable $\mathbf{E}(t, x_i)$ has only to be known at the end points (entrance and exit of the volume considered) of the x-interval, Fig. 15.1a. This constitutes the well-known fact that the volume integral of the divergence about the volume G is equal to the surface integral of the argument functions of the divergence about the boundary C of G (Gauss' theorem), Fig. 15.1b. In our case, the argument function is given by the fluxes $\mathbf{E}(t, x_i)$ across the boundary. This implies directly that the integral value is also correct if a discontinuity exists inside the volume G.

Fig. 15.2 Sketch of the Crank-Nicholson discretisation scheme

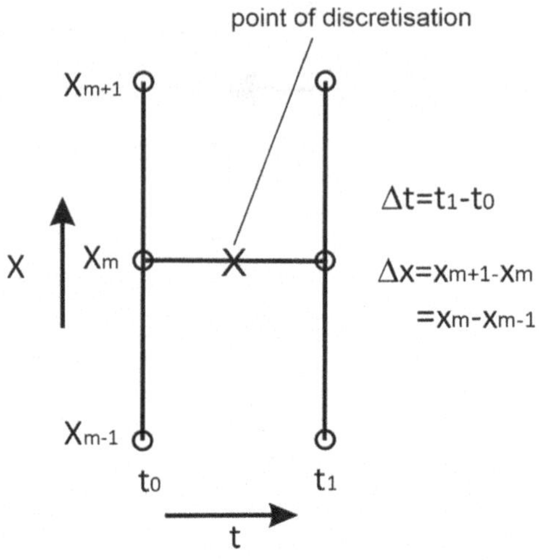

The integral version of the non-conservative form (Eq. (15.9)) shows besides the integrable term $\mathbb{A}\mathbf{X}$ the integral about the space coordinate $\int_{x_1}^{x_2} \frac{\partial \mathbb{A}}{\partial x} \mathbf{X} dx$, which in general is not equal to zero because the integrand does not disappear. The necessary pointwise knowledge of $\mathbf{X}$ and $\frac{\partial \mathbb{A}}{\partial x}$ interdicts the appearance of discontinuities in the x-interval because $\frac{\partial \mathbb{A}}{\partial x}$ is not existing there. Further, the quantity $\mathbb{A}\mathbf{X}$ cannot be construed as a conservative flux.

Notwithstanding, that Eq. (15.4) is formally similar to Eq. (15.3), the integral version Eq. (15.10) yields exactly the same formulation as Eq. (15.8) because, with respect to the homogeneous relation of Euler both, $\frac{\partial \mathbb{J}}{\partial x} \mathbf{Q} = 0$ and $\mathbb{J}\mathbf{Q} = \mathbf{E}$ hold.

15.3 Numerical Approximations of the Differential and Integral Versions of the Euler Equations

The differential Eqs. (15.1) to (15.4) are discretized with a Crank-Nicholson scheme, Fig. 15.2. Summation about an appropriate spatial x-interval leads to an approximation of the integral versions of the Euler equations, Eqs. (15.8), (15.9), (15.10). These approximations provide information how exactly the Gauss theorem of the continuum equations is satisfied, which is at the same time also a statement about the quality of the conservative properties of the discretised equations.

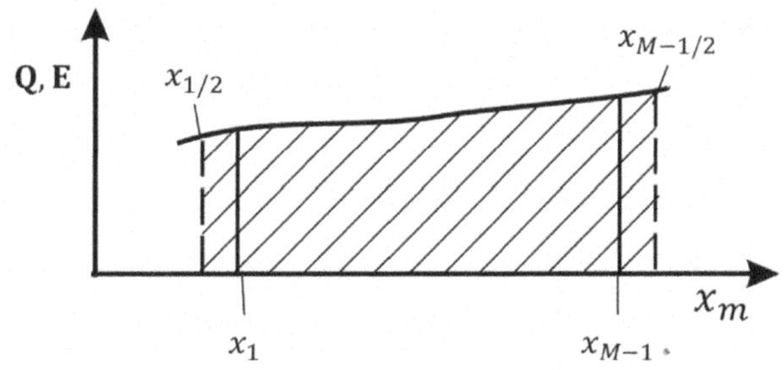

Fig. 15.3 Integration area of Crank-Nicholson discretisation scheme

Conservative form

The x-interval for the cases considered is given by $x_1 \leq x_m \leq x_{M-1}$ (see Fig. 15.3) with $x_m = m\Delta x$ and $\kappa = \Delta t/\Delta x$, $\Delta t = t_n - t_{n-1}$. The discretisation of Eq. (15.1) with a subsequent summation about m results in

$$\sum_{m=1}^{M-1} (\mathbf{Q}_m^{n+1} - \mathbf{Q}_m^n) + \frac{\kappa}{4} \sum_{m=1}^{M-1} (\mathbf{E}_{m+1}^{n+1} - \mathbf{E}_{m-1}^{n+1} + \mathbf{E}_{m+1}^n - \mathbf{E}_{m-1}^n) =$$

$$= \sum_{m=1}^{M-1} (\mathbf{Q}_m^{n+1} - \mathbf{Q}_m^n) +$$

$$+ \frac{\kappa}{4} \left[(\mathbf{E}_M^{n+1} + \mathbf{E}_{M-1}^{n+1} + \mathbf{E}_M^n + \mathbf{E}_{M-1}^n) - (\mathbf{E}_1^{n+1} + \mathbf{E}_0^{n+1} + \mathbf{E}_1^n + \mathbf{E}_0^n) \right] = 0. \quad (15.11)$$

By application of the trapezium formula the first part of Eq. (15.11) approximates the spatial integral. The second part of Eq. (15.11) constitutes the flux $\mathbf{E}$ through the considered interval whose entrance and exit boundaries are shifted outwards by $\Delta x/2$, Fig. 15.3. This part approximates the time integral of Eq. (15.8). Indeed, the approximation Eq. (15.11) shows that the fluxes $\mathbf{E}$ inside the x-interval are cancelled out by each other and have no influence on the value of the integral as Gauss' theorem requires.

Non-conservative form

The discretisation of the non-conservative Eq. (15.3) with the subsequent summation along m yields

$$\sum_{m=1}^{M-1}(\mathbf{X}_m^{n+1}-\mathbf{X}_m^n)+\frac{\kappa}{4}\sum_{m=1}^{M-1}\mathbb{A}_m^{n+1/2}(\mathbf{X}_{m+1}^{n+1}-\mathbf{X}_{m-1}^{n+1}+\mathbf{X}_{m+1}^n-\mathbf{X}_{m-1}^n)=$$

$$\sum_{m=1}^{M-1}(\mathbf{X}_m^{n+1}-\mathbf{X}_m^n)+$$

$$+\frac{\kappa}{4}\left\{\underbrace{(\mathbb{A}_M^{n+1}\mathbf{X}_M^{n+1}+\mathbb{A}_{M-1}^{n+1}\mathbf{X}_{M-1}^{n+1}+\mathbb{A}_M^n\mathbf{X}_M^n+\mathbb{A}_{M-1}^n\mathbf{X}_{M-1}^n)}_{(*)}-\right.$$

$$-\underbrace{(\mathbb{A}_1^{n+1}\mathbf{X}_1^{n+1}+\mathbb{A}_0^{n+1}\mathbf{X}_0^{n+1}+\mathbb{A}_1^n\mathbf{X}_1^n+\mathbb{A}_0^n\mathbf{X}_0^n)}_{(*)}-$$

$$\left.\underbrace{-2\Delta x\sum_{m=1}^{M-1}\left[\left(\frac{\partial\mathbb{A}}{\partial x}\right)_m^{n+1}\mathbf{X}_m^{n+1}+\left(\frac{\partial\mathbb{A}}{\partial x}\right)_m^n\mathbf{X}_m^n\right]}_{(**)}+\mathcal{O}(\Delta x^2,\Delta t^2,\Delta x\Delta t)\right\}=0. \tag{15.12}$$

The time integral in Eq. (15.9) is split into two parts by partial integration. If the integral $\int_{x_1}^{x_2}\frac{\partial\mathbb{A}}{\partial x}\mathbf{X}dx$ would be equal to zero, the formulation with respect to Eq. (15.3) would also have conservative properties. But in general this integral is nonzero. The discrete approximation of the first part of the time integral, given by (*) in Eq. (15.12), again leaves only the fluxes "$\mathbb{A}\mathbf{X}$" through the entrance and exit boundaries of the x-interval, whereas the discrete approximation of the second part of the time integral (**) requires the knowledge of $\frac{\partial\mathbb{A}}{\partial x}$ inside the x-interval and would lead to wrong results if discontinuities appear.[3]

Quasi-conservative form

Finally, considering the discrete approximation of Eq. (15.4), which is also summed up along the x-interval, one finds

$$\sum_{m=1}^{M-1}(\mathbf{Q}_m^{n+1}-\mathbf{Q}_m^n)+\frac{\kappa}{4}\sum_{m=1}^{M-1}\mathbb{J}_m^{n+1/2}(\mathbf{Q}_{m+1}^{n+1}-\mathbf{Q}_{m-1}^{n+1}+\mathbf{Q}_{m+1}^n-\mathbf{Q}_{m-1}^n)=0. \tag{15.13}$$

[3] Many terms of $\mathcal{O}(\Delta x^2,\Delta t^2,\Delta x\Delta t)$ are generated by the discretisation procedure due to the quasi-linearity of Eq. (15.3). These terms are of the same order as the truncation error of the discretized equation itself.

After some mathematical operations, one obtains

$$\sum_{m=1}^{M-1}(\mathbf{Q}_m^{n+1} - \mathbf{Q}_m^n) + \frac{\kappa}{4}\Big\{(\mathbb{J}_M^{n+1}\mathbf{Q}_M^{n+1} + \mathbb{J}_{M-1}^{n+1}\mathbf{Q}_{M-1}^{n+1} + \mathbb{J}_M^n\mathbf{Q}_M^n + \mathbb{J}_{M-1}^n\mathbf{Q}_{M-1}^n) - (\mathbb{J}_1^{n+1}\mathbf{Q}_1^{n+1} + \mathbb{J}_0^{n+1}\mathbf{Q}_0^{n+1} + \mathbb{J}_1^n\mathbf{Q}_1^n + \mathbb{J}_0^n\mathbf{Q}_0^n) + \mathcal{O}\left(\frac{\Delta x^2}{2}, \frac{\Delta t^2}{4}\right) + \underbrace{\Delta x \Delta t \sum_{m=1}^{M-1}\left[\left(\frac{\partial^2 \mathbb{J}}{\partial x \partial t}\right)_m^{n+1}\mathbf{Q}_m^{n+1} - \left(\frac{\partial^2 \mathbb{J}}{\partial x \partial t}\right)_m^{n}\mathbf{Q}_m^{n}\right]}_{(***)}\Big\} = 0. \tag{15.14}$$

The conservative properties of the differential Eq. (15.4) are already addressed, and also the ones of the integral version Eq. (15.9). The discrete approximation Eq. (15.14) has strong conservative properties, too. The fluxes $\mathbb{J}\mathbf{Q} = \mathbf{E}$ through the respective boundaries of the x-interval are approximated by the difference formula up to $\mathcal{O}(\Delta x^2/2, \Delta t^2/4)$. In addition, the term $(***)$ is $\mathcal{O}(\Delta x \Delta t^2)$ and can be interpreted as a measure of the conservative characteristics of the difference approximation.

15.4 Numerical Test Case

The test case[4] chosen is the well known initial value problem given in [8], which can be understood as a one-dimensional shock wave running into a gas at rest, Fig. 15.4.

The initial conditions are

$$\mathbf{X}(x,0) = \begin{cases} \mathbf{X}_1 = \begin{pmatrix} \rho_1 = 50 \\ v_{x,1} = 2 \\ p_1 = 50 \end{pmatrix} & x < 0 \\ \mathbf{X}_2 = \begin{pmatrix} \rho_2 = 10 \\ v_{x,2} = 0 \\ p_2 = 0 \end{pmatrix} & x > 0 \end{cases} \tag{15.15}$$

[4] The test case serves only as a benchmark for checking the capacity of numerical discretisation schemes. Therefore the variables do not have dimensions and the "gas" has no physical relevance.

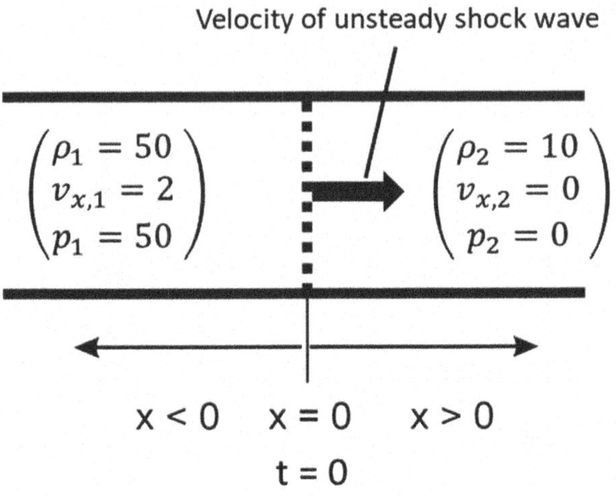

Fig. 15.4 Definition of test case conditions. Index 1: gas behind the shock wave. Index 2: gas ahead of the shock wave. Position of shock wave $x = 0$ at $t = 0$

The analytical solution of this problem is

$$\mathbf{X}(x,t) = \begin{cases} \mathbf{X}_1, & x - c_{in}\,t < 0 \\ \mathbf{X}_2, & x - c_{in}\,t > 0 \end{cases} \tag{15.16}$$

with c_{in} the unsteady shock speed and $\gamma = 1.5$ the ratio of specific heats.

The pressure, the velocity and the density of the test case are evaluated for 25 and 300 time steps and are compared with the analytical solution. The numerical solution of the conservative discretisation (Eq. (15.11)) of the Euler equations approximates the analytical solution of the shock position, the sharpness of the shock and the magnitude of the variables fairly well, Fig. 15.5.

The results of the non-conservative discretisation (Eq. (15.12)) of the Euler equations are shown in Fig. 15.6. It can clearly be seen that this approximation fails because all variables obviously tend to another solution. Pressure and density are too high and the velocity is too low, which applies also for the shock wave velocity. The latter leads to a shorter shock wave position.

The solution of the quasi-conservative discretisation scheme (Eq. (15.14)) meets all requirements on the accuracy for capturing shock waves, Fig. 15.7. This solution at least has the quality of the conservative one, and the predicted functions, for example the density, seem a bit smoother in comparison to Fig. 15.5.

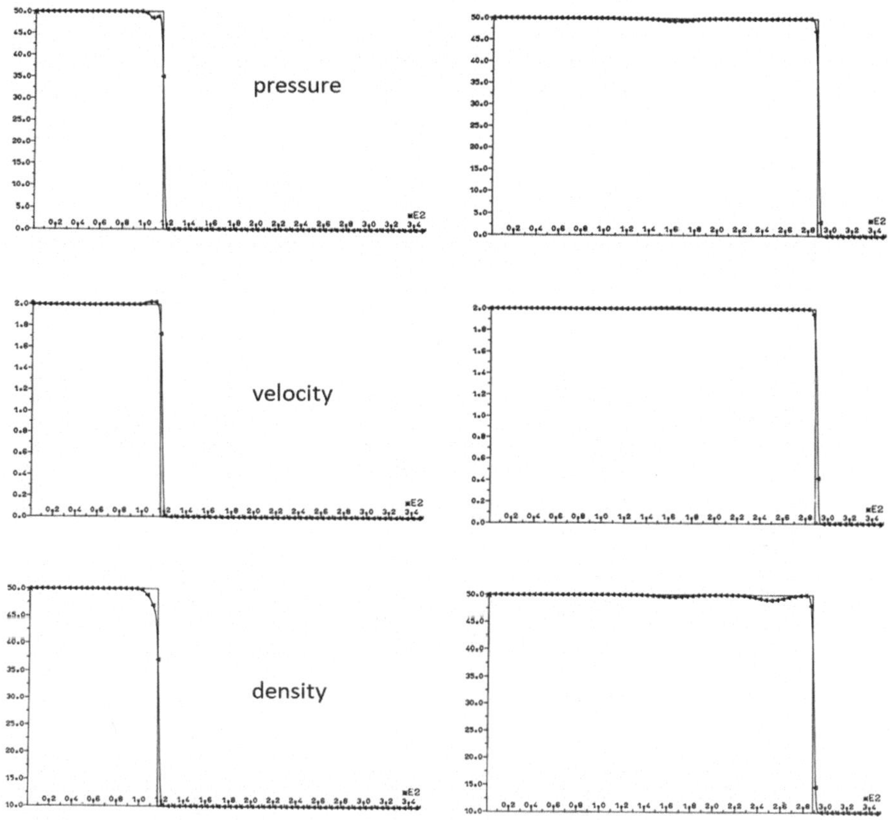

Fig. 15.5 Numerical solution of the test case with conditions Eq. (15.15) using the conservative approximation Eq. (15.11). Left column: result after 25 time steps, right column: result after 300 time steps

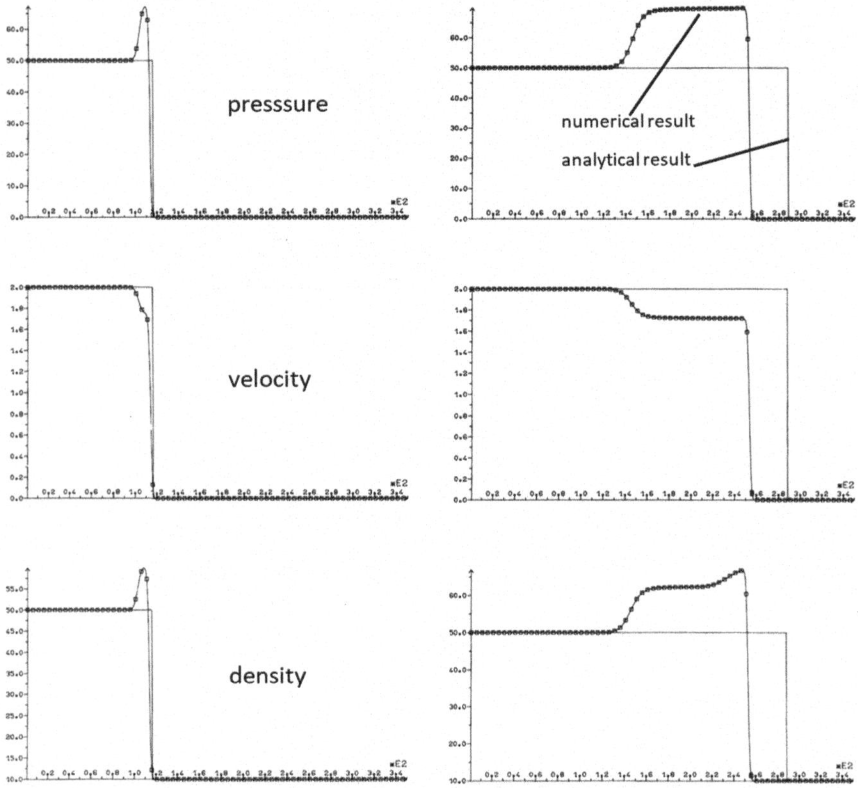

Fig. 15.6 Numerical solution of the test case with conditions Eq. (15.15) using the non-conservative approximation Eq. (15.12). Left column: result after 25 time steps, right column: result after 300 time steps

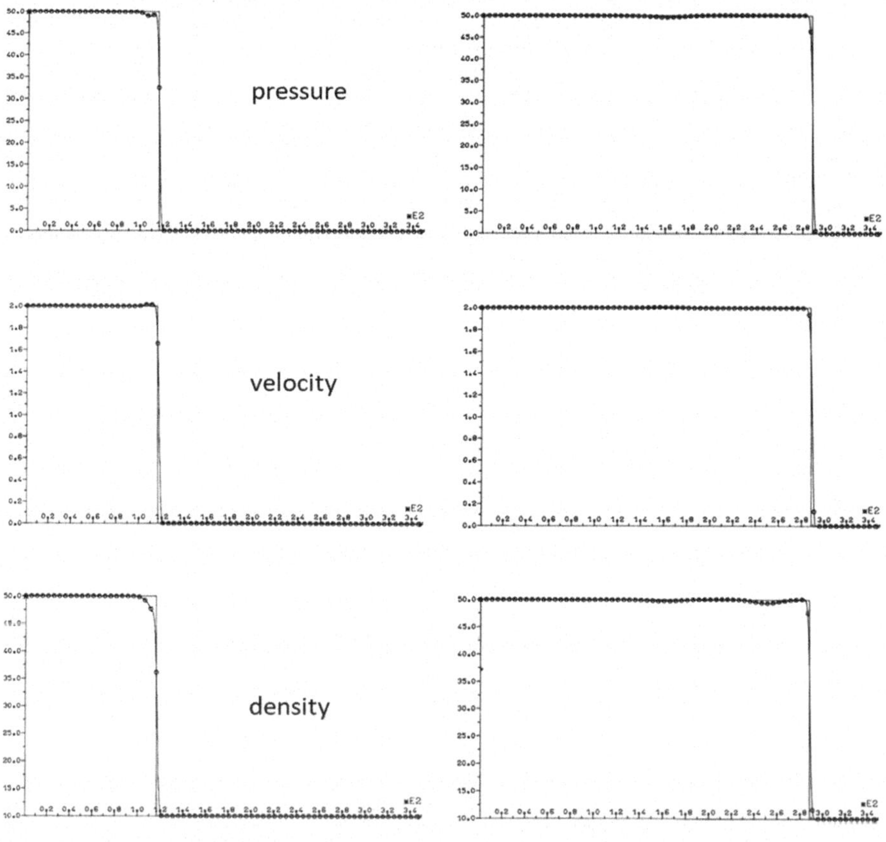

Fig. 15.7 Numerical solution of the test case with conditions Eq. (15.15) using the quasi-conservative approximation Eq. (15.14). Left column: result after 25 time steps, right column: result after 300 time steps

References

1. Mach, E., Salcher, P.: Photographische Fixierung der durch Projectile in der Luft eingeleiteten Vorgänge. Sitzungsberichte der Kaiserlichen Akademie der Wissenschaften **95**(2), 764–781 (1887)
2. Mach, E.: Über die Abbildung der von Projectilen mitgeführten Luftmasse durch Momentphotographie. Anzeiger der Kaiserlichen Akademie der Wissenschaften **23**, 136 (1886)
3. Chorin, A.J., Marsden, J.E.: A Mathematical Introduction to Fluid Mechanics. Universitext, Springer Verlag Berlin/Heidelberg (1979)
4. Mach, E., Salcher, P.: Optische Untersuchung der Luftstrahlen. Sitzungsberichte der Akademie der Wissenschaften, XCVIII Band, Wien (1889)
5. Lyubimow, A.N., Rusanow, V.V.: Gas Flows Past Blunt Bodies. (NASA TT F-714 (1973)), Isdatel'stoo "Nauka", Moskau (1970)
6. Moretti, G., Abbett, M.: A time-dependent computational method for blunt body flows. AIAA J. **4**(12), 2136–2141 (1966)
7. Weiland, C.: Zwei und dreidimensionale Umströmungen stumpfer Körper unter Berücksichtigung schallnaher Überschallströmungen. ZfW **24**, 237–245 (1976)
8. Lax, P.D.: Weak solutions of nonlinear hyperbolic equations and their numerical computations. Com. Pure Appl. Math., **7** (1954)
9. Lax, P.D.: Hyperbolic systems of conservative laws II. Com. Pure Appl. Math., **10** (1957)

Chapter 16
Flow Discontinuities

Abstract Since the beginning of the nineteenth century many well known scientists among them Jean-Baptiste le Rond d'Alembert, Simèon Denis Poisson, George Gabriel Stokes, Lord Rayleigh, Lord Kelvin, Bernhard Riemann, William John Macquorn Rankine and Pierre Henri Hugoniot worked on the theory of sound, in particular the propagation of sound waves of finite amplitude. In this view it is interesting to comprehend the evolution of the theory of discontinuous surfaces in a flow field [1].

- Siméon Denis Poisson delivered his lecture "On the theory of sound" in 1808, which outlined the propagation of sound waves, but did not give any advice on discontinuous functions,
- George Gabriel Stokes discussed the problem of the motion of sound waves and find theoretical indications for the existence of discontinuous solutions by considering the continuity and momentum equations (jump conditions), but ignores the influence of the energy equation. Stokes did discuss the problem over the time period 1848–1880 with Lord Rayleigh and Lord Kelvin with the conclusion that discontinuous surfaces violates the principle of energy conservation and could therefore not exist in nature.
- Bernhard Riemann presented in his paper on the propagation of sound waves of finite amplitude (1860) a solution of moving compression waves where he derived the jumps in mass and momentum for an isentropic gas (This means an adiabatic process in the shock transition.) and failed also like Stokes before him in the understanding of the true nature of a shock wave.
- It was William John Macquorn Rankine (1870) who has first recognized that during a shock transition the temperature and the conduction of heat must change which implies a non-adiabatic process, see also Sect. 18.17.
- The transition of a gas across a shock wave is not an adiabatic process and therefore the usual relation $p \propto \rho^\gamma$ is not valid. Pierre Henri Hugoniot has distinguished this and found by evaluation of the equations for the conservation of mass, momentum and energy the famous relation between the internal energy, the pressure and the density $e' - e = \frac{p'+p}{2}(\frac{1}{\rho} - \frac{1}{\rho'})$ which bears his name. This equation is the basic principle of all the work in the field of shock waves.

We address in this chapter strong, weak and contact discontinuities.

C. Weiland, *The Hierarchy of Fluid Dynamic Equations*,
https://doi.org/10.1007/978-3-031-71198-5_16

The Variables and Quantities Used in this Chapter

$\mathbf{E}$	vector: conservatively formulated Euler equations
$\mathbf{Q}$	vector: conservative variables
$\mathbb{J}(\mathbf{Q})$	Jacobian of conservatively formulated Euler equations
$\rho, m, \bar{e}$	dependent variables
ρ	density
p	pressure
e	total energy per unit mass
e_i	internal energy per unit mass
h	enthalpy per unit mass
c	speed of sound
γ	ratio of specific heats
q_1, q_2, c_1, c_2	elements of advection equation
$\lambda_{1,2,3}$	eigenvalues of Jacobian $\mathbb{J}(\mathbf{Q})$
$n_{1,2,3}$	eigenvectors of Jacobian $\mathbb{J}(\mathbf{Q})$
$\Gamma_{1,2,3}$	Riemann invariants
u	one-dimensional velocity component
$\mathbf{U}_1, \mathbf{U}_2$	vector: velocity in discontinuous flow
$U_{1,n}, U_{2,n}$	normal velocity components of $\mathbf{U}_1, \mathbf{U}_2$
$U_{1,t}, U_{2,t}$	tangential velocity components of $\mathbf{U}_1, \mathbf{U}_2$
$V_{1,n}, V_{2,n}$	normal velocity components for the system at rest
D	velocity of discontinuous surface
$\Sigma(\xi, \eta, \zeta, t)$	discontinuous surface in arbitrary coordinates (ξ, η, ζ, t)

16.1 General Considerations

With the seminal work of B. Riemann (1826–1866) a theory became available to understand discontinuous solutions of hyperbolic partial differential equations, [2–4], despite the fact that also Riemann has failed in the understanding of the relevance of the energy equation for shock waves ($\Rightarrow$ strong discontinuity).

The classical Riemann problem consists of two regions $x < 0$ and $x \geq 0$, where at $t = 0$ the dependent variables have different values. The question is how to develop the solution for $t > 0$. For linear equations like the advection equation

$$\frac{\partial}{\partial t}\begin{pmatrix} q_1 \\ q_2 \end{pmatrix} + \begin{pmatrix} c_1 & 0 \\ 0 & c_2 \end{pmatrix} \frac{\partial}{\partial x}\begin{pmatrix} q_1 \\ q_2 \end{pmatrix} = 0,$$

with constant advection velocities or a constant Jacobi matrix, the solution is simply that the component q_1 travels to the right with velocity $c_1 > 0$ and the component q_2 travels to the left with velocity $c_2 < 0$.

For non-linear problems like the one-dimensional Euler equations the situation is much more complicated and strongly depends on the initial conditions. Considering the one-dimensional Euler equations in the form of Eqs. (15.2), (15.4), (15.7)

$$\frac{\partial \mathbf{Q}}{\partial t} + \mathbb{J}(\mathbf{Q})\frac{\partial \mathbf{Q}}{\partial x} = 0, \tag{16.1}$$

$$\mathbb{J}(\mathbf{Q}) = \begin{pmatrix} 0 & 1 & 0 \\ \frac{\gamma - 3}{2}\frac{m^2}{\rho^2} & (3-\gamma)\frac{m}{\rho} & (\gamma - 1) \\ \left[-\gamma\frac{m\bar{e}}{\rho^2} + (\gamma - 1)\frac{m^3}{\rho^3}\right] & \left[\frac{\gamma\bar{e}}{\rho} - \frac{3(\gamma-1)}{2}\frac{m^2}{\rho^2}\right] & \gamma\frac{m}{\rho} \end{pmatrix},$$

$$\mathbf{Q} = \begin{pmatrix} \rho \\ m \\ \bar{e} \end{pmatrix}, \quad \mathbf{E} = \begin{pmatrix} m \\ m^2/\rho + p \\ (\bar{e} + p)m/\rho \end{pmatrix}, \quad J_{lm} = \frac{\partial E_l}{\partial Q_m},$$

with $m = \rho u$ and[1] $\bar{e} = \rho e$. The eigenvalues of the Jacobian $\mathbb{J}(\mathbf{Q}) \doteqdot J_{lm} = \frac{\partial E_l}{\partial Q_m}$ are

$$\begin{aligned} \lambda_1 &= u - c, \\ \lambda_2 &= u, \\ \lambda_3 &= u + c, \end{aligned} \tag{16.2}$$

where $c = \sqrt{\gamma \frac{p}{\rho}}$ is the speed of sound of an adiabatic gas. A corresponding set of eigenvectors for the Jacobian $\mathbb{J}(\mathbf{Q})$ is given by[2]

[1] In the whole book the velocity component in x-direction is denoted by v_x. Instead, in this chapter we use the letter u for this velocity component, because the presentation is then closer to other literature.

[2] Here and in the following perfect gas is assumed.

$$n_1 = \begin{pmatrix} 1 \\ u - c \\ \frac{u^2}{2} + \frac{c^2}{\gamma - 1} - cu \end{pmatrix}, \; n_2 = \begin{pmatrix} 1 \\ u \\ \frac{u^2}{2} \end{pmatrix}, \; n_3 = \begin{pmatrix} 1 \\ u + c \\ \frac{u^2}{2} + \frac{c^2}{\gamma - 1} + cu \end{pmatrix}. \quad (16.3)$$

Remark A set of eigenvectors for the matrix $\mathbb{A}$ of the non-conservative formulation is given by the matrix $\mathbb{P}$ (here for the one-dimensional case) of Eq. (14.16) and reads

$$m_1 = \begin{pmatrix} 1 \\ 0 \\ 0 \end{pmatrix}, \; m_2 = \begin{pmatrix} \frac{1}{c^2} \\ \frac{1}{\rho c} \\ 1 \end{pmatrix}, \; m_3 = \begin{pmatrix} \frac{1}{c^2} \\ -\frac{1}{\rho c} \\ 1 \end{pmatrix}.$$

The eigenvalues can be interpreted as characteristics in spacetime with $\frac{dx}{dt} = \lambda_{1,2,3}$. Waves traveling along these characteristics are for $\lambda_{1,3}$ sound waves and for λ_2 entropy waves. Further it has turned out that scalars exist, which are constant along these waves. These scalars are the so-called Riemann invariants, which have the general form

$$\Gamma_{1,2,3} = u + \int \frac{c(\rho)}{\rho} d\rho, \; u, \; u - \int \frac{c(\rho)}{\rho} d\rho, \quad (16.4)$$

and for a perfect gas read

$$\Gamma_{1,2,3} = u + \frac{2c}{\gamma - 1}, \; u, \; u - \frac{2c}{\gamma - 1}, \quad (16.5)$$

A closer look at the theory reveals that the entropy s is constant along λ_1 and λ_3 and the pressure p along λ_2. Thus, they are also Riemann invariants. Section 16.4 discusses the Riemann problem for a practical case, namely a shock tube, where an analytical solution including the contact discontinuity, the rarefaction wave and the shock wave (strong discontinuity) is worked out with Riemann's theory.

16.2 Strong Discontinuities

A strong discontinuity is characterized by the mass that is transported across the surface of discontinuity and by the fact that the fluxes of mass, momentum and energy are conserved. Figure 16.1 shows a discontinuous surface positioned in a flow with the freestream velocity $\mathbf{U}_1$. On a point of this surface a coordinate system

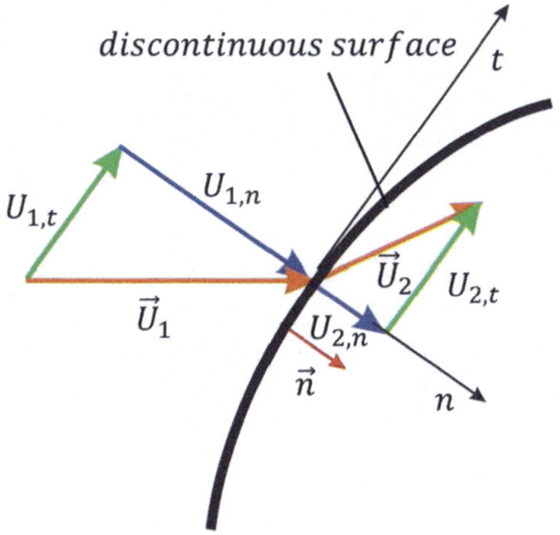

Fig. 16.1 Sketch of the jump conditions with respect to a steady discontinuous surface (shock front) in a flow with the free stream velocity $\mathbf{U}_1$

is defined with the components n normal and t tangential to the surface.[3] Thus, the normal components of the mass flux, Eq. (16.6a), of the momentum flux, Eq. (16.6d), and of the energy, Eq. (16.6e), are conserved. From vector Eq. (16.6b) together with Eq. (16.6a), Eq. (16.6c) is found, which says that the tangential velocity is the same on both sides of the discontinuity and hence is conserved, too. Physically, the discontinuous surface can be identified as a compression shock front (shortly called shock wave).

$$\rho_1 U_{1,n} = \rho_2 U_{2,n}, \tag{16.6a}$$

$$[\mathbf{U}\rho\mathbf{U}\cdot\mathbf{n} + p\,\mathbf{n}]_1 = [\mathbf{U}\rho\mathbf{U}\cdot\mathbf{n} + p\,\mathbf{n}]_2 \quad \text{or}$$

$$\left[\begin{Bmatrix} U_t \\ U_n \end{Bmatrix} \rho \begin{Bmatrix} U_t \\ U_n \end{Bmatrix} \cdot \begin{Bmatrix} 0 \\ 1 \end{Bmatrix} + p \begin{Bmatrix} 0 \\ 1 \end{Bmatrix}\right]_1 = \left[\begin{Bmatrix} U_t \\ U_n \end{Bmatrix} \rho \begin{Bmatrix} U_t \\ U_n \end{Bmatrix} \cdot \begin{Bmatrix} 0 \\ 1 \end{Bmatrix} + p \begin{Bmatrix} 0 \\ 1 \end{Bmatrix}\right]_2, \tag{16.6b}$$

$$U_{1,t} = U_{2,t}, \tag{16.6c}$$

$$\rho_1 U_{1,n}^2 + p_1 = \rho_2 U_{2,n}^2 + p_2, \tag{16.6d}$$

$$h_1 + \frac{1}{2} U_{1,n}^2 = h_2 + \frac{1}{2} U_{2,n}^2, \tag{16.6e}$$

where the index 1 denotes the variables in front of the discontinuity and index 2 behind the discontinuity.

[3] In three dimensions the third coordinate lies also in the tangential plane perpendicular to n and t, and the relation $\mathbf{U} \times \mathbf{n}$ holds instead of Eq. (16.6c).

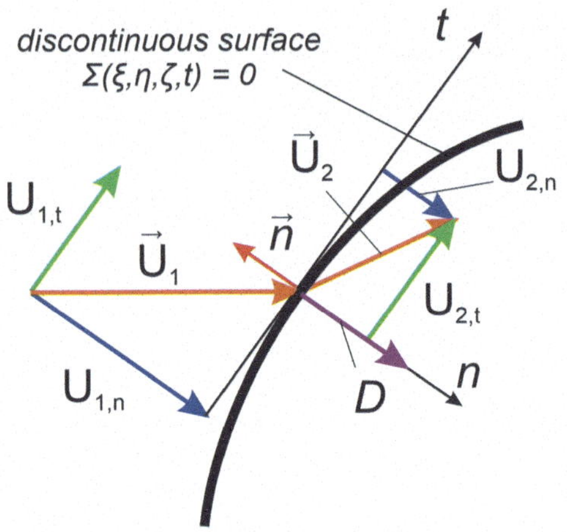

Fig. 16.2 Sketch of the jump conditions with respect to a moving discontinuity described by the time dependent function $\Sigma(\xi, \eta, \zeta, t) = 0$ (unsteady shock front)

A simple relation between the thermodynamic state variables enthalpy, pressure and density h, p, ρ can be obtained by eliminating the normal velocity components $U_{1,n}$ and $U_{2,n}$ in Eqs. (16.6a) and (16.6e), namely

$$h_2 - h_1 = \frac{1}{2}\left(\frac{1}{\rho_1} + \frac{1}{\rho_2}\right)(p_2 - p_1), \tag{16.7}$$

or with the internal energy e_i, defined by $h = e_i + p/\rho$

$$e_{i,2} - e_{i,1} = \frac{1}{2}\left(\frac{1}{\rho_1} - \frac{1}{\rho_2}\right)(p_2 + p_1). \tag{16.8}$$

Originally, the above Eqs. (16.7) and (16.8) are called Rankine-Hugoniot relations.[4] At later times the equation system (16.6a) to (16.6e) was also termed Rankine-Hugoniot relations in the literature.

Considering a perfect gas with $h = \dfrac{\gamma}{(\gamma - 1)}\dfrac{p}{\rho}$ and $e_i = \dfrac{1}{(\gamma - 1)}\dfrac{p}{\rho}$, one obtains from Eqs. (16.7) and (16.8)

$$\frac{p_2}{p_1} = \frac{(\gamma + 1)\rho_2 - (\gamma - 1)\rho_1}{(\gamma + 1)\rho_1 - (\gamma - 1)\rho_2} \quad \text{or} \quad \frac{p_2 - p_1}{\rho_2 - \rho_1} = \gamma\frac{p_2 + p_1}{\rho_2 + \rho_1}, \tag{16.9}$$

which is called von Kármán relation.

Flight vehicles like hypersonic airbreakers and winged re-entry vehicles generate a lot of shocks in the flow field around their configurations. When these vehicles are in an unsteady operation phase, like acceleration and deceleration or in flight maneuvers, the shocks change their shape, strength and position with time. In particular, the

[4] Equation (16.8) was first published by P.H. Hugoniot in his memoir *On the Propagation of Motion in Bodies* 1887 (In the year of his death!).

bow shock is of interest in this regard. Such a bow shock can be determined using numerical approximation methods, besides others, by the shock fitting technique, where the time-dependent shock surface is transformed into a coordinate surface given by the function $\Sigma(\xi, \eta, \zeta, t) = 0$, Fig. 16.2, [5, 6]. The coordinates ξ, η, ζ are curvilinear and non-orthogonal.

It is reasonable to transform the shock surface into a system at rest. If a local point of the shock surface moves with the velocity D, the equation system (16.6a)–(16.6e) gets the form

$$\rho_1 V_{1,n} = \rho_2 V_{2,n}, \tag{16.10a}$$

$$U_{1,t} = U_{2,t}, \tag{16.10b}$$

$$\rho_1 V_{1,n}^2 + p_1 = \rho_2 V_{2,n}^2 + p_2, \tag{16.10c}$$

$$h_1 + \frac{1}{2} V_{1,n}^2 = h_2 + \frac{1}{2} V_{2,n}^2, \tag{16.10d}$$

where

$$V_{1,n} = U_{1,n} - D, \quad V_{2,n} = U_{2,n} - D, \quad D = -\frac{\partial \Sigma / \partial t}{|\mathrm{grad}\,\Sigma|}.$$

The equation system (16.10a)–(16.10d) describes the situation when the moving shock is transformed into the system of rest.

A Typical Example

Figure 16.3 exhibits a prominent example of a strong discontinuity, namely the bow shock. The left figure a) presents the numerical solution of a flow field past the Hermes configuration. The three-dimensional Euler equations including an equilibrium real gas approach are applied. The bow shock was handled with the shock fitting technique, whereas the embedded shocks are computed by the shock capturing procedure, [5, 8]. In the right figure b) we see the supersonic flow past the Phoenix demonstrator. The experimental investigation shown in the colored shadowgraph was made in the TWG windtunnel of the German Aerospace Center (DLR), Goettingen.

16.3 Weak Discontinuities

Shear layers or vortex sheets are discontinuities without mass transfer $U_{1,n} = U_{2,n} = 0$ (weak discontinuities), Fig. 16.4. Therefore, by applying the equation system (16.6a) to (16.6e) to the shear layer situation one finds

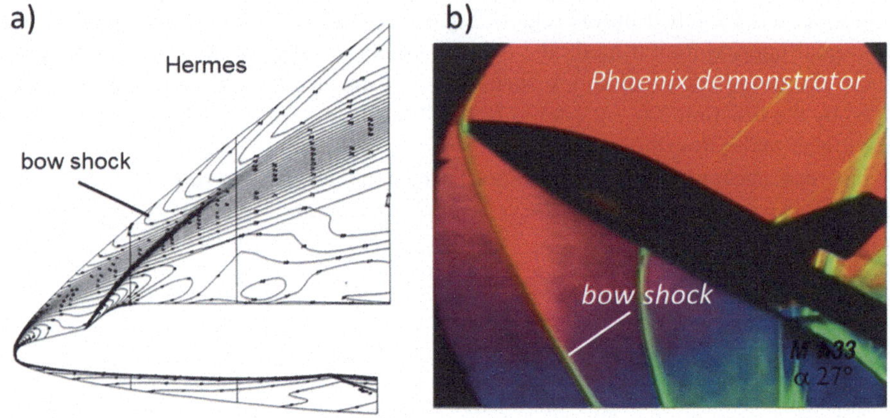

Fig. 16.3 Hypersonic flow field past the Hermes configuration, $M_\infty = 10, \alpha = 30^0$; **a** numerical prediction by solving the Euler equations with a chemical equilibrium real gas approach, [7]; **b** colored shadowgraph of a supersonic flow past the Phoenix demonstrator. Photograph made by DLR in the TWG windtunnel, $M_\infty = 2.21, \alpha = 27^0$

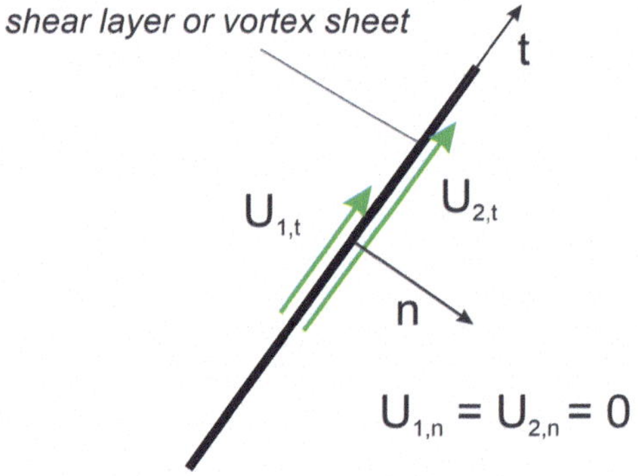

Fig. 16.4 Sketch of the situation in a shear layer or vortex sheet

$$\rho_1 U_{1,n} = \rho_2 U_{2,n} = 0 \quad \Longrightarrow \quad \rho_1 \neq \rho_2, \tag{16.11a}$$

$$\rho_1 U_{1,n}^2 + p_1 = \rho_2 U_{2,n}^2 + p_2 \quad \Longrightarrow \quad p_1 = p_2\,, \tag{16.11b}$$

$$\rho_1 U_{1,n} U_{1,t} = \rho_2 U_{2,n} U_{2,t} = 0 \quad \Longrightarrow \quad U_{1,t} \neq U_{2,t}, \tag{16.11c}$$

$$h_1 + \frac{1}{2} U_{1,t}^2 = h_2 + \frac{1}{2} U_{2,t}^2, \tag{16.11d}$$

Thermodynamics and Crocco's law (Eq. (3.23)) state that the entropy due to Eq. (16.11c) is discontinuous across the shear layer, which means $s_1 \neq s_2$.

Weak discontinuities in form of boundary layers and free shear layers exist in aircraft wing flow. The concept of the 'local vorticity content of a shear layer' [9] permits to connect the viscous and vortical flow phenomena of reality with the singularities and discontinuities of inviscid (potential) flow theory.

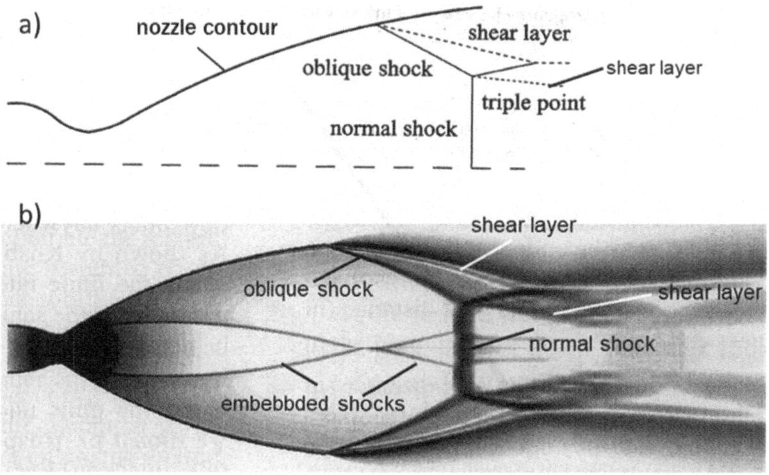

Fig. 16.5 Flow in a rocket motor test nozzle. Sketch of the structure of discontinuities, **a** Numerical solution of the non-equilibrium flow with a H_2/O_2 gas composition. The structure of the discontinuities is visualized with the pseudo-Schlieren technique, (**b**) [10, 11]

A Typical Example

The flow in rocket motor nozzles generates several discontinuous surfaces. Most of them are shocks and their interactions, see the sketch in Fig. 16.5a). The figure is also an impressive example of the generation of shear layers. One of these shear layers arises when the nozzle flow gets in touch with the ambient air. Another one develops behind the interaction of shocks of different strength, for example normal shocks and oblique shocks, in the sketch of figure a) in the triple point. Figure 16.5b) shows the result of a turbulent Navier-Stokes computation with a non-equilibrium real gas approach for a H_2/O_2 gas composition, [10],[11], where all the shocks, interactions and shear layers are present in the realistic flow field.

16.4 Contact Discontinuities

Again no mass transfer across the discontinuity is present: $U_{1,n} = U_{2,n} = 0$. Further, the tangential velocity component is given by $U_{1,t} = U_{2,t}$. With these definitions, one obtains from evaluating of the equation system (16.6a)–(16.6e) (See Fig. 16.6)

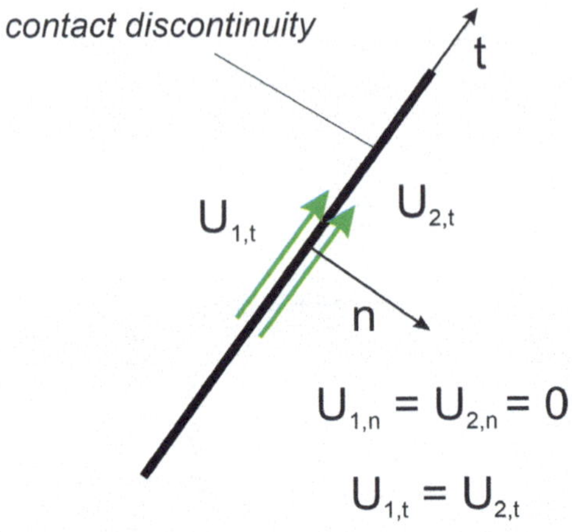

Fig. 16.6 Sketch of a contact discontinuity

$$\begin{aligned}
\rho_1 U_{1,n} &= \rho_2 U_{2,n} = 0 &&\Longrightarrow \quad \rho_1 \neq \rho_2, &\text{(16.12a)}\\
\rho_1 U_{1,n}^2 + p_1 &= \rho_2 U_{2,n}^2 + p_2 &&\Longrightarrow \quad p_1 = p_2\,, &\text{(16.12b)}\\
\rho_1 U_{1,n} U_{1,t} &= \rho_2 U_{2,n} U_{2,t} = 0 &&\Longrightarrow \quad U_{1,t} = U_{2,t}, &\text{(16.12c)}\\
h_1 + \frac{1}{2} U_{1,t}^2 &= h_2 + \frac{1}{2} U_{2,t}^2, && &\text{(16.12d)}
\end{aligned}$$

Different values of the entropy exist across the discontinuous surface: $s_1 \neq s_2$ (Fig. 16.6).

A Typical Example

In the following, the flow situation in a shock tube is discussed, which is a typical example on the one hand for the appearance of a contact discontinuity and on the other hand for the application of the theory of Riemann. The shock tube problem is a special form of a Riemann problem [2]. The physical aspects of shock tubes are described in detail in the book of H. Oertel, [12]. The shock tube problem described and discussed below, had the initial conditions first formulated in [13] and, since then, has been a classical benchmark problem of numerical methods [14].

The initial conditions at $t = 0$ are:

$$\begin{aligned}
\rho_1 &= 1.0, & p_1 &= 1.0,\ u_1 = 0 && \text{left of the membrane,}\\
\rho_5 &= 0.125, & p_5 &= 0.1,\ u_5 = 0 && \text{right of the membrane.} \qquad \text{(16.13)}
\end{aligned}$$

At a specific time interval Δt after the burst of the membrane, located at $x = 0$ for $t = 0$, the follwing flow situation appears, shown in the $x - t$ plane in Fig. 16.7 and in a scenario regarding the states of the gases in the shock tube in Fig. 16.8.

A right running characteristic $dx/dt = u + c$ divides the undisturbed region 5 with ρ_5, p_5, u_5 from the disturbed region 4 with ρ_4, p_4, $u_4 \Longrightarrow$ a shock wave. Then

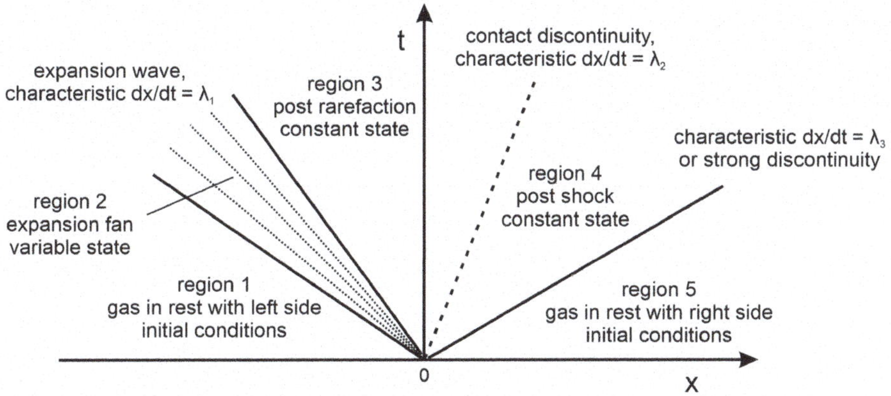

Fig. 16.7 Shock tube representation in a x-t plane with a right-going shock, a left-going expansion fan as well as a contact discontinuity in between of the two

Fig. 16.8 State of gases in a shock tube. Blue: driver gas; green: running gas. Shock tube flow at a small time interval after the burst of the membrane (membrane located at $x = 0$ for $t = 0$)

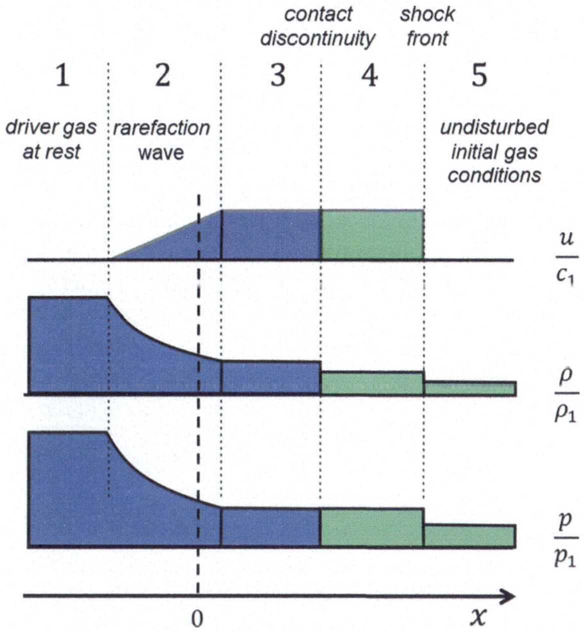

the characteristic $dx/dt = u$ separates the region 4 with ρ_4, p_4, u_4 from the region 3 with $\rho_3, p_3, u_3 \Longrightarrow$ a contact discontinuity. The left running characteristic $dx/dt = u - c$ starts the rarefaction wave at the junction between region 3 with ρ_3, p_3, u_3 and region 2. Region 2 is the only non-constant region in the solution. Finally, the last characteristic of the expansion fan $dx/dt = u(x,t) - c(x,t)$ separates region 2 with $\rho_2(x,t), p_2(x,t), u_2(x,t)$ from region 1 with the undisturbed values ρ_1, p_1, u_1.

Analytical Solution

We start with the determination of the values ρ_4, p_4, u_4 in region 4. With Eqs. (16.10a) and (16.10c) for the unsteady shock motion ($u_s \equiv D$) and the von Kármán relation Eq. (16.9) we obtain with $a = (\gamma - 1)/(\gamma + 1)$

$$u_s = u_4 \frac{\rho_4}{\rho_4 - \rho_5} \tag{16.14}$$

$$\frac{\rho_4}{\rho_5} = \frac{p_4 + a\, p_5}{a\, p_4 + p_5} \tag{16.15}$$

$$u_4 = (p_4 - p_5)\sqrt{\frac{1 - a}{\rho_5(p_4 + a\, p_5)}} \,. \tag{16.16}$$

We know, that across a contact discontinuity, only the density has a jump, which means that $u_3 = u_4$ and $p_3 = p_4$. Further, the velocity of the contact discontinuity is in agreement with Eqs. (16.12a) and (16.12c) $u_c = u_4 = u_3$. The gas velocity u_2 at the beginning of the expansion fan is with $b = (\gamma - 1)/2\gamma$

$$u_2 = \sqrt{\frac{(1 - a^2)p_1^{\frac{1}{\gamma}}}{a^2\, \rho_1}\left(p_1^b - p_2^b\right)} \,. \tag{16.17}$$

Therefore, it holds $u_4 = u_2$ and with Eqs. (16.16) and (16.17), the quantity $p_3 = p_4$ ($= p_2$ at the beginning of the expansion fan) can be iteratively determined. If $p_3 = p_4$ is known, u_4 can be calculated by Eq. (16.16) and ρ_4 by Eq. (16.15).
Note:
An alternative to Eqs. (16.16), (16.17) is given by the relations [12]

$$p_1 = p_4\left(1 - \frac{(\gamma - 1)(c_5/c_1)(p_4/p_5 - 1)}{\sqrt{2\gamma(2\gamma + (\gamma + 1)(p_4/p_5 - 1))}}\right), \tag{16.18}$$

$$u_4 = \frac{c_5}{\gamma}\,(p_4/p_5 - 1)\sqrt{(\frac{1/b}{p_4/p_5 + a}}, \tag{16.19}$$

where Eq. (16.18) must also be iteratively solved for p_4. The shock velocity u_s is then known by Eq. (16.14) and the shock position by $x_s = u_s\, t$. The gas in the regions 1, 2, 3 has not crossed any shock with a change of entropy and therefore can be considered an adiabatic gas with $p/\rho^\gamma = const$. The constant is valid in all three regions. With this, one finds

$$\rho_3 = \rho_1\left(\frac{p_3}{p_1}\right)^{1/\gamma} \,.$$

The leftward characteristic in region 2 is described by the relation

$$\frac{dx}{dt} = u - c \quad \Longrightarrow \quad c = u - \frac{x}{t}, \tag{16.20}$$

with the Riemann invariant $\Gamma = u - \frac{2c}{\gamma - 1}$. From the rightward characteristic and the Riemann invariant $\Gamma = u + \frac{2c}{\gamma - 1}$, one finds, due to the fact that at $t = 0$ all characteristics emerge from the origin

$$u = \frac{2(c_1 - c)}{\gamma - 1}. \tag{16.21}$$

Combining Eq. (16.20) with Eq. (16.21) yields

$$u_2(x, t) = \frac{2}{\gamma + 1}\left(c_1 + \frac{x}{t}\right). \tag{16.22}$$

The speed of sound in region 2 follows by evaluation of the left running characteristic

$$u_2(x, t) - c_2(x, t) = \frac{x}{t} \quad \Longrightarrow \quad c_2^2(x, t) = \left(u_2(x, t) - \frac{x}{t}\right)^2 = \gamma \frac{p_2(x, t)}{\rho_2(x, t)}. \tag{16.23}$$

Finally, with the relations

$$\begin{aligned} \rho_2(x, t) &= \left[\frac{\rho_1^\gamma}{\gamma p_1}\left(u_2(x, t) - \frac{x}{t}\right)^2\right]^{\frac{1}{\gamma - 1}}, \\ p_2(x, t) &= p_1\left(\frac{\rho_2(x, t)}{\rho_1}\right)^\gamma, \end{aligned} \tag{16.24}$$

the system of equations is complete.

Note that an alternative for the pressure $p_2(x, t)$ is given in [12] with

$$p_2(x, t) = p_1\left[1 - \frac{(\gamma - 1)}{2}\frac{u_2(x, t)}{c_1}\right]^{\frac{1}{b}}. \tag{16.25}$$

The analytical solution for the above-defined initial conditions is determined with the system of Eqs. (16.14)–(16.25). The time after the burst of the membrane was taken to be $t = 1.0$. The results are plotted in Fig. 16.9. The x-positions of the surfaces separating the different regions, also listed in the figure, are

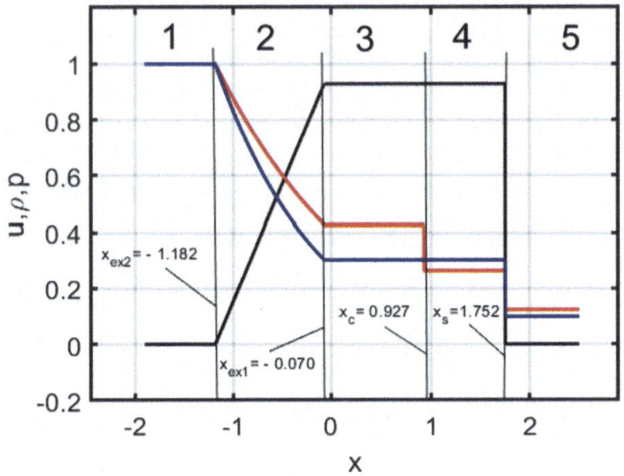

Fig. 16.9 Analytical solution of the shock tube problem with the initial conditions $\rho_1 = 1.0, p_1 = 1.0, u_1 = 0.0, \rho_5 = 0.125, p_5 = 0.1, u_5 = 0.0, \gamma = 1.4$. The solution was calculated for $t = 1.0$. Black line: velocity u, red line: density ρ, blue line: pressure p

shock position	$x_s = 1.7522,$
contact discontinuity	$x_c = 0.9275,$
onset of the expansion fan	$x_{ex1} = -0.0703,$
end of the expansion fan	$x_{ex2} = -1.1828\,.$

The constant values in regions 3 and 4 are

$$\begin{aligned}&\text{velocity } u_3 = u_4 = 0.9275,\\&\text{pressure } p_3 = p_4 = 0.3031,\\&\text{density } \rho_3 = 0.4263, \;\; \rho_4 = 0.2656,\end{aligned}$$

and finally the velocities of the separation surfaces are

velocity of the shock front	$u_s = 1.7522,$
velocity of the contact discontinuity	$u_c = 0.9275,$
velocity of the dividing surface 2-3	$u_{2,3} = u_3 - c_3 = -0.0703,$
velocity of the dividing surface 1-2	$u_{1,2} = c_1 = -1.1828.$

References

1. Salas, M.D.: The Curious Events Leading to the Theory of Shock Waves. Invited lecture at the 17th Shock Interaction Symposium, Rome (2006)
2. Chorin, A.J., Marsden, J.E.: A Mathematical Introduction to Fluid Mechanics. Springer Verlag, Berlin/Heidelberg (1979)

3. LeVeque, R.J.: Numerical Methods for Conservation Laws. Birkhäuser Verlag, Berlin (1990)
4. LeVeque, R.J.: Finite-Volume Methods for Hyperbolic Problems. Cambridge University Press, Cambridge (2004)
5. Weiland, C.: Zwei- und dreidimensionale Umströmungen stumpfer Körper unter Berücksichtigung schallnaher Überschallströmungen. ZfW **24**(5), 237–245 (1976)
6. Weiland, C.: A contribution to the computation of transonic supersonic flows over blunt bodies. Comp. and Fluids **9**, 143–162 (1981)
7. Hartmann, G., Menne, S., Schröder, W.: HERMES uncertainties analysis/critical points. HERMES Report H-NT-1-0329-DASA (1994)
8. Weiland, C., Pfitzner, M.: 3-D and 2-D solutions of the quasi-conservative euler equations. Lecture Notes in Physics vol. 264, pp. 654–659. Springer (1986)
9. Hirschel, E.H., Rizzi, A., Breitsamter, Ch., Staudacher, W.: Separated and Vortical Flow in Aircraft Wing Aerodynamics. Springer Verlag, Berlin/Heidelberg (2021)
10. Gross, A., Weiland, C.: Numerical simulation of separated cold gas Nozzle flows. J. Propulsion Power **20**(3), 509–519 (2004)
11. Gross, A., Weiland, C.: Numerical simulation of hot gas nozzle flows. J. Propulsion Power **20**(5), 879–891 (2004)
12. Oertel, H.: Stossrohre (Shock Tubes). Springer Verlag, Wien (1966)
13. Sod, G.A.: A survey of several finite difference methods for systems on nonlinear hyperbolic conservation Laws. J. Comp. Phys. **27** (1978)
14. https://www.ita.uni-heidelberg.de/~dullemond/lectures/num_fluid_2012/Chapter_6.pdf

Chapter 17
Boundary Conditions

Abstract Flow fields around configurations like airplanes, space planes, missiles, helicopters etc. or through technical devices like pipes, pumps, inlets of propulsion systems, turbo machinery, ram- and scramjets etc. are defined by their individual initial and boundary conditions. Generally, the system of Navier-Stokes equations is parabolic-hyperbolic in space and time and elliptic-parabolic for steady flow. The time dependent Euler equations are purely hyperbolic in time. Typically, for hyperbolic and parabolic problems initial boundary value conditions are imposed, whereas for elliptic problems, values on the boundaries of the flow domain determine the solution.

The Variables and Quantities Used in this Chapter

q_{gw}	heat flux of the gas
q_w	heat flux into the wall
$q_{rad,w}$	heat flux due to surface radiation
T, T_w, T_{ra}	temperature, wall temperature, radiation adiabatic temperature
κ	thermal conductivity
$F(x, y, z)$	description of body contour
$\mathbf{X}$	vector: non-conservative variables
A^ξ, A^η, A^ζ	Jacobians of non-conservatively formulated Euler equations in arbitrary coordinates
$\mathbf{Q}$	vector: conservative variables
K^ξ, K^η, K^ζ	Jacobians of conservatively formulated Euler equations in arbitrary coordinates
$\mathbb{P}_k$	matrix of eigenvectors, non-conservative variables
$\mathbb{T}_k$	matrix of eigenvectors, conservative variables
Λ_k	eigenvalue matrix of $\mathsf{A}^k(\mathbf{X})$ and $\mathsf{K}^k(\mathbf{Q})$
Γ_ξ^n	tangential terms of boundary condition equation
$\mathbf{r}_\xi$	$= \mathbb{P}_{\xi,0}^{-1}\mathbf{X}$ characteristic vector, non-conservative
$\mathbf{q}_\xi$	$= \mathbb{T}_{\xi,0}^{-1}\mathbf{Q}$ characteristic vector, conservative
v_x, v_y, v_z	components of velocity vector in Cartesian coordinates

C. Weiland, *The Hierarchy of Fluid Dynamic Equations*,
https://doi.org/10.1007/978-3-031-71198-5_17

ξ, η, ζ	arbitrary coordinates
ρ, ρ_0	density, reference density
c, c_0	speed of sound, reference speed of sound
p	pressure
e	total energy per unit mass
h	enthalpy per unit mass
Θ_ξ	$= v_x\xi_x + v_y\xi_y + v_z\xi_z$ abbreviation
$\psi_0, \Theta_{\xi,0}, \kappa_{2,0}$	abbreviations of reference values
$\mathbf{U}_1, \mathbf{U}_2$	vector: velocity in discontinuous flow
$U_{1,n}, U_{2,n}$	normal velocity components of $\mathbf{U}_1, \mathbf{U}_2$
$U_{1,t}, U_{2,t}$	tangential velocity components of $\mathbf{U}_1, \mathbf{U}_2$
D	velocity of discontinuous surface
Ω	abbreviation of iteration system
$\phi(D)$	iteration function
ϕ	$= \frac{\gamma-1}{2}\lvert\mathbf{v}\rvert^2$ abbreviation of iteration system
A, B, C, χ, q_n	abbreviations of iteration system
I	identity matrix
$\mathbf{Y}$	$= \frac{1}{\lvert\nabla\xi\rvert}(0, \xi_x, \xi_y, \xi_z, 0)^T$ metric vector
τ_w	tangential component of shear stress vector
μ	dynamic viscosity
κ	thermal conductivity
λ	molecular mean free path
γ	ratio of specific heats
σ_v, σ_T	accommodation coefficients
Pr	Prandlt number
v_{slip}	slip velocity tangentially to the wall

17.1 Physical Boundary Conditions

It is an experimental matter of fact that in viscous flows at solid surfaces, the velocity relative to the surface is zero (no slip condition, $\mathbf{v} = 0$). Further, in viscous flows where all sorts of heat are involved (due to dissipation, diffusion, conduction and radiation), several thermal boundary conditions can be imposed.

We consider the general case, including high supersonic and hypersonic flows, where heat fluxes of the gas (q_{gw}), heat fluxes into the wall (q_w) and heat fluxes due to wall radiation ($q_{rad,w}$) play a role and follow the argumentation in [1] and [2]. Figure 17.1 provides an insight into the schematic of the heat flux structures as well as the temperatures at a solid wall.

The following thermal boundary conditions are possible:

(1) Assumed, that the radiation heat flux from the wall is zero ($q_{rad,w} = 0$), then at the solid wall boundary a temperature T_w can be prescribed, where $q_w = q_{gw}$.

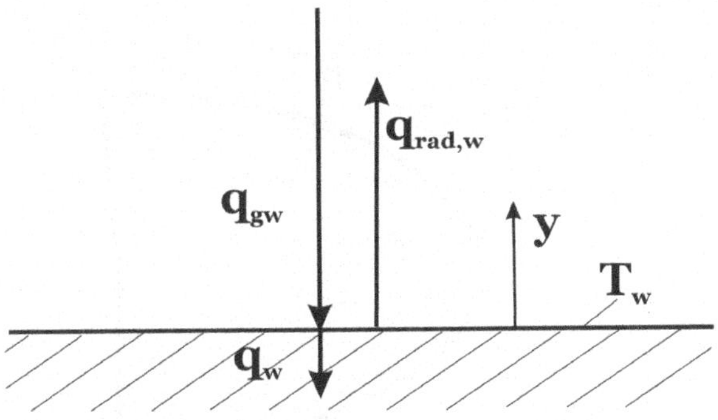

Fig. 17.1 Schematic description of the thermal state of solid surfaces. Heat flux in the gas at the wall $\Longrightarrow q_{gw}$; heat flux into the the wall $\Longrightarrow q_w$; heat flux due to surface radiation $\Longrightarrow q_{rad,w}$. For wall boundary condition 2) the wall temperature is $T_w = T_r$ and for condition 3) $T_w = T_{ra}$ (see above)

(2) Assumed, that the radiation heat flux from the wall is zero ($q_{rad,w} = 0$), then at the solid wall boundary a zero heat flux $q_w = q_{gw} = -\kappa \left(\frac{\partial T}{\partial y}\right)_w = 0$ can be prescribed, which would lead to $T_w = T_r$ (T_r the recovery temperature).
(3) Assumed, that the radiation heat flux from the wall is $q_{rad,w} = -q_{gw}$, which means $q_w = 0$. Again, at the solid wall boundary a zero heat flux $q_w = -\kappa \left(\frac{\partial T}{\partial y}\right)_w = 0$ can be prescribed, which would lead to $T_w = T_{ra}$, T_{ra} being the radiation adiabatic temperature.
(4) Assumed, that the radiation heat flux from the wall is non-zero $q_{rad,w} > 0$, then at the solid wall boundary, a temperature T_w can be prescribed, where q_w is the result of the balance of q_{gw} and $q_{rad,w}$.
(5) Assumed, that all three heat fluxes are non-zero and the heat flux into the wall q_w is prescribed, then the wall temperature T_w is variable and a function of the balance of all three heat fluxes.

The boundary condition for inviscid flow consists of the fact that the flow velocity vector has to be tangential to the body surface (or: no inflow into the body), Fig. 17.2. This is valid for the Euler equations and the full potential equation as well.

$$\mathbf{v} \cdot \mathbf{n} = 0, \quad \text{with} \quad \mathbf{n} = \frac{\nabla F}{|\nabla F|} \quad \text{and} \quad F(x, y, z) = 0\,, \tag{17.1}$$

where $\mathbf{n}$ is the body normal, and $F(x, y, z)$ describes the body contour.

For inviscid flow, no physical mechanism is available for a relation between the temperature field and the solid surface. Thus, no boundary condition can be formulated in this regard.

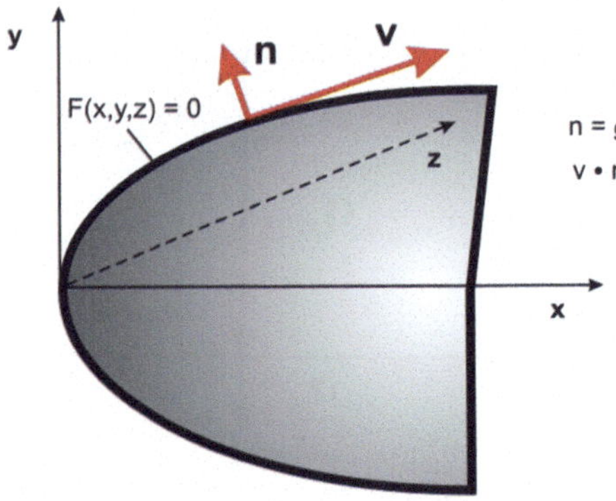

Fig. 17.2 Boundary condition for inviscid flow using the function of the surface contour $F(x, y, z) = 0$

17.2 Characteristic Variable Boundary Conditions for Euler Equations

In Sect. 14.1.5 "Diagonalization of Quasi-Linear Forms of the Euler Equations" we have introduced the determination of the eigenvalues and eigenvectors for the three-dimensional Euler equations when formulated in arbitrary coordinates (ξ, η, ζ, τ). It was shown that the Jacobians of the non-conservative or primitive variables $\mathbf{X} = (\rho, v_x, v_y, v_z, p)^T$ and of the conservative variables $\mathbf{Q} = (\rho, \rho v_x, \rho v_y, \rho v_z, \rho e)^T$ have the same eigenvalues because the Jacobians in both systems are similar matrices but have different eigenvectors. We will use these results when below we discuss the boundary conditions based on the characteristic variables.

The result of the theory of characteristics for hyperbolic partial differential equations is based on the fact that inside a corresponding flow domain the information travels along characteristic directions. Some of these characteristics point from inside the domain to the boundary and can therefore be utilized to determine the quantities at the boundary. Other characteristics point from outside the flow domain to the boundary and can not be used in this regard. Physical boundary conditions must replace them. The sign of the eigenvalues specifies which characteristic equations can be used to calculate the boundary values and which can not [3–5].

Let us consider the Euler equations in non-conservative form Eq. (14.17) with the eigenvalues Eqs. (14.22), the eigenvector matrices Eqs. (14.26) and the relation Eq. (14.25). By multiplying Eq. (14.17) from the left by $\mathbb{P}_\xi^{-1}$ we obtain[1]

$$\mathbb{P}_\xi^{-1}\frac{\partial \mathbf{X}}{\partial \tau} + \mathbb{P}_\xi^{-1}\mathbb{P}_\xi\Lambda_\xi\mathbb{P}_\xi^{-1}\frac{\partial \mathbf{X}}{\partial \xi} + \underbrace{\mathbb{P}_\xi^{-1}\left(\mathsf{A}^\eta(\mathbf{X})\frac{\partial \mathbf{X}}{\partial \eta} + \mathsf{A}^\zeta(\mathbf{X})\frac{\partial \mathbf{X}}{\partial \zeta}\right)}_{\Gamma_\xi^n} =$$

$$= \mathbb{P}_\xi^{-1}\frac{\partial \mathbf{X}}{\partial \tau} + \Lambda_\xi\mathbb{P}_\xi^{-1}\frac{\partial \mathbf{X}}{\partial \xi} + \Gamma_\xi^n = 0\,. \tag{17.2}$$

[1] In this case we assume that the boundary considered coincides with a coordinate surface $\xi = \text{const}$.

If we define $\mathbb{P}_{\xi}^{-1}$ to be a constant matrix, which means to linearize Eq. (17.2), we can write

$$\frac{\partial(\mathbb{P}_{\xi,o}^{-1}\mathbf{X})}{\partial\tau}+\Lambda_{\xi}\frac{\partial(\mathbb{P}_{\xi,o}^{-1}\mathbf{X})}{\partial\xi}+\Gamma_{\xi}^{n}=0\,, \tag{17.3}$$

where $\mathbb{P}_{\xi,o}^{-1}\mathbf{X}=\mathbf{r}_{\xi}$ can be denoted as characteristic vector with the characteristic variable elements $r_{\xi,i}$ and it follows

$$\frac{\partial\mathbf{r}_{\xi}}{\partial\tau}+\Lambda_{\xi}\frac{\partial\mathbf{r}_{\xi}}{\partial\xi}+\Gamma_{\xi}^{n}=0\,, \tag{17.4}$$

with $\Lambda_{\xi}=\frac{d\xi}{d\tau}$ being the eigenvalues of the matrix $\mathsf{A}^{\xi}(\mathbf{X})$, which determine the direction of the information fluxes.

To demonstrate the idea behind the characteristic variable boundary conditions we drop Γ_{ξ}^{n} in Eq. (17.4) and find

$$\frac{d\mathbf{r}_{\xi}}{d\tau}=\frac{\partial\mathbf{r}_{\xi}}{\partial\tau}+\Lambda_{\xi}\frac{\partial\mathbf{r}_{\xi}}{\partial\xi}=\frac{\partial\mathbf{r}_{\xi}}{\partial\tau}+\frac{d\xi}{d\tau}\frac{\partial\mathbf{r}_{\xi}}{\partial\xi}=0, \tag{17.5}$$

which can be considered as a locally one-dimensional approach. The elements of $\mathbf{r}_{\xi}$ are then given by

$$\begin{aligned}
r_{\xi,1}&=\frac{1}{|\nabla\xi|}\{\rho\xi_x+v_y\xi_z-v_z\xi_y-\frac{p}{c_0^2}\xi_x\},\\
r_{\xi,2}&=\frac{1}{|\nabla\xi|}\{\rho\xi_y-v_x\xi_z+v_z\xi_x-\frac{p}{c_0^2}\xi_y\},\\
r_{\xi,3}&=\frac{1}{|\nabla\xi|}\{\rho\xi_z+v_x\xi_y-v_y\xi_x-\frac{p}{c_0^2}\xi_z\},\\
r_{\xi,4}&=\frac{1}{\sqrt{2}|\nabla\xi|}\{+v_x\xi_x+v_y\xi_y+v_z\xi_z+\frac{p|\nabla\xi|}{\rho_0c_0}\},\\
r_{\xi,5}&=\frac{1}{\sqrt{2}|\nabla\xi|}\{-v_x\xi_x-v_y\xi_y-v_z\xi_z+\frac{p|\nabla\xi|}{\rho_0c_0}\}.
\end{aligned} \tag{17.6}$$

Now, we look at the Euler equations in quasi-conservative form Eq. (14.13), where the eigenvalues of the corresponding Jacobians, eqs. (14.22), are the same as for the non-conservative form, and where the eigenvector matrices are given by (14.29). Multiplying Eq. (14.13) from the left by $\mathbb{T}_{\xi}^{-1}$ and notice of Eq. (14.24) leads to[2]

[2] The same can be done with the eigenvector matrix $\mathbb{T}_{\xi}^{-1}$ for real gas Eq. (14.30).

$$\mathbb{T}_\xi^{-1}\frac{\partial \mathbf{Q}}{\partial \tau} + \mathbb{T}_\xi^{-1}\mathbb{T}_\xi\Lambda_\xi\mathbb{T}_\xi^{-1}\frac{\partial \mathbf{Q}}{\partial \xi} + \underbrace{\mathbb{T}_\xi^{-1}\left(\mathsf{K}^\eta(\mathbf{Q})\frac{\partial \mathbf{Q}}{\partial \eta} + \mathsf{K}^\zeta(\mathbf{Q})\frac{\partial \mathbf{Q}}{\partial \zeta}\right)}_{\Gamma_\xi^c} =$$

$$= \mathbb{T}_\xi^{-1}\frac{\partial \mathbf{Q}}{\partial \tau} + \Lambda_\xi\mathbb{T}_\xi^{-1}\frac{\partial \mathbf{Q}}{\partial \xi} + \Gamma_\xi^c = 0\,. \tag{17.7}$$

By linearization we have

$$\frac{\partial(\mathbb{T}_{\xi,o}^{-1}\mathbf{Q})}{\partial \tau} + \Lambda_\xi\frac{\partial(\mathbb{T}_{\xi,o}^{-1}\mathbf{Q})}{\partial \xi} + \Gamma_\xi^c = 0\,. \tag{17.8}$$

Then, we introduce the characteristic vector $\mathbb{T}_{\xi,o}^{-1}\mathbf{Q} = \mathbf{q}_\xi$, drop the term Γ_ξ^c and find the relation

$$\frac{d\mathbf{q}_\xi}{d\tau} = \frac{\partial \mathbf{q}_\xi}{\partial \tau} + \Lambda_\xi\frac{\partial \mathbf{q}_\xi}{\partial \xi} = \frac{\partial \mathbf{q}_\xi}{\partial \tau} + \frac{d\xi}{d\tau}\frac{\partial \mathbf{q}_\xi}{\partial \xi} = 0\,. \tag{17.9}$$

In components we obtain

$$\begin{aligned}
q_{\xi,1} &= \tfrac{1}{|\nabla\xi|}\{\rho[\xi_x(1-\tfrac{\psi_0}{c_0^2}) + \tfrac{1}{\rho_0}(v_{z,0}\xi_y - v_{y,0}\xi_z)] + \rho v_x[\xi_x\tfrac{v_{x,0}(\gamma-1)}{c_0^2}]+\\
&+\rho v_y[\xi_z\tfrac{1}{\rho_0} + \xi_x\tfrac{v_{y,0}(\gamma-1)}{c_0^2}] + \rho v_z[-\xi_y\tfrac{1}{\rho_0} + \xi_x\tfrac{v_{z,0}(\gamma-1)}{c_0^2}] + \rho e[-\xi_x\tfrac{\gamma-1}{c_0^2}]\}\,,\\
q_{\xi,2} &= \tfrac{1}{|\nabla\xi|}\{\rho[\xi_y(1-\tfrac{\psi_0}{c_0^2}) + \tfrac{1}{\rho_0}(v_{x,0}\xi_z - v_{z,0}\xi_x)] + \rho v_x[-\xi_z\tfrac{1}{\rho_0} + \xi_y\tfrac{v_{x,0}(\gamma-1)}{c_0^2}]+\\
&+\rho v_y[\xi_y\tfrac{v_{y,0}(\gamma-1)}{c_0^2}] + \rho v_z[\xi_x\tfrac{1}{\rho_0} + \xi_y\tfrac{v_{z,0}(\gamma-1)}{c_0^2}] + \rho e[-\xi_y\tfrac{\gamma-1}{c_0^2}]\}\,,\\
q_{\xi,3} &= \tfrac{1}{|\nabla\xi|}\{\rho[\xi_z(1-\tfrac{\psi_0}{c_0^2}) + \tfrac{1}{\rho_0}(v_{y,0}\xi_x - v_{x,0}\xi_y)] + \rho v_x[\xi_y\tfrac{1}{\rho_0} + \xi_z\tfrac{v_{x,0}(\gamma-1)}{c_0^2}]+\\
&+\rho v_y[-\xi_x\tfrac{1}{\rho_0} + \xi_z\tfrac{v_{y,0}(\gamma-1)}{c_0^2}] + \rho v_z[\xi_z\tfrac{v_{z,0}(\gamma-1)}{c_0^2}] + \rho e[-\xi_z\tfrac{\gamma-1}{c_0^2}]\}\,,\\
q_{\xi,4} &= \{\rho\kappa_{2,0}(\psi_0 - c_0\Theta_{\xi,0}) + \rho v_x\kappa_{2,0}(c_0\hat{\xi}_x - v_{x,0}(\gamma-1))+\\
&+\rho v_y\kappa_{2,0}(c_0\hat{\xi}_y - v_{y,0}(\gamma-1)) + \rho v_z\kappa_{2,0}(c_0\hat{\xi}_z - v_{z,0}(\gamma-1))+\\
&+\rho e\kappa_{2,0}(\gamma-1)\}\,,\\
q_{\xi,5} &= \{\rho\kappa_{2,0}(\psi_0 + c_0\Theta_{\xi,0}) - \rho v_x\kappa_{2,0}(c_0\hat{\xi}_x + v_{x,0}(\gamma-1))-\\
&-\rho v_y\kappa_{2,0}(c_0\hat{\xi}_y + v_{y,0}(\gamma-1)) - \rho v_z\kappa_{2,0}(c_0\hat{\xi}_z + v_{z,0}(\gamma-1))+\\
&+\rho e\kappa_{2,0}(\gamma-1)\}\,,
\end{aligned} \tag{17.10}$$

where

$$\psi_0 = \frac{\gamma-1}{2}(v_{x,0}^2 + v_{y,0}^2 + v_{z,0}^2)\,, \qquad \Theta_{\xi,0} = \xi_x v_{x,0} + \xi_y v_{y,0} + \xi_z v_{z,0}\,,$$

$$\kappa_{2,0} = \frac{1}{\sqrt{2}\rho_0 c_0}\,, \quad \hat{\xi}_{x_i} = \frac{\xi_{x_i}}{|\nabla\xi|}\,, \quad x_i = x, y, z\,, \quad |\nabla\xi| = (\xi_x^2 + \xi_y^2 + \xi_z^2)^{1/2}\,.$$

Remark Equations (17.3) and (17.5) are formally identical to Eqs. (17.8) and (17.9) because

$$\mathbb{T}_{\xi,0}^{-1}\mathbf{Q} = \mathbf{q}_\xi \quad \text{and} \quad \mathbb{T}_{\xi,0}^{-1} = P_{\xi,0}^{-1}\mathbb{M}^{-1} \quad \text{lead to} \quad P_{\xi,0}^{-1}\mathbb{M}^{-1}\mathbf{Q} = P_{\xi,0}^{-1}\mathbf{X}_0$$

with $\mathbb{M}^{-1}\mathbf{Q} = \mathbf{X}_0$ as long as linearization is assumed, $\mathbb{M}, \mathbb{M}^{-1}$ are given by Eq. (14.19), and one obtains

$$\mathbf{X}_0 = \mathbb{P}_{\xi,0}\mathbf{q}_\xi \ .$$

But by evaluation of $\mathbf{X}_0 = \mathbb{P}_{\xi,0}\mathbf{q}_\xi$ one finds the components

$$\mathbf{X}_0 = \begin{pmatrix} \rho \\ \frac{\rho}{\rho_0} v_x \\ \frac{\rho}{\rho_0} v_y \\ \frac{\rho}{\rho_0} v_z \\ p + \frac{\gamma-1}{2}\rho(|\mathbf{v}| - |\mathbf{v}_0|)^2 \end{pmatrix},$$

which indeed transfers the vector $\mathbf{X}_0$ to $\mathbf{X}$ by dropping the linearization assumptions $\rho_0 \Rightarrow \rho$, $|\mathbf{v}_0| \Rightarrow |\mathbf{v}|$.

17.3 Boundary Conditions for Problems Solved by Numerical Methods

17.3.1 Euler Equations

The task of theoretical fluid dynamicists consists in solving the governing equations around configurations or through flow devices and instruments. In modern fluid dynamics, in particular aerodynamics and aerothermodynamics, highly sophisticated numerical approximation methods are used to meet this challenge. For external flows, the flow domain (form and enlargement of the computational area) and the computational grid must be specified and designed by the user. For internal flows the shape of the devices typically gives the boundaries, and only the positions of the inflow and outflow boundaries have to be specified.

Generally we identify the following boundaries:

- subsonic inflow,
- subsonic outflow,
- supersonic inflow,
- supersonic ouflow,
- impermeable surface (solid surface),
- permeable surface (fitted shock surface).

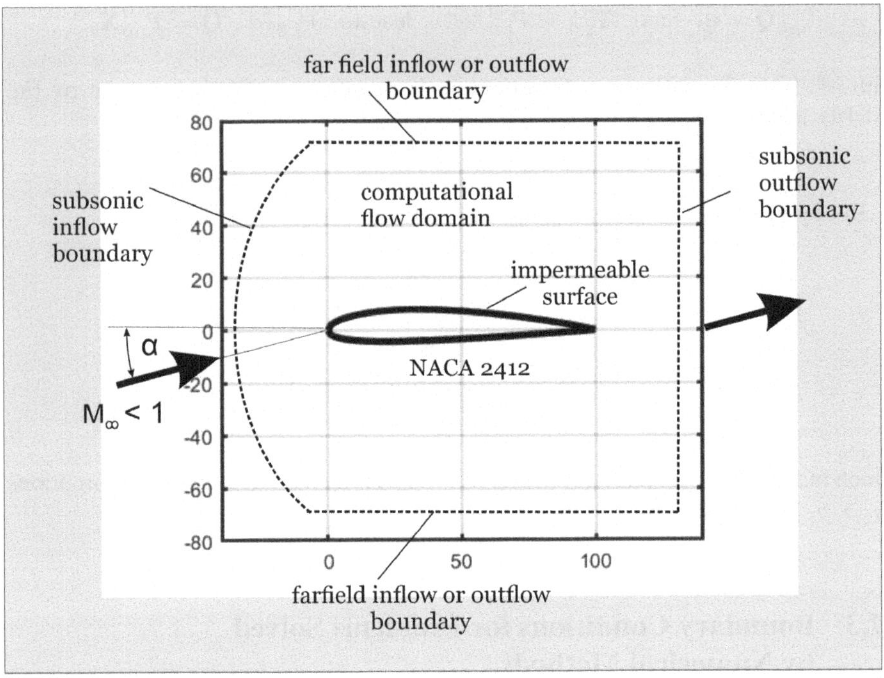

Fig. 17.3 Sketch of the boundary conditions for subsonic flow with respect to a computational domain around a NACA 2412 profile

External Flows

Below we discuss the various boundary conditions present with external flows. Figure 17.3 shows the typical situation of a flow case with subsonic inflow, subsonic outflow and impermeable surface boundaries.

Subsonic Inflow

The eigenvalues Eq. (14.22) for the subsonic inflow case have different signs. In the case that the increasing coordinate ξ points in the opposite direction compared to the flow, Fig. 17.4 a), the first three eigenvalues and the fifth eigenvalue are negative, whereas the fourth eigenvalue is positive. In the other case, when the coordinate ξ points in the same direction as the flow, Fig. 17.4 b), the first four eigenvalues are positive, and the fifth is negative.

Considering the elements of the characteristic variables, Eq. (17.6), one obtains

$$(\rho\xi_x + v_y\xi_z - v_z\xi_y - \frac{p}{c_0^2}\xi_x)_e = (\rho\xi_x + v_y\xi_z - v_z\xi_y - \frac{p}{c_0^2}\xi_x)_b\,, \tag{17.11a}$$

$$(\rho\xi_y - v_x\xi_z + v_z\xi_x - \frac{p}{c_0^2}\xi_y)_e = (\rho\xi_y - v_x\xi_z + v_z\xi_x - \frac{p}{c_0^2}\xi_y)_b\,, \tag{17.11b}$$

$$(\rho\xi_z + v_x\xi_y - v_y\xi_x - \frac{p}{c_0^2}\xi_z)_e = (\rho\xi_z + v_x\xi_y - v_y\xi_x - \frac{p}{c_0^2}\xi_z)_b\,, \tag{17.11c}$$

$$\left(\frac{p|\nabla\xi|}{\rho_0 c_0} \pm \Theta_\xi\right)_e = \left(\frac{p|\nabla\xi|}{\rho_o c_0} \pm \Theta_\xi\right)_b , \tag{17.11d}$$

$$\left(\frac{p|\nabla\xi|}{\rho_0 c_0} \mp \Theta_\xi\right)_i = \left(\frac{p|\nabla\xi|}{\rho_0 c_0} \mp \Theta_\xi\right)_b , \tag{17.11e}$$

with $\Theta_\xi = v_x\xi_x + v_y\xi_y + v_z\xi_z$, where the upper signs in Eqs. (17.11d), (17.11e) refer to Fig. 17.4 b) and the lower signs to Fig. 17.4 a). This means that four characteristic variables are prescribed by the freestream conditions and one is determined by the computational domain. The index b denotes the boundary, the index i the internal, and the index e the external part of the flow field.

Evaluation of the equation system (17.11) leads to the relations for the five non-conservative flow variables at the boundary

$$p_b = \frac{1}{2}(p_e + p_i \pm \rho_0 c_0(\hat{\xi}_x(v_{x,e} - v_{x,i}) + \hat{\xi}_y(v_{y,e} - v_{y,i}) + \hat{\xi}_z(v_{z,e} - v_{z,i}))) , \tag{17.12a}$$

$$\rho_b = \rho_e + \frac{p_b - p_e}{c_0^2} , \tag{17.12b}$$

$$v_{x,b} = v_{x,e} \pm \hat{\xi}_x \frac{p_e - p_b}{\rho_0 c_0} , \tag{17.12c}$$

$$v_{y,b} = v_{y,e} \pm \hat{\xi}_y \frac{p_e - p_b}{\rho_0 c_0} , \tag{17.12d}$$

$$v_{z,b} = v_{z,e} \pm \hat{\xi}_z \frac{p_e - p_b}{\rho_0 c_0} , \tag{17.12e}$$

where the minus sign refers to the situation in Fig. 17.4 a) and the plus sign to Fig. 17.4 b).

Subsonic Outflow

For the subsonic outflow boundary conditions, the relations regarding the elements of the characteristic variables, Eqs. (17.11), are formally the same as for the subsonic inflow conditions. Only the index e has the be replaced by the index i and vice versa, see Fig. 17.5. This leads to the fact that in contrast to the subsonic inflow conditions, four of the characteristic variables are determined by the information from inside the computational domain and one by information from outside the computational domain. Therefore, the result of the evaluation of this equation system for determining the five flow variables differs slightly from that of the subsonic inflow conditions because from the adjusted fifth component of Eqs. (17.11)

$$\left(\frac{p|\nabla\xi|}{\rho_0 c_0} \mp \Theta_\xi\right)_e = \left(\frac{p|\nabla\xi|}{\rho_0 c_0} \mp \Theta_\xi\right)_b ,$$

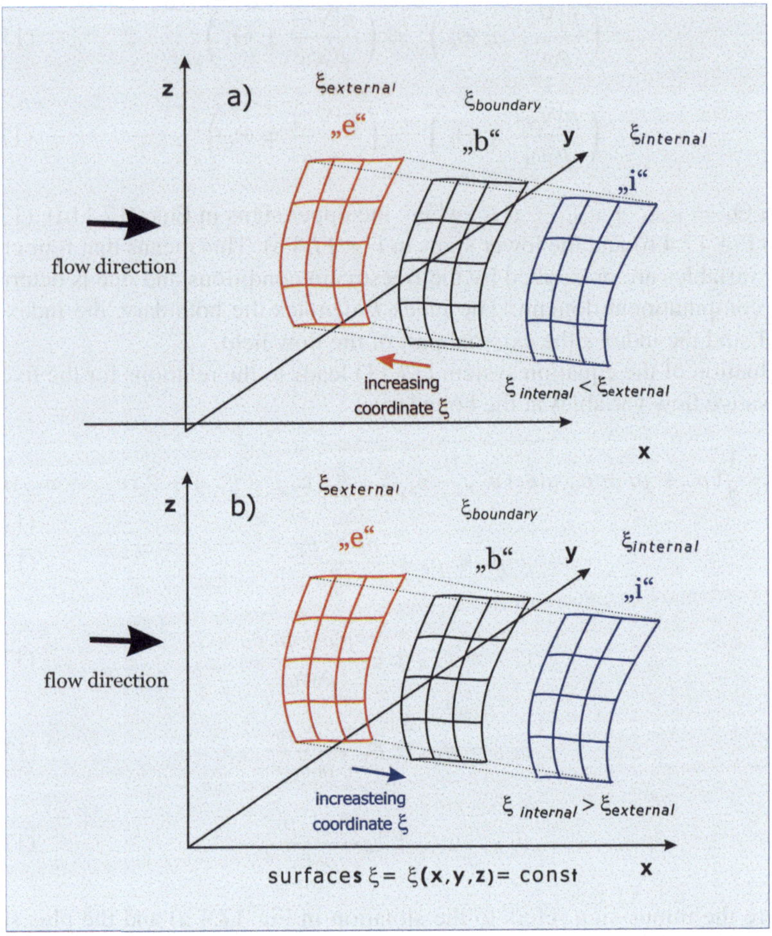

Fig. 17.4 Characteristic variable boundary conditions for subsonic inflow: definition of computational surfaces at boundary (index b), internally (index i) and externally (index e) of the flow domain. **a** explains the situation when the flow direction is in contrast to the direction of increasing values of the ξ coordinate. **b** constitutes the behavior when the flow and the increasing values of the ξ coordinate have the same direction

it follows $p_b = p_e$, since the outflow is assumed to be uniform. In total we obtain

$$p_b = p_e, \tag{17.13a}$$

$$\rho_b = \rho_i + \frac{p_b - p_i}{c_0^2}, \tag{17.13b}$$

$$v_{x,b} = v_{x,i} \pm \hat{\xi}_x \frac{p_i - p_b}{\rho_0 c_0}, \tag{17.13c}$$

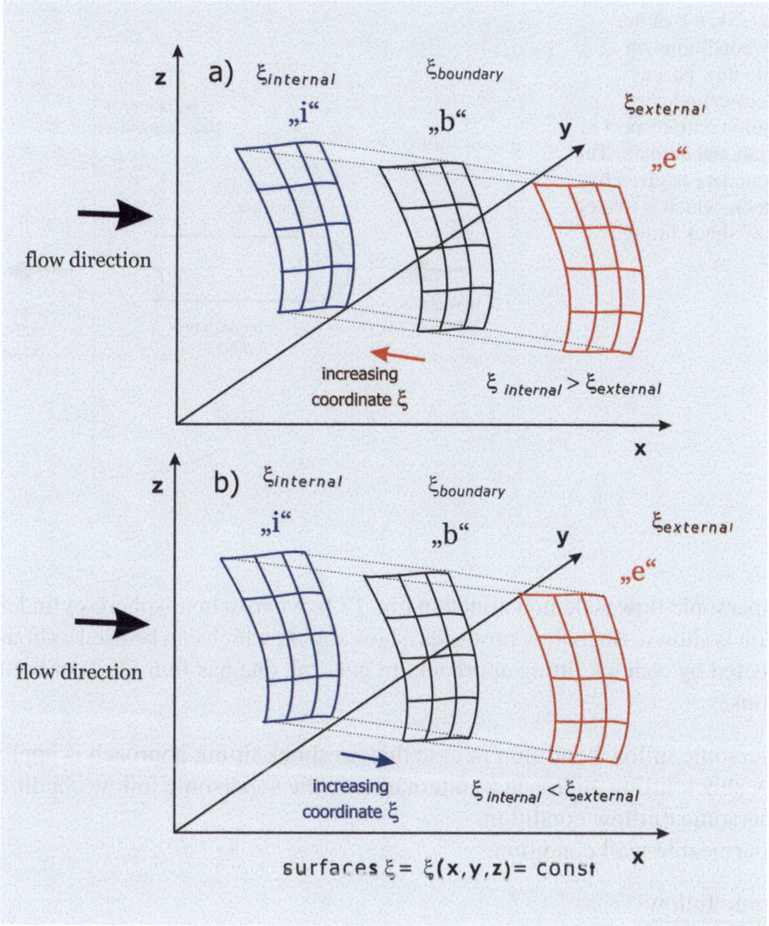

Fig. 17.5 Characteristic variable boundary conditions for subsonic outflow: Definition of computational surfaces at boundary (index b), internally (index i) and externally (index e) of the flow domain. **a** explains the situation when the flow direction is opposite to the direction of increasing values of the ξ coordinate. **b** constitutes the behavior when the flow and the increasing values of the ξ coordinate have the same direction

$$v_{y,b} = v_{y,i} \pm \hat{\xi}_y \frac{p_i - p_b}{\rho_0 c_0}, \tag{17.13d}$$

$$v_{z,b} = v_{z,i} \pm \hat{\xi}_z \frac{p_i - p_b}{\rho_0 c_0}. \tag{17.13e}$$

If the flow direction is opposite to the direction of the increasing ξ coordinate, Fig. 17.5a, the minus sign is valid, whereas the plus sign holds for the situation given in Fig. 17.5b.

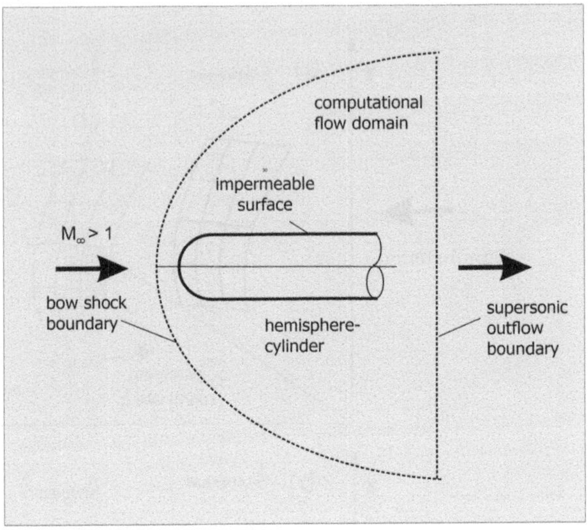

Fig. 17.6 Sketch of the boundary conditions for supersonic flow past a hemisphere-cylinder configuration with respect to a computational domain. The inflow boundary is given by a bow shock, which is treated with a bow shock fitting approach

A supersonic flow is demonstrated in Fig. 17.6, where a hemisphere-cylinder configuration is shown. Such flow produces a bow shock, which can be used as boundary to be treated by a shock fitting approach. In general, one has four kinds of boundary conditions.

- supersonic inflow condition in case that no shock fitting approach is applied,
- bow shock fitting approach as alternative to the supersonic inflow condition,
- supersonic outflow condition,
- impermeable wall condition.

Supersonic Inflow

All the eigenvalues, Eq. (14.22), have the same sign in this case, and therefore all variables are prescribed (or given by the free-stream conditions).

Supersonic Outflow

This situation is also distinguished by the same sign for the set of eigenvalues. The flow variables at the boundary are determined by the values from inside the computational domain, which means that they are extrapolated from inside.

Shock Fitting Approach

The shock fitting approach requires the five Rankine-Hugoniot equations for moving shocks in the form of the equation system (16.10a) to (16.10d) and additionally, one equation based on a characteristic variable, which is typically given by the fourth component of Eq. (17.2) for non-conservative variables as well as Eq. (17.7) for conservative variables.

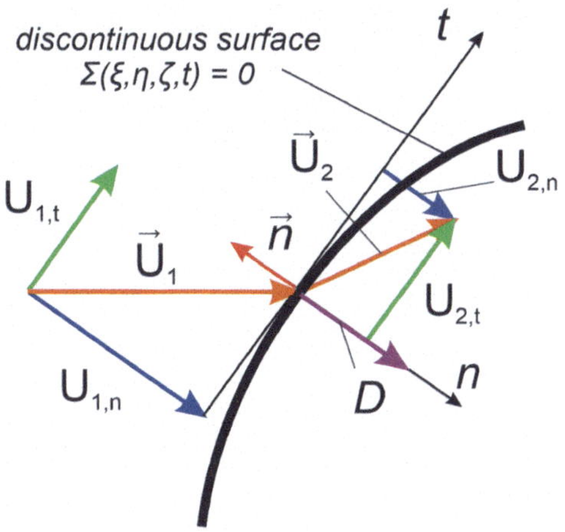

Fig. 17.7 See Fig. 16.2

Non-Conservative Variables

Let us first consider the situation if non-conservative variables are used. We present for distinctness Fig. 16.2 again and compose the set of equations in the following form (Fig. 17.7).

$$
\begin{aligned}
\rho_1(U_{1,n} - D) &= \rho_2(U_{2,n} - D)\,, \\
U_{1,t} &= U_{2,t}\,, \\
\rho_1(U_{1,n} - D)^2 + p_1 &= \rho_2(U_{2,n} - D)^2 + p_2\,, \\
h_1 + \frac{1}{2}(U_{1,n} - D)^2 &= h_2 + \frac{1}{2}(U_{2,n} - D)^2\,, \\
\frac{p_1}{\rho_0 c_0} + U_{1,n} &= \frac{p_2}{\rho_0 c_0} + U_{2,n}\,.
\end{aligned}
\tag{17.14}
$$

The latter equation is the fourth component of Eq. (17.6), where the assumptions of Eqs. (17.3)–(17.5) are utilized. Index 2 refers to the conditions downstream of the shock, and index 1 to the conditions upstream of the shock, which is given for a bow shock by the freestream. If it is assumed that the shock surface coincides with a coordinate surface $\xi = \xi(x, y, z, t) = \text{const}$, one can define

$$
D = -\frac{\xi_t}{|\nabla \xi|}, \qquad \mathbf{n} = \frac{\nabla \xi}{|\nabla \xi|}\,.
$$

Resolution of the equation system (17.14) leads to

$$\begin{aligned}
\rho_2 &= \rho_1 \frac{(U_{1,n} - D)}{(U_{1,n} - D) + \Omega}\,,\\
p_2 &= p_1 - \rho_1 \Omega (U_{1,n} - D)\,,\\
v_{x,2} &= \Omega \frac{\xi_x}{|\nabla \xi|} + v_{x,1}\,,\\
v_{y,2} &= \Omega \frac{\xi_y}{|\nabla \xi|} + v_{y,1}\,,\\
v_{z,2} &= \Omega \frac{\xi_z}{|\nabla \xi|} + v_{z,1}\,,\\
D &= U_{1,n} - \frac{\rho_0 c_0}{\rho_1}\,,
\end{aligned} \tag{17.15}$$

where the equation for the variable $\Omega = U_{2,n} - U_{1,n}$

$$\Omega = -(U_{1,n} - D) \pm \sqrt{(U_{1,n} - D)^2 - 2(h_2 - h_1)} \tag{17.16}$$

has to be solved by an iteration procedure with $h_2 = h_2(p_2(\Omega), \rho_2(\Omega))$ and $h_1 = h_\infty$. Further, we need the relation

$$[\xi_x(v_{x,2} - v_{x,1}) + \xi_y(v_{y,2} - v_{y,1}) + \xi_z(v_{z,2} - v_{z,1})]\frac{1}{|\nabla \xi|} - \Omega = 0\,, \tag{17.17}$$

and the components of $U_{2,t} = U_{1,t}$

$$\begin{aligned}
v_{y,2}\xi_z - v_{z,2}\xi_y &= v_{y,1}\xi_z - v_{z,1}\xi_y\,,\\
v_{z,2}\xi_x - v_{x,2}\xi_z &= v_{z,1}\xi_x - v_{x,1}\xi_z\,,\\
v_{x,2}\xi_y - v_{y,2}\xi_x &= v_{x,1}\xi_y - v_{y,1}\xi_x\,,
\end{aligned} \tag{17.18}$$

which provide two linearly independent correlations.The constant values ρ_0 and c_0 must be chosen with respect to the iteration process, for example as an averaging between state 1 and state 2.

Remark The determination of the velocity components $v_{x,2}$, $v_{y,2}$, $v_{z,2}$ is performed by the evaluation of eqs. (17.17) and (17.18). We demonstrate this for the $v_{x,2}$ component. With

$$\Omega \frac{\nabla \xi}{|\nabla \xi|} = (U_{2,n} - U_{1,n}) \frac{\nabla \xi}{|\nabla \xi|}$$

the $v_{x,2}$ component reads

$$\begin{aligned}
&\Omega \frac{\xi_x}{|\nabla \xi|} \\
&= \left[(v_{x,2}\xi_x + v_{y,2}\xi_y + v_{z,2}\xi_z) - (v_{x,1}\xi_x + v_{y,1}\xi_y + v_{z,1}\xi_z)\right] \frac{\xi_x}{|\nabla \xi|^2} = \\
&= [(v_{x,2}\xi_x\xi_x + v_{y,2}\xi_x\xi_y - v_{x,2}\xi_y\xi_y + v_{x,2}\xi_y\xi_y + v_{z,2}\xi_x\xi_z - v_{x,2}\xi_z\xi_z + v_{x,2}\xi_z\xi_z) - \\
&- (v_{x,1}\xi_x\xi_x + v_{y,1}\xi_x\xi_y - v_{x,1}\xi_y\xi_y + v_{x,1}\xi_y\xi_y + v_{z,1}\xi_x\xi_z - v_{x,1}\xi_z\xi_z + v_{x,1}\xi_z\xi_z)]\frac{1}{|\nabla \xi|^2} \\
&= [v_{x,2}(\xi_x^2 + \xi_y^2 + \xi_z^2) + \xi_y(v_{y,2}\xi_x - v_{x,2}\xi_y) + \xi_z(v_{z,2}\xi_x - v_{x,2}\xi_z) - \\
&- (v_{x,1}(\xi_x^2 + \xi_y^2 + \xi_z^2) + \xi_y(v_{y,1}\xi_x - v_{x,1}\xi_y) + \xi_z(v_{z,1}\xi_x - v_{x,1}\xi_z))]\frac{1}{|\nabla \xi|^2} \\
&= v_{x,2} - v_{x,1}\,.
\end{aligned}$$

Another possibility to close the equation system Eq. (17.14) with a characteristic variable relation consists in applying the fourth component of the non-linear equation[3] (17.2). This means to replace the last equation of Eq. (17.14) by

$$\begin{aligned}
&\frac{1}{\sqrt{2}}\left[\frac{1}{|\nabla \xi|}\left(\xi_x v_{x,2} + \xi_y v_{y,2} + \xi_z v_{z,2}\right) + \frac{1}{\rho' c'}\, p_2\right] + \Gamma^n_{\xi,4;2} = \\
&\frac{1}{\sqrt{2}}\left[\frac{1}{|\nabla \xi|}\left(\xi_x v_{x,1} + \xi_y v_{y,1} + \xi_z v_{z,1}\right) + \frac{1}{\rho' c'}\, p_1\right] + \Gamma^n_{\xi,4;1}\,, \\
\Longrightarrow\quad &\underbrace{\frac{1}{|\nabla \xi|}\left(\xi_x \bar{v}_x + \xi_y \bar{v}_y + \xi_z \bar{v}_z\right)}_{\Omega(D)} + \frac{1}{\rho' c'}\, p = -\sqrt{2}(\Gamma^n_{\xi,4;2} - \Gamma^n_{\xi,4;1}) - \bar{\Gamma}^n\,, \\
\Longrightarrow\quad &\Omega(D) = \frac{\bar{\Gamma}^n}{\frac{\rho_1}{\rho' c'}(U_{1,n} - D) - 1}\,,
\end{aligned} \qquad (17.19)$$

where $\bar{v}_x = v_{x,2} - v_{x,1}$, $\bar{v}_y = v_{y,2} - v_{y,1}$, $\bar{v}_z = v_{z,2} - v_{z,1}$, $\bar{p} = p_2 - p_1$ and the values for ρ', c' are taken from a previous time step.

Then we obtain

$$\Phi(D) = \frac{h_2(\rho_2(D), p_2(D)) - h_1 + \frac{1}{2}\Omega(D)^2 + \Omega(D)U_{1,n}}{\Omega(D)}\,, \qquad (17.20)$$

which has to be solved iteratively, for example by a Newton iteration, with s the iteration number

[3] For real gas flow Eq. (17.2) has to be built with the matrix $\mathbb{P}_\xi^{-1}$ for real gas, Eq. (14.27), where the assumption is referred to that the pressure p and the speed of sound c_R are functions of the density ρ and the internal energy e_i. Usually both variables are determined by evaluation of data systems like the one given in Ref. [6].

$$D^{(s+1)} = D^{(s)} - \frac{\Phi(D^{(s)}) - D^{(s)}}{\Phi'(D^{(s)}) - 1} .$$

Conservative Variables

Today, modern numerical methods are formulated predominantly in conservative variables owing to the capability of such formulations to deal with all sorts of discontinuities, in particular shocks, shear layers and contact discontinuities, see Chap. 16. Therefore we present the shock fitting algorithm also with the conservative variables $\rho,\ \rho v_x,\ \rho v_y,\ \rho v_z,\ \rho e$, [7–9].

In analogy to Eq. (17.14) we can compose

$$\begin{aligned}\rho_1(U_{1,n} - D) &= \rho_2(U_{2,n} - D) ,\\ U_{1,t} &= U_{2,t} ,\\ \rho_1(U_{1,n} - D)^2 + p_1 &= \rho_2(U_{2,n} - D)^2 + p_2 ,\\ e_{i,1} + \frac{p_1}{\rho_1} + \frac{1}{2}(U_{1,n} - D)^2 &= e_{i,2} + \frac{p_2}{\rho_2} + \frac{1}{2}(U_{2,n} - D)^2 ,\end{aligned}$$

$$t_{41}^{-1}\rho_2 + t_{42}^{-1}\rho_2 v_{x,2} + t_{43}^{-1}\rho_2 v_{y,2} + t_{44}^{-1}\rho_2 v_{z,2} + t_{45}^{-1}\rho_2 e_2 = -\Gamma^c_{\xi,4;2} , \tag{17.21}$$

where the latter equation is given by the fourth component of Eq. (17.7). In the case of real gas[4] Eq. (17.7) is built with $\mathbb{T}_\xi^{-1}$ for real gas, Eq. (14.30). The enthalpy is replaced by the relation $h = e_i + p/\rho$ to render possible that the thermodynamic variables p and T are functions of the chosen independent thermodynamic variables e_i and ρ, which then allows considering flows with real gas effects in thermodynamic equilibrium.

The resolution of the equation system (17.21) yields for the conservative variables $\rho_2,\ \rho_2 v_{x,2},\ \rho_2 v_{y,2},\ \rho_2 v_{z,2},\ \rho_2 e_2$

$$\begin{aligned}\frac{\rho_2}{\rho_1} &= \frac{1}{2A}\left(-B \pm \sqrt{B^2 - 4AC}\right) ,\\ e_2 &= e_1 + \frac{1}{t_{45}^{-1}}\left(\frac{1}{\rho_1}\Gamma^c_{\xi,4;1} - \frac{1}{\rho_2}\Gamma^c_{\xi,4;2} - \left(\frac{1}{\rho_2} - \frac{1}{\rho_1}\right)\chi\, q_n\right) ,\\ \begin{pmatrix}\rho_2 v_{x,2}\\ \rho_2 v_{y,2}\\ \rho_2 v_{z,2}\end{pmatrix} &= \frac{\rho_2}{\rho_1}\begin{pmatrix}\rho_1 v_{x,1}\\ \rho_1 v_{y,1}\\ \rho_1 v_{z,1}\end{pmatrix} + \begin{pmatrix}\xi_x\\ \xi_y\\ \xi_z\end{pmatrix}\left(1 - \frac{\rho_2}{\rho_1}\right)\chi\,\frac{1}{|\nabla\xi|} ,\end{aligned} \tag{17.22}$$

and the equation for D, which has to be solved by a numerical iteration process, is

$$p_2(\rho_2(D), e_{i,2}(D)) - p_1 - \frac{\chi(D)^2}{\rho_1}\left(1 - \frac{\rho_1}{\rho_2(D)}\right) = 0 . \tag{17.23}$$

[4] Real gas means changes of the thermodynamic properties of gases due to hot temperatures.

The following relations and abbreviations are used in eqs. (17.22) and (17.23)

$$
\begin{aligned}
\chi &= \rho_1(U_{1,n} - D)\,,\\
q_n &= (t_{42}^{-1}\xi_x + t_{43}^{-1}\xi_y + t_{44}^{-1}\xi_z)\frac{1}{|\nabla\xi|}\,,\\
\Gamma^c_{\xi,4;1} &= -(t_{41}^{-1}\rho_1 + t_{42}^{-1}\rho_1 v_{x,1} + t_{43}^{-1}\rho_1 v_{y,1} + t_{44}^{-1}\rho_1 v_{z,1} + t_{45}^{-1}\rho_1 e_1)\,,\\
A &= \frac{1}{t_{45}^{-1}}\left(\chi\, q_n + \Gamma^c_{\xi,4;1}\right) + \chi\, D - p_1\,,\\
B &= \frac{1}{t_{45}^{-1}}\left(-\Gamma^c_{\xi,4;2} - \chi\, q_n\right) - \chi\, D + p_1 + \chi^2\frac{1}{\rho_1}\,,\\
C &= -\frac{\chi^2}{\rho_1}\,.
\end{aligned}
\tag{17.24}
$$

Impermeable Wall Boundary Conditions

There are many possibilities for treating impermeable wall boundary conditions by the idea of characteristic variables. In the following we present two of them. Generally, at an impermeable wall -no flow across the boundary-, the first three eigenvalues of Eqs. (14.22) are zero, the fourth is positive and the fifth is negative.

- A boundary condition procedure based on the equation system (17.6).

The four characteristic variable relations (1, 2, 3, 5) of Eqs. (17.6) point towards the boundary and can be used. In contrast, the fourth component points away from the boundary and has to be replaced by the condition Eq. (17.1) in the form $\mathbf{v} \cdot \mathbf{n} = 0$ with $\mathbf{v} = (v_x, v_y, v_z)^T$ and $\mathbf{n} = \dfrac{\nabla\xi}{|\nabla\xi|}$, if $\xi = \xi(x, y, z) = \text{const}$ denotes the body contour. One obtains

$$(\rho\xi_x + v_y\xi_z - v_z\xi_y - \frac{p}{c_0^2}\xi_x)_b = (\rho\xi_x + v_y\xi_z - v_z\xi_y - \frac{p}{c_0^2}\xi_x)_i\,, \tag{17.25a}$$

$$(\rho\xi_y - v_x\xi_z + v_z\xi_x - \frac{p}{c_0^2}\xi_y)_b = (\rho\xi_y - v_x\xi_z + v_z\xi_x - \frac{p}{c_0^2}\xi_y)_i\,, \tag{17.25b}$$

$$(\rho\xi_z + v_x\xi_y - v_y\xi_x - \frac{p}{c_0^2}\xi_z)_b = (\rho\xi_z + v_x\xi_y - v_y\xi_x - \frac{p}{c_0^2}\xi_z)_i\,, \tag{17.25c}$$

$$(\xi_x v_x + \xi_y v_y + \xi_z v_z)_b = \Theta_\xi = 0\,, \tag{17.25d}$$

$$\left(\frac{p\,|\nabla\xi|}{\rho_0 c_0} \mp \Theta_\xi\right)_b = \left(\frac{p\,|\nabla\xi|}{\rho_0 c_0} \mp \Theta_\xi\right)_i\,. \tag{17.25e}$$

The subscript b denotes the boundary point, whereas the subscript i refers to values inside the computational domain, which are typically chosen a half cell width from the wall boundary (or halfway from the related boundary point). The pressure at the wall can easily be determined by Eqs. (17.25d) and (17.25e), and one obtains[5]

$$p_b = p_i \mp \rho_0 c_0 \left(\frac{\Theta_\xi}{|\nabla \xi|} \right) . \qquad (17.26a)$$

The evaluation of the relations (17.25a)–(17.25d) yields for the variables

ρ, v_x, v_y, v_z at the boundary b

$$\rho_b = \rho_i + \frac{p_b - p_i}{c_0^2} , \qquad (17.26b)$$

$$v_{x,b} = v_{x,i} - \xi_x \Theta_{\xi,i} \frac{1}{|\nabla \xi|^2} , \qquad (17.26c)$$

$$v_{y,b} = v_{y,i} - \xi_y \Theta_{\xi,i} \frac{1}{|\nabla \xi|^2} , \qquad (17.26d)$$

$$v_{z,b} = v_{z,i} - \xi_z \Theta_{\xi,i} \frac{1}{|\nabla \xi|^2} . \qquad (17.26e)$$

- A boundary condition procedure based on the post correction technique, [10, 11]

As already mentioned above, at the wall boundary, four characteristic variable relations point towards the boundary and therefore are included in the predictor step performed by the computer code.[6] But the result does not satisfy the boundary condition $\mathbf{v} \cdot \mathbf{n} = 0$, which reads when formulated with the conservative solution vector $\mathbf{Q} = (\rho, \rho v_x, \rho v_y, \rho v_z, \rho e)^T$

$$\mathbf{Y} \cdot \mathbf{Q} = 0 \qquad (17.27)$$

with $\mathbf{Y} = \frac{1}{|\nabla \xi|}(0, \xi_x, \xi_y, \xi_z, 0)^T$. The correction is performed using the eigenvector $\mathbf{t}$, which corresponds to the characteristic pointing away from the wall, taken from the matrix $\mathbb{T}_\xi$, Eq. (14.29) for a perfect gas or Eq. (14.30) for a real gas. As an example,[7] the fourth right eigenvector $\mathbf{t}_4$ (for perfect gas), that is the fourth column of $\mathbb{T}_\xi$, is applied, and the correction is then given by the relation

$$\mathbf{Q} = \mathbf{Q}^* - \frac{\mathbf{Y} \cdot \mathbf{Q}^*}{\mathbf{Y} \cdot \mathbf{t}_4} \mathbf{t}_4 , \qquad (17.28)$$

---

[5] The minus sign holds if the point i (internal) is in the positive ξ direction from the boundary and vice versa for the plus sign.

[6] The computer code may solve the Euler equations either in non-conservative or in conservative variables. In the case considered here, conservative variables are examined.

[7] This example is valid if the ξ coordinate is increasing from the wall. Otherwise, the fifth right eigenvector comes into play.

where $\mathbf{Q}^*$ is the conservative solution vector of the predictor step. Finally one obtains

$$\mathbf{Q} = \mathbf{Q}^* - \frac{\rho\hat{\Theta}_\xi}{\hat{\Theta}_\xi + c}\begin{pmatrix} 1 \\ v_x + c\,\hat{\xi}_x \\ v_y + c\,\hat{\xi}_y \\ v_z + c\,\hat{\xi}_z \\ \frac{\phi + c^2}{\gamma - 1} + c\,\hat{\Theta}_\xi \end{pmatrix} \tag{17.29}$$

with $\phi = \dfrac{\gamma-1}{2}|\mathbf{v}|^2$, $\hat{\xi}_i = \dfrac{\xi_i}{|\nabla\xi|}$, $i = x, y, z$, $\hat{\Theta}_\xi = \dfrac{\Theta_\xi}{|\nabla\xi|}$, c speed of sound.

Remark To better understand the post correction technique, we present below a derivation based on the matrix calculus, [9]. Assumed, that the characteristic variables defined by $W = \mathbb{T}_\xi^{-1}\mathbf{Q} = (w_1, w_2, w_3, w_4, w_5)^T$, see also Eq. (17.10), are preliminarily calculated at the wall by the numerical field algorithm with

$$\mathbf{Q}^* = (\rho, \rho v_x, \rho v_y, \rho v_z, \rho e)^{*T} \qquad \text{and} \qquad W^* = (w_1, w_2, w_3, w_4, w_5)^{*T}\,.$$

By replacing the fourth component of W with the wall boundary condition ($\rho\Theta_\xi = 0$) one obtains

$$\begin{pmatrix} w_1 \\ w_2 \\ w_3 \\ \rho\Theta_\xi \\ w_5 \end{pmatrix} = \begin{pmatrix} w_1 \\ w_2 \\ w_3 \\ 0 \\ w_5 \end{pmatrix}^* = (I - I_4)\mathbb{T}_\xi^{-1}\mathbf{Q}^*, \qquad I = \begin{pmatrix} 1\,0\,0\,0\,0 \\ 0\,1\,0\,0\,0 \\ 0\,0\,1\,0\,0 \\ 0\,0\,0\,1\,0 \\ 0\,0\,0\,0\,1 \end{pmatrix}, \quad I_4 = \begin{pmatrix} 0\,0\,0\,0\,0 \\ 0\,0\,0\,0\,0 \\ 0\,0\,0\,0\,0 \\ 0\,0\,0\,1\,0 \\ 0\,0\,0\,0\,0 \end{pmatrix}. \tag{17.30}$$

Let

$$\tilde{\mathbb{T}}^{-1} = (I - I_4)\mathbb{T}_\xi^{-1} + I_4\begin{pmatrix} 0 & 0 & 0 & 0 & 0 \\ 0 & 0 & 0 & 0 & 0 \\ 0 & 0 & 0 & 0 & 0 \\ 0 & \xi_x & \xi_y & \xi_z & 0 \\ 0 & 0 & 0 & 0 & 0 \end{pmatrix}, \tag{17.31}$$

then from Eq. (17.30) follows $\tilde{\mathbb{T}}^{-1}\mathbf{Q} = (w_1, w_2, w_3, \rho\Theta_\xi, w_5)^T$ and with $(I - I_4)I_4 = 0$, $I_4^2 = I_4$ one finds

$$\mathbf{Q} = \tilde{\mathbb{T}}(I - I_4)\tilde{\mathbb{T}}^{-1}\mathbf{Q}^* = \mathbf{Q}^* - \tilde{\mathbb{T}}I_4\tilde{\mathbb{T}}^{-1}\mathbf{Q}^*. \tag{17.32}$$

Evaluating the term $\tilde{\mathbb{T}}I_4\tilde{\mathbb{T}}^{-1}\mathbf{Q}^*$ gives exactly the same result as the evaluation of the term $\dfrac{\mathbf{Y}\cdot\mathbf{Q}^*}{\mathbf{Y}\cdot\mathbf{t}_4}\mathbf{t}_4$, eqs. (17.28), (17.29), but is much more tedious.

Internal Flows

As an example of internal flow a rocket motor nozzle is chosen. In the late eighties and the nineties of the last century in Europe the main engines for the ARIANE V rocket,

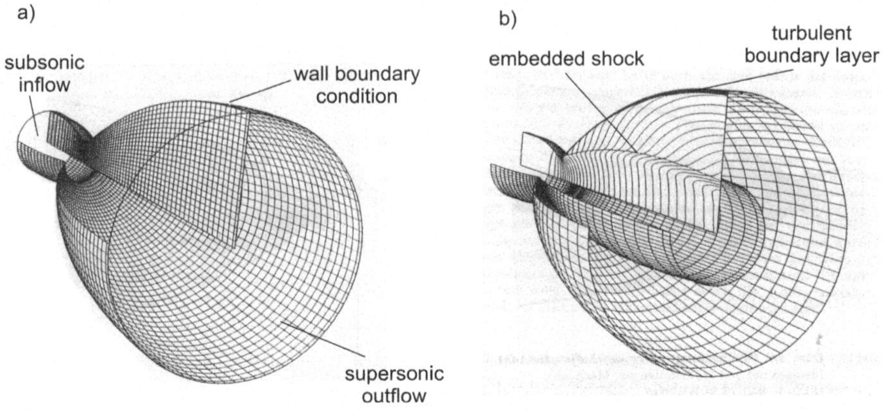

Fig. 17.8 Boundary conditions for internal flows demonstrated by a rocket motor nozzle, **a** Navier-Stokes solution with turbulence modeling and equilibrium real gas effects, **b** [13]. Isolines of Mach number are plotted

the Vulcain motor, and for the ARIANE V Extension rocket, the Vulcain 2 motor, were developed, [12]. The design and the performance of the nozzles were determined using numerical methods either by Euler or Navier-Stokes solutions. In the most sophisticated approach, the modeling of turbulence and of the thermodynamic non-equilibrium of a hydrogen-oxygen gas mixture was included.

In Fig. 17.8a the boundary conditions are drawn, which must be met for nozzle flows. They consist of a subsonic inflow, a supersonic outflow and an impermeable wall. The subsonic inflow condition must satisfy, using one-dimensional theory, the area ratio of inflow to the throat cross section. At the exit position, the pure supersonic outflow condition is sufficient as long as Euler solutions are considered. For viscous flow exists due to the boundary layer a small region at the exit with subsonic outflow conditions, where one characteristic points from the outside and must be replaced by a variable to be prescribed. Such a variable mostly is the pressure, which is not known a priori, but can be extrapolated from the supersonic regime.[8] Figure 17.8b shows the result of a Navier-Stokes solution, where turbulence and equilibrium real gas effects are simulated, [13]. The embedded shock is clearly detected by the isolines of the Mach number.

17.3.2 Boundary Conditions for the Navier-Stokes Equations

In flow around configurations, the viscous effects decay fast with increasing distance from the wall. Therefore, at far field boundaries viscous effects play only a minor role and the boundary conditions for subsonic inflow-, subsonic outflow-, supersonic

[8] Sometimes the influence from outside is simply ignored.

inflow- and supersonic outflow, derived for the Euler equations, Sect. 17.3.1, can be used. Even though in shock layers large gradients occur, the shock fitting approach treats the shock as a singularity and it makes no sense to resolve the strong viscous effects inside the very thin shock layer. Hence, the shock fitting approach derived in Sect. 17.3.1 can also be applied to Navier-Stokes flows.

At the wall the no-slip condition is used, which means that all velocity components are zero. Further, one of the five thermal boundary conditions, discussed in Sect. 17.1, is imposed and often a zero normal pressure gradient is prescribed. Another possibility for the determination of the pressure at the wall is discussed in [14]. Because flow problems are usually treated as unsteady with the steady result found asymptotically for $t \to \infty$, a compatibility condition for the density can be formulated from the continuity equation. Assuming a perfect gas a relation for the time derivative of the pressure at the wall, $\partial p/\partial t$ can be derived, which allows the prediction of the wall pressure.

When the flow leaves the computational domain at outflow recirculation regions may occur, which cannot be handled by the above described boundary condition processes. The only possibility to avoid this problem consists of an enlargement of the computational domain to such an extent that the recirculation region will be captured.

17.3.3 Boundary Conditions for Rarefied Gases

The above defined boundary conditions for the Navier-Stokes equations are valid, if the behavior of the flow can be described as a continuum. During space flight, which means the ascent and the re-entry through the Earth atmosphere, space vehicles encounter regions with strongly varying densities, see [1], Chap. 2 and [15], Chap. 7. By the decrease of the density, or the other way around, the increase of the Knudsen number, at first the assumptions of the no-slip condition and the continuous behavior of the temperature at the wall fail and next the continuum hypothesis itself cannot be kept. In the Knudsen number range $0.01 \leq Kn \leq 0.1$ the flow can still be considered as a continuum, but the tangential velocity at the wall has a slip and the temperature of the gas at the wall differs from the wall temperature (temperature jump). This is the regime of rarefied gases.

The mathematical formulation of the boundary conditions of rarefied gases requires a look into the kinetic theory of gases, [16, 17]. In 1879, Maxwell has presented a relation for the velocity slip at walls depending on the tangential shear stress vector $\boldsymbol{\tau}$ and the heat flux vector $\mathbf{q}$ (effect of thermal creep), which reads[9]

[9] The tangential shear stress vector $\boldsymbol{\tau}_w$ is evaluated from the tensor product $S \cdot (\mathbf{n} \cdot \bar{\bar{\tau}}) = -\boldsymbol{\tau}_w$, where the normal components of the shear stress tensor $\bar{\bar{\tau}}$ are removed by the tensor $S = I - \mathbf{nn}$, with I the identity matrix, $\mathbf{n}$ the wall normal.

$$\mathbf{v}_{slip} = \frac{2-\sigma_v}{\sigma_v \mu}\lambda \boldsymbol{\tau}_w - \frac{3}{4}\frac{Pr(\gamma-1)}{\gamma p}\mathbf{q}\,, \tag{17.33}$$

with $\boldsymbol{\tau}_w$ being the tangential shear stress vector at the wall, σ_v the momentum accommodation coefficient ($0 \le \sigma_v \le 1$), μ the gas viscosity, λ the molecular mean free path, Pr the Prandtl number, γ the ratio of specific heats, p the pressure at the wall. The value of the momentum accommodation coefficient is equal to one ($\sigma_v = 1$), if all molecules are diffusely reflected from the wall and zero ($\sigma_v = 0$) for purely specular reflection, [18, 19].

The temperature jump condition first formulated by Smoluchowski [20] can be written

$$T_w = T - \frac{2-\sigma_T}{\sigma_T}\frac{2\gamma\lambda}{(\gamma+1)Pr}\mathbf{n}\cdot\mathbf{q} = T - \frac{2-\sigma_T}{\sigma_T}\frac{2\gamma\lambda}{(\gamma+1)Pr}q_w\,, \tag{17.34}$$

with σ_T being the thermal accommodation coefficient ($0 \le \sigma_T \le 1$), $\mathbf{q} = -\kappa \operatorname{grad} T$ the heat flux vector, κ the thermal conductivity, q_w the heat flux into the wall, T_w the wall temperature, T the temperature of the gas at the wall, $\mathbf{n}$ the wall normal.

For a one-dimensional wall the Eqs. (17.33),(17.34) can be reduced to the scalar forms

$$v_{slip} = \frac{2-\sigma_v}{\sigma_v \mu}\lambda\frac{\partial v_x}{\partial n} + \frac{3}{4}\frac{\mu}{\rho T}\frac{\partial T}{\partial x}\,, \tag{17.35}$$

$$T_w = T + \frac{2-\sigma_T}{\sigma_T}\frac{2\gamma\lambda}{(\gamma+1)Pr}\kappa\frac{\partial T}{\partial y}\,, \tag{17.36}$$

with $Pr = \frac{\mu c_p}{\kappa}$ being the Prandtl number and $q_w = -\kappa\frac{\partial T}{\partial y}$. These equations are valid for the flow computation along a flat plate. An example is given in [21]. In order to deal with curved walls several extensions of the Eqs. (17.35), (17.36) are proposed, for example the one reported in [22]. But all this must be checked and sometimes calibrated by experimental data.

References

1. Hirschel, E.H.: Basics of Aerothermodynamics, 2nd edn. Springer, Berlin (2015)
2. Hirschel, E.H., Weiland, C.: Selected Aerothermodynamic Design Problems of Hypersonic Flight Vehicles. Springer, Berlin/Heidelberg, and vol. 229, Progress in Astronautics and Aeronautics, AIAA, Reston, Va. (2009)
3. Chakravarthy, S.R., Anderson, D.A., Salas, M.D.: The split-coefficient matrix method for hyperbolic systems of gas dynamic equations. AIAA Paper 80-0268 (1980)
4. Whitfield, D.L., Janus, J.M.: Three-dimensional unsteady Euler equations solution using flux vector splitting. AIAA Paper 84-1552 (1984)
5. Chakravarthy, S.R.: Euler equations–implicit schemes and boundary conditions. AIAA J. **21**, 699–706 (1983)

6. Gordon, S., McBride, B.J.: Computer program for calculation of complex chemical equilibrium compositions, rocket performance, incident and reflected shocks, and Chapman-Jouguet detonations. NASA TN D-7056, Lewis Research Center (1973)
7. Pfitzner, M., Weiland, C.: 3-D Euler Solutions for Hypersonic Freestream Mach Numbers. AGARD-CP-428, Paper 22, Bristol, UK (1987)
8. Riedelbauch, S., Weiland, C.: Inviscid Laval-nozzle flowfield calculation. J. Spacecraft and Rockets **25**, 88–90 (1988)
9. Weiland, C.: A split-matrix Runge-Kutta type space marching procedure. J. Comp. Phys **102**, 319–335 (1992)
10. Pandolfi, M.: Upwind formulations for the Euler equations. Lecture Series 1987-04, v. Karman Institute for Fluid Dynamics, Brussels, Belgium (1987)
11. Pfitzner, M.: A 3-D Non-equilibrium shock-fitting algorithm using effective Rankine-Hugoniot relations. AIAA Paper 91-1467 (1991)
12. Caporicci, M., Eriksson, L.-E., Pekkari, L.O., Onofri, M., Popp, M., Weiland, C.: Advanced nozzle technologies for the Ariane 5 Vulcain Engine. AIAA Paper-94-3263 (1994)
13. Hartmann, G., Menne, S., Schröder, W., Weiland, C.: Strömungsfeldberechnungen und Leistungsvorhersage von fortgeschrittenen Raketenmotordüsen. DGLR Jahrestagung, Bremen Germany, Paper 92-03-047 (1992)
14. Hirschel, E.H., Groh, A.: Wall-compatibility condition for the solution of the Navier-Stokes equations. Note: J. of Comp. Phys. **53**(2) (1984)
15. Weiland, C.: Computational Space Flight Mechanics. Springer, Berlin/Heidelberg (2010)
16. Hänel, D.: Molekulare Gasdynamik. Springer, Berlin/Heidelberg (2004)
17. Frohn, A.: Einführung in die kinetische Gastheorie. Akademische Verlagsgesellschaft, Wiesbaden (1979)
18. Maxwell, J.C.: Phil. Trans. Roy. Soc., Part 1, **170**, 231ff (1879)
19. Koppenwallner, G.: Rarefied gas dynamics. In: Hypersonics, Volume 1, Eds. Bertins, J.J., Glowinski, R., Periaux, J., Birkhäuser, Boston, pp. 511–547 (1989)
20. von Smolan, S.: M. Annalen der Physik **300**, 101–130 (1898)
21. Hirschel, E.H.: The Flow of a Rarefied, Dissociated Gas over a Sharp Flat Plate. Deutsche Luft- und Raumfahrt, Cologne, Germany, DLR FB 70-60 (1970)
22. Lockerby, D.A., Reese, J.M., Emerson, D.R., Barber, R.W.: The velocity boundary condition at solid walls in rarefied gas simulation. Phys. Rev. E, **300**(1), Am. Phys. Soc. (2004)

Chapter 18
Significant Scientists

Abstract From the beginning of the human race, some individuals always requested to understand the experiences and observations they had of intellectual and natural incidents. This led to the origination of philosophy, theology, law and in particular the natural sciences. In earlier times, persons wanted to know for example the laws of fluids and the laws of moving bodies as well as the structure and the motion of stars, suns, planets and other celestial bodies. Sometimes the explanation they had for certain observations were not accurate. The most famous example is the *Ptolemaic View of the World*, which was the basis of thinking for centuries. During the *Scientific Revolution* in the second half of the sixteenth century, the science of nature became an autonomous discipline, distinct from philosophy and replacing the *Christianity*. The shift from the geocentric system to the heliocentric system, strongly supported by the invention of the telescope, was a significant milestone in the evolution of natural science. From that time on, the influence of any religious faith on observations of nature and physical processes and their scientific explanations has steadily decreased. In the last centuries, natural sciences, in particular physics have experienced an extreme expansion, for example in optics, acoustics, fluid dynamics, thermodynamics, electricity and magnetism as well as in modern physics with quantum mechanics, the theory of elementary particles, the theory of relativity, etc. Present-day advancements are based predominantly on rationalism. There are often a few individuals, who are responsible for developments having seminal character. In this chapter a selection of 25 scientists, who made significant contributions to fluid mechanics, the new world view (heliocentric system) and the mathematics of field theory, including infinitesimal calculus, vector-tensor algebra and analysis, continuum approach, are portrayed. The lifetime and periods of appearances of these scientists are listed in Fig. 18.1.

C. Weiland, *The Hierarchy of Fluid Dynamic Equations*,
https://doi.org/10.1007/978-3-031-71198-5_18

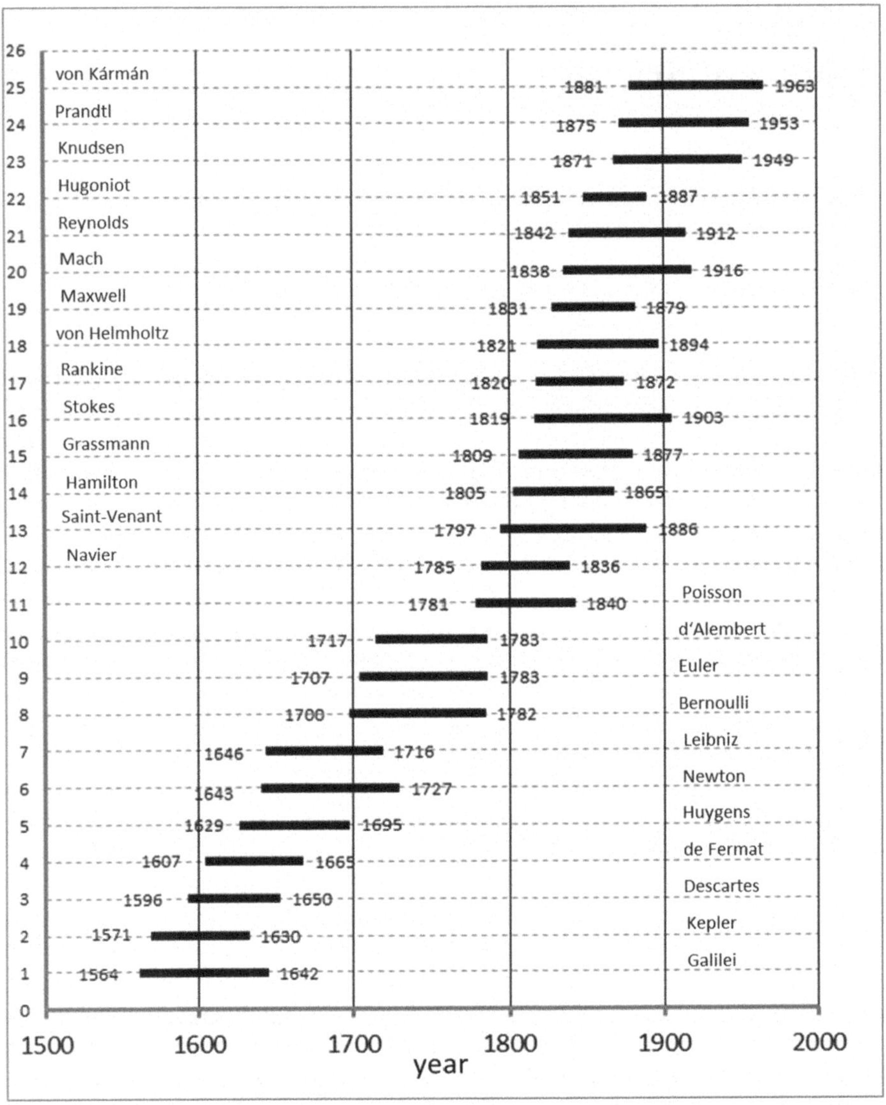

Fig. 18.1 Chronological table and lifetimes of the scientists with impacts on the considerations in this book

18.1 Galileo Galilei

Galileo Galilei

Galileo Galilei was an Italian polymath working in the fields of astronomy, cosmology, mathematics, physics, engineering and philosophy. He was born in Pisa on February 15, 1564, and died on January 8, 1642, in Arcetri near Florence. Due to the influence of his father he began to study medicine at the University of Pisa in 1581. In 1585, Galilei left the university without obtaining a medical degree, because his genuine interest was mathematics. In the years 1582–1583, he attended courses in mathematics at the University of Pisa, studying the works of Euclid and Archimedes. During 1585–1586, he began teaching mathematics at the Accademia Fiorentina and in Siena. Subsequently (1587) he tried to secure a position as a mathematics teacher at the University of Bologna, but with no success. Two years later, he was appointed to the chair of mathematics at the University of Pisa due to his excellent reputation through his lectures at the Accademia Fiorentina. In 1592, be became professor of mathematics at the University of Padua, where he stayed for the next 18 years. Hans Lipperhey, a Dutchman, has constructed a spyglass in 1608 and this information reached Galilei in 1609. He was so impressed with this device that he constructed a spyglass by himself whose optical performance was much better than that of the Dutchman. With this telescope Galilei made several groundbreaking observations: the surface of the moon is mountainous, the Milky Way is composed of a huge number of single stars and the discovery of four moons of the planet Jupiter and others [1]. He then resigned from his post at Padua and became Chief Mathematician at the University of Pisa.

In a letter he wrote in 1616, Galilei stated that his discoveries are much more in agreement with the Copernican theory (heliocentric system) and that these refute the Ptolemaic theory (geocentric system). This position eventually resulted in an accusation at the Inquisition in 1633. In addition, Galilei made major contributions to kinematics and mechanics with his formulation of inertia, the famous law of falling bodies, the parabolic trajectory of projectiles and the mechanics of materials.

18.2 Johannes Kepler

Johannes Kepler

Johannes Kepler was a German astronomer and mathematician. He was born in Weil der Stadt, Germany on December 27, 1571, and died in Regensburg, Germany on November 15, 1630. He studied theology at the University of Tübingen, Germany, where a lecture in mathematics was part of the course. This mathematical course also encompassed the discipline of astronomy. There, he learned about the peculiarities of the old geocentric system of Ptolemy and Aristotle and the ideas of the new heliocentric system of Copernicus. He was more inclined towards the latter as this theory was considerably better in agreement with the observations available in these days. In 1594, he accepted a post as a mathematician at a college in Graz, Austria where he stayed with his family until 1600. Following that, he relocated to Prague, Bohemia now Czech Republic, accepting a position as assistant of Tycho Brahe the foremost observational astronomer at that time. After Tycho Brahe's passing in 1601, he succeeded him as Imperial Mathematician. Johannes Kepler analyzed the extensive number of observations Tycho Brahe has executed from the path of the planet Mars and concluded that its orbit is an ellipse with the sun at one of its foci. Further exploitation for the other planets, known at that time, had shown that this was true for all the planets. This was the origin of the formulation of the two laws, later named Kepler's first and second law, published in 1609 [2]. Because he was a profoundly religious man (Protestant), he regarded these laws as celestial harmonies that reflected God's design of the universe. After Emperor Rudolf abdicated in favour of his brother Matthias (1611), who was not tolerant against Protestants, he was obliged to leave Prague moving to Linz, Austria. There he published the work *"Harmonice Mundi"* (The Harmony of the World) in 1619, where he presented a more elaborated mathematical model than the previous one. In this work he also added the discovery that for any two planets the ratio of the squares of their periods will be the same as the ratio of the cubes of their semi-major axis, now known as Kepler's third law [3]. He also found that using logarithms could rigorously simplify the intricate calculations in astronomy.

18.3 René Descartes

Renè Descartes

René Descartes was a French philosopher, natural scientist and mathematician of the first half of the seventeenth century. Birth on March 31, 1596, in La Haye en Touraine, France and death on February 11, 1650, in Stockholm, Sweden. He studied at the University of Poitiers receiving a law degree in 1616. In 1618, he enrolled in the military school of Breda, becoming a volunteer in the army of Maurice of Nassau. After three years he left the service of Maurice of Nassau and traveled throughout Europe, eventually joining the Bavarian army of Maximilian of Bavaria. One year later, he left the army and spent eight years (1620–1628) traveling across Europe with stays in Bohemia, Hungary, Germany, The Netherlands, France, Switzerland and Italy. Then he sought a place to settle down, where he could work in solitude. He found it in The Netherlands which was a haven of tolerance for him and lived there for the next 18 years.

In his publication "*Discourse of Methods*" (1637), he propagates the method for utilizing reason in the search for truth in the sciences. To illustrate these ideas he attached three appendices to this publication, where the first is concerned with the law of light refraction, the second focuses on meteorology, explaining, for example, the rainbow and the third, the most interesting one, the founding of algebraic geometry, which has made the understanding and handling of the geometry easier. René Descartes is credited with the introduction of the Cartesian coordinate system. Further, he introduced specific letters for the knowns and the unknowns, respectively, which have simplified the arithmetic. Known are also his contributions to mechanics, with his laws of impact and the conservation of momentum and energy. However, it was later discovered, that most of his impact laws are not true [4].

The treatises "*Meditations of First Philosophy, in which is Proved the Existence of God and the Immortality of the Soul*", published in 1641, and "*Principles of Philosophy*" (1644), placed him the dignity of the founder of the modern philosophy [5].

18.4 Pierre de Fermat

Pierre de Fermat

Pierre de Fermat was a French lawyer and government official and shared interests for mathematical questions with a lot of mathematicians at that time. His day of birth is not truly known, but historians conjecture that he was probably born in the year 1607 in Beaumont-de-Lomagne, France. He died on January 12, 1665, in Castres, France. He attended the University of Orléans (1623–1626) and graduated with a degree in civil law. After a short stay in Bordeaux, he purchased the office of councillor at the parliament in Toulouse (1631), where he remained for the rest of his life. Fermat was appointed to a lower chamber of the parliament in 1631, was promoted to a higher chamber in 1638 and finally reached 1652 the highest level at the criminal court.

The beginning of his mathematical engagement can be dated to the late 1620 s, presumable when he lived in Bordeaux. In these days, he produced important work on methods to determine extreme values of functions (maxima-minima) and the tangent problem. These activities can be considered as the precursor for the development of the infinitesimal calculus. It was René Descartes who criticized Fermat's methods regarding the treatment of the maxima-minima problem and for finding tangents to curved lines. However after a heated debate Descartes eventually wrote to Fermat that he now appraised his method as "very good". Fermat also did intensive work in number theory (see "Fermat's Last Theorem") and the theory of probability. More than 300 years later the British mathematician Andrew Wiles succeeded in 1994 a correct proof of Fermat's Last Theorem. It should be mentioned that Pierre de Fermat published nearly nothing of his mathematical work. We know from his work through the correspondence with other mathematicians and by the notes he made in the publications and books of other researchers. In optics he formulated a theory regarding the path of rays which now bears his name: The Fermat Principle, [6].

18.5 Christiaan Huygens

Christiaan Huygens

Christiaan Huygens was a Dutch mathematician, physicist and astronomer. Born on April 24, 1629, and death on July 8, 1695, The Hague, The Netherlands. The family of Christiaan Huygens had many contacts with scholarly persons at that time including René Descartes, who undoubtedly influenced his early mathematical education. Later Christiaan Huygens studied law and mathematics at the University of Leiden (1645–1647) and at the College of Orange at Breda (1647–1649).

Since some of his greatest interests were optics and the construction of telescopes he detected in 1655 the moon Titan of the planet Saturn using lenses he had ground himself, [7]. Later, he discovered the ring of Saturn and explained the phases and changes in the shape of the ring by his ring theory (1656). Not everyone was pleased with his ring theory including the Jesuit Fabri who attacked not only the theory but also the observations. However, after improved telescopes had confirmed the previous observations, Fabri was finally persuaded to accept Huygens' ring theory in 1665. Work in astronomy required accurate timekeeping, which led Huygens to tackle this problem. In 1656, he invented the first pendulum clock, which greatly increased the accuracy of time measurement. He also built several pendulum clocks to determine the degree of longitude at sea and they underwent sea trials in 1662 and again in 1686. Further, he formulated the law of centrifugal force for uniform circular motion, described the theory of a moving pendulum and developed the wave theory of light. The latter enabled him to grind lenses with lower aberrations. By his work on the collision of elastic bodies and the reformulation of the laws of impact he revealed the errors René Descartes made in this regard. In 1661, Huygens visited London and was interested there in the newly formed Royal Society, where he was elected as a member in 1663. Huygens met Leibniz 1672 in Paris and also Newton 1689 in London. Although he had great admiration for Newton he was not convinced of his theory of gravitation.

18.6 Isaac Newton

Isaac Newton

Issac Newton was one of the most famous English natural scientists. He was born on January 4, 1643, in Woolsthorpe-by-Colsterworth and died on March 3, 1727, in Kensington. Isaac Newton was educated at the Free Grammar School in Grantham from 1655 to 1659 and was admitted to Trinity College at the University of Cambridge in June 1661. Newton's aim at Cambridge was a law degree. It seems that his interest in mathematics began in the autumn of 1663 when he read books regarding geometry and algebra by the leading mathematicians of this time, like R. Descartes (1596–1650), F. Viéte (1540–1603), F. van Schooten (1615–1660), J. Wallis (1616–1703) and I. Barrow (1630–1677). In April 1665, he received a bachelor's degree but then returned to Woolsthorpe when the university was closed due to the plague. After university's reopening in 1667, Newton was elected to a fellowship in 1668 when being awarded his master's degree. During this period he developed widespread interests in physics, mathematics, celestial mechanics and astronomy. In 1669, he was appointed Lucasian professor for mathematics in Cambridge. The Lucasian Chair of Mathematics was endowed by Henry Lucas in 1663 and is one of the most prestigious academic posts in the world. In the early years of his career Newton primarily focused on optics, which was part of the Scientific Revolution of the 17th century. He was elected in 1672 to the Royal Society. At this time he discovered that white light is a composite of all the colors of the spectrum. He later assumed the role as President of the Royal Society from 1703 on until his death. During the 1660 s, Newton developed the infinitesimal calculus concept independently of G.W. Leibniz. However, he did not published it at the time, which led to a fierce controversy between these two scholars.

His greatest achievements were the formulation of the three laws of motion and the development of the law of universal gravitation, which were described in his famous book *"Philosophiae Naturalis Principia Mathematica"* (Mathematical Principles of Natural Philosophy) published in 1687, [8]. This was the origin of classical mechanics, which are still in force to this day.

18.7 Gottfried Wilhelm Leibniz

Gottfried Wilhelm Leibniz

The German philosopher Gottfried Wilhelm Leibniz was born on July 1, 1646, in Leipzig and died on November 14, 1716, in Hannover. He was regarded as one of the last universal geniuses, [9, 10]. As a child, he was educated at the Nicolai school and later self-taught in the library of his father. During his studies of law at the University of Leipzig, he came in contact with the toughts of scientists and philosophers of that time, which had a strong impact on his further scientific life. His interests covered philosophy, mathematics, physics, theology, law, history and other disciplines. From a mathematical perspective he was the originator of the binary number system (1679), which is the foundation of every computer system today and was the founder of the infinitesimal calculus (integral and differential calculus, 1675), independently of the work of Isaac Newton.

Johann Christian Freiherr von Boyneburg took him into his service and introduced him to the court of the prince-elector, the archbishop of Mainz, Johann Philipp von Schönborn, where he was concerned with matters related to law and politics. After the deaths of Boyneburg 1672 and of the prince-elector 1673 he has left these places. In 1678, he was appointed as a councillor by John Frederick, the Duke of Braunschweig-Lüneburg. Some years later (1685), he acted as historian for the house of Brunswick and was named Hofrat (court advisor), a position he held until the end of his life. In addition he was also named librarian in Wolfenbüttel since the year 1691.

When the Duke of Brunswick, Ernest Augustus, died in 1698, G.W. Leibniz reduced his activities for the house of Brunswick. In the meantime his fame as a philosopher and scientist had spread already throughout Europe. Therefore, he was in correspondence with most of the European scholars. The Russian Tsar Peter the Great promoted him 1714 to the position of Reichshofrat (adviser to the empire), which entailed the title Freiherr (Baron).

Throughout his entire lifetime, beginning with the completion of his legal studies in 1666, he published numerous books and treatises concerning the aforementioned disciplines.

18.8 Daniel Bernoulli

Daniel Bernoulli

Daniel Bernoulli was a Swiss mathematician and physicist, who was born on February 8, 1700, in Groningen, The Netherland, and died on March 17, 1782, in Basel, Switzerland. He was the second son of the mathematician Johann Bernoulli (1667–1748). In the seventeenth and eighteenth centuries the Bernoulli family has spawned many renowned mathematicians, physicists and astronomers. Although his father was a mathematician filling the chair of mathematics at the Basel University, Daniel had to study medicine and he completed his doctorate in medicine at the Basel University in 1720. Subsequently, he applied for two chairs at the Basel University: the first in anatomy and botany and the second in logic. The final decision for both chairs were made by drawings of lots and in both cases he was unlucky in these games of chance. As a consequence he went to Venice, worked on mathematics for the first time and performed medical work on the flow of blood indicating a shift in his academic interests.

In 1725, he went with his older brother Nicolaus to St. Petersburg, where they both took up a chair in mathematics. Upon his recommendation Leonhard Euler arrived at St. Petersburg in 1727 to fill a position in the mathematical department. This period in St. Petersburg, which he left in 1733, was his most productive time [11]. When Daniel Bernoulli returned to Basel in 1734, he applied for some posts in mathematics and physics, but with no success. Therefore, for the first time, he accepted a position in botany and later, in 1750, he was appointed to the chair of physics where he taught until 1776.

His most important work was on hydrodynamics. He studied the principle of energy conservation (Bernoulli equation, Bernoulli principle) and discussed a first model on the kinetic theory of gases. All these derivations and progressions were published in 1738 in his famous book *"Hydrodynamica"* [12]. Further he performed work on other mathematical topics like the solution of the Riccati differential equation and discussed many mechanical questions like the behavior of vibrating strings. For these activities, he won the Grand Prize of the Paris Academy ten times.

18.9 Leonhard Euler

Leonard Euler

Leonard Euler born April 15, 1707, in Basel, Switzerland and death in St. Petersburg, Russia on September 18, 1783, was a mathematician, physicist and astronomer. The family of Euler and the Bernoulli family were acquainted with each other. In particular Johann Bernoulli, the father of Daniel Bernoulli, a professor of mathematics at the University of Basel, discovered Leonhard's great mathematical capacity and provided him some private tuition. When he entered the University of Basel in 1720, he first studied philosophy and later mathematics and physics and completed his studies in 1726. A recommendation of Daniel Bernoulli led him to the Imperial Russian Academy of Sciences in Saint Petersburg in 1727, where he accepted a position in the mathematical department, which had become vacant after the death of Nicolaus Bernoulli, a brother of Daniel, in July 1726. In 1730, he became professor for physics at the Academy. Daniel Bernoulli was unhappy in Russia and left St. Petersburg in 1733 to return to Basel. As he held the senior chair at the Academy, he suggested that Euler should be the successor to this position. Due to the turmoil situation in Russia Euler departed from St. Petersburg in 1741 to take over a post at the Berlin Academy, where he stayed for 25 years. In 1766, he accepted an invitation to the St. Petersburg Academy again, mainly driven by the increasingly shattered relationship with Frederick the Great of Prussia. He remained there until his death in 1783, [13]. Leonard Euler was an extremely productive scientist. He wrote hundreds of scientific articles (approximately 900) and more than 20 books on a large variety of mathematical and physical problems. He was the first who formulated the hydromechanical differential equations in three dimensions. We know these equations as the convective part (including the pressure) respectively the inviscid part of the fundamental fluid dynamic momentum equations, today called the Euler equations. Further, he made important contributions to various branches of mathematics including geometry, trigonometry, graph theory, number theory, topology, infinitesimal calculus, algebra. In the field of physics, he worked on continuum mechanics, optics, lunar theory, stability theory, ballistics and gyroscopic theory.

18.10 Jean-Baptiste le Rond d'Alembert

Jean-Baptiste le Rond d'Alembert

Jean-Baptiste le Rond d'Alembert was a French mathematician, physicist and philosopher. He was born on November 16, 1717, and died on October 29, 1783, both in Paris. d'Alembert was a pioneer in studying differential equations describing physical processes like the mechanics of fluids [14]. His education began when he entered the Jansenist Collège des Quatre Nations, where he graduated in 1735. Afterwards, he studied law and qualified as a advocate in 1738. In the following year, he started studies in medicine, but he decided that both topics were not the right career for him. Instead, his real enthusiasm had prevailed for mathematics, which he studied almost exclusively on his own. In May 1741, d'Alembert was admitted to the Paris Academy of Science due to some published papers on infinitesimal calculus. In another paper (1743), he formulated, with respect to Newton's third law of motion, an improved principle of mechanics, called the d'Alembert Principle [15]. d'Alembert, who was a mathematician, believed that mechanics were based on metaphysical principles rather than experimental evidence and he clearly stated his position on this matter. This was in contrast to Newton, who has based his laws of motion on experimental experience. With this attitude he frequently disagreed with other natural scientists at that time among them Daniel Bernoulli, Alexis-Claude Clairaut and Leonhard Euler. His famous article about the first solution of the wave equation, demonstrated on a vibrating string, was published by him in 1747. But this work also has suffered from the defect that he used mathematically pleasing simplifications, which led to results that were at odds with observations.

Between 1751–1780, he worked with the philosopher Denis Diderot as an editor, on the *"Encyclopédie"*, which was denominated as the quintessence of human knowledge at that time. d'Alembert became a fellow of the French Academy and was elected to the Russian Academy of Sciences, the American Academy of Arts and Sciences as well as the Prussia Academy of Sciences.

18.11 Siméon-Denis Poisson

Siméon-Denis Poisson

Siméon-Denis Poisson was a French mathematician and physicist. He was born on June 21, 1781, in Pithiviers and died on April 25, 1840, in Sceaux, France. He began the study of mathematics in 1798 at the École Polytechnique in Paris. Pierre-Simon Laplace and Joseph-Louis Lagrange were his teachers. In his early days (1802), he became a deputy professor at the École Polytechnique, a position he held until 1806, when he was appointed to the professorship at the École Polytechnique, which Fourier had vacated when Napoleon sent him to Grenoble. In 1809, he was appointed a professor of pure mathematics at the newly instituted Faculty of Sciences. He was a member and delegate of many international academies and scientific societies, among them the American Academy of Arts and Sciences and the Russian Academy of Sciences in St. Petersburg. We know from him in the field of mathematics his contributions to potential theory, probability theory $\Longrightarrow$ Poisson distribution, Fourier series and integrals among others. In the field of physics, he worked on electricity and magnetism, optics and analytical mechanics, where he had a share on the evolution of the calculus of variations [16]. It is worth noting that Poisson did not share the chauvinistic attitude of many French scientists of his day to give credit to Fermat as the inventor of the infinitesimal calculus. He stated that Leibniz was the creator of the algorithm of the calculus and of the notation that has generally prevailed.

His contribution to fluid mechanics was the extension of Euler's momentum equations by the friction force derived on the basis of interactions between molecules (1831) as C.L.M.H. Navier 1822 did (Navier-Stokes equations). Today, we know, that this interpretation was not true.

18.12 Claude Louis Marie Henri Navier

Claude Louis Marie
Henri Navier

Claude Louis Marie Henri Navier was a French mathematician, physicist and mechanical engineer. He was born in Dijon, France on February 10, 1785, and died in Paris on August 21, 1836. When his father died, 1793 in Paris, his mother returned to her hometown and left Navier in Paris under the care of his granduncle Emiland Gauthey a leading civil engineer in France working at the Corps des Ponts et Chaussées. He received his education at the École Polytechnique and the École Nationale des Ponts et Chaussées, where he graduated in 1806. During Navier's first year at the École Polytechnique, he was taught analysis by Fourier who had a remarkable influence on the young man. Fourier became a lifelong friend of Navier and took an active interest in his career. After Navier's granduncle Gauthey passed away, he was asked by the Corps des Ponts et Chaussées to edit the granduncle's work. Gauthey's work was based on traditional empirical approaches to numerous applications in civil engineering. During the editing process, Navier supplemented some of the problems with analytical solutions in the sense of his own research in theoretical mechanics.

In 1824, he was elected into the French Academy of Science. Already in 1819, he took charge of the applied mechanics courses at the École Nationale des Ponts et Chaussées and later became a post as professor there (1830). A year later he succeeded Augustin Louis Chauchy at the chair for calculus and mechanics at the École Polytechnique. Navier specialized in the field of structural analysis and discovered, among other things, a mathematical usable form for the theory of elasticity (1821). Therefore, he is often regarded as the originator of modern structural analysis. However, his greatest contribution to physics was the extension of the hydrodynamic momentum equations of L. Euler to viscous fluids in 1822 (Navier-Stokes equations). Despite the fact that he had no clear understanding about shear stress in fluids the equations arrived at the proper form we know today.

18.13 Jean Claude Barré de Saint-Venant

Jean Claude Barré de Saint-Venant

Jean Claude Barré de Saint-Venant was a French engineer with interests in physics and mathematics. He was born on August 23, 1797, in Villiers-en Biére and died on January 6, 1886, in Saint-Quen, France. His education has taken place at the École Polytechnique, where he graduated in 1816. Then he spent many years working as an engineer for the Corps des Ponts et Chaussées. He taught mathematics at the École des Ponts et Chaussées, where he succeeded G.G. de Coriolis. Mechanics, elasticity and hydrodynamics including hydrostatics were the disciplines in which he was engaged [17]. Among other contributions, he is known for the shallow water equation and the elements of the structural mechanics: Saint-Venant principle, Saint-Venant theorem. For the consideration of field theoretical aspects vector calculus is indispensable. Saint-Venant published in 1845 his development of the vector calculus he used in his lectures. This was one year after H.G. Grassmann (1844) presented his foundation of vector calculus, see 18.15. Saint-Venant claimed that he had already developed these ideas in 1832.

His most significant contribution to the physics of fluids was the extension of Euler's momentum equations to viscous flows in 1843, seven years after Navier's demise ($\Longrightarrow$ Navier-Stokes equations). He was the first, who gave the correct explanation of the viscous terms, in the sense that these were the products of the viscosity and the velocity gradients (viscous stresses). In contrast, Navier's explanation of the viscous terms was based on a molecular approach, which has been found to be inaccurate. Two years later (1845) G.G. Stokes published the same approach. It is one of the historical curiosities that Saint-Venant's name was not associated with these equations, [18].

18.14 William Rowan Hamilton

Sir William Rowan Hamilton

Sir William Rowan Hamilton was an Irish mathematician and physicist. Born on August 4, 1805, and death on September 2, 1865, in Dublin, Ireland. He was regarded as a child prodigy. Starting his education at Trinity College in Dublin at the age of 18 (1823), he already published as an undergraduate the memoir *"Theory of Systems of Rays"* in 1826 [19]. He completed this work in 1832 with the publication of a third supplement. In 1827, while still an undergraduate aged twenty-one years, Hamilton was appointed as professor of Astronomy at Trinity College. Afterwards, he took up residence at Dunsink Observatory where he spent the rest of his life. Up to the middle of the 1830s, he mainly worked in the field of optics. Hamilton was knighted in 1835. Then, he encountered the challenge to extend his discovery of algebraic couples to a theory of triplets. This work finally let to the formulation of the theory of quaternions in 1844, [20], which states the mathematical connectivity of four quantities in the sense of complex numbers. The mathematical rules and laws of the quaternions have aspects similar to the vector calculus. In 1853, he published the book "Lectures on Quaternions" but he soon realized that it lacked of the necessary qualities to effectively teach the theory of quaternions. Therefore, he began to write another book, "Elements of Quaternions", which was not fully completed when he died. Not everyone was enthused about Hamilton's quaternions. Thomson (1. Baron Kelvin) wrote: "Quaternions came from Hamilton after his really good work had been done, and though beautifully ingenious, have been an unmixed evil to those who have touched them in any way."

Furthermore, Hamilton conceived a mathematical formalism wherein a mechanical event is considered in a two dimensional phase space described by a geometrical and a momentum coordinate. This is today known as Hamilton mechanics and is often exerted in classical and relativistic quantum mechanics.

18.15 Hermann Günther Grassmann

Hermann Günther
Grassmann

Hermann Günther Grassmann was a German mathematician and physicist. He was born on April 15, 1809, in Stettin (now Poland) and died there on September 26, 1877. He started studies of theology at the University of Berlin in 1827 and took courses on theology, classical languages, philosophy and literature, but did not appear to have taken any courses on mathematics or physics. After completing his studies in 1830 and passing several exams to receive permission to qualify as gymnasium teacher, he was appointed in 1832 to the gymnasium and later the Friedrich Wilhelm Schule in Stettin as an assistant teacher. It was during this year that Grassmann began to develop vector methods. He applied these first in a scientific work required for a further examination regarding an advanced qualification for teaching at higher levels at gymnasiums. In an essay (1840) for this examination with the title "Theorie der Ebbe und Flut" (Theory of low and high tide) he applied his vector methods for the first time and produced an original and simplified approach. All of these new ideas including vector summation, vector differentiation, vector function theory as well as the associated linear algebra was summarized by him in the book *"Die lineale Ausdehnungslehre, ein neuer Zweig der Mathematik"* (The straight extension theory, a new branch of mathematics) published in 1844 [21]. However, his work was not understood at that time by his colleagues. As a consequence, the book was largely ignored. Even a revised version of the book published in 1862 fared no better than the first version.

He wrote to the Prussian Ministry of Education to be considered for a university position due to his innovative mathematics, which he believed to be significant. But this was prevented by a negative comment of a reviewer. Therefore, Grassmann could never achieve a post at an university.

It is interesting to see how many leading mathematicians failed to recognise the importance of Grassmann's new mathematical subject [23]. Only in 1866 the worthiness of Grassmann's work was appreciated by Hermann Hankel and Felix Klein and later by Josiah Willard Gibbs. Since then, he prevails as the man who was the founder of the vector and tensor analysis.

18.16 George Gabriel Stokes

Sir George Gabriel Stokes

George Gabriel Stokes, 1st Baronet, was an Irish mathematician and physicist. He was born on August 13, 1819, in Skreen, Ireland, and died on February 1, 1903, in Cambridge, England. He was educated and taught by his father, a Protestant minister of the parish of Skreen, along with a clerk of his father's parish until he was 13. He left Skreen in 1832 and attended a school in Dublin for three years. At the age of 16, George Gabriel Stokes moved to England and entered Bristol College in Bristol. His stay in Bristol was very important for him in preparing for the studies in Cambridge, which he began at Pembroke College in 1837. At Pembroke College Stokes was trained by William Hopkins, a famous Cambridge tutor who played a more significant role than the lecturers. It was Hopkins, who advised Stokes to undertake research into hydrodynamics. Four years later, he graduated in mathematics and was the first Smith's prizeman.

George Gabriel Stokes became the Lucasian Professor of Mathematics at Cambridge University in 1849. He was the secretary of the Royal Society since 1854 and its president from 1885–1890. In 1889, he was made a Baronet by Queen Victoria and served as a representative of his university in the British Parliament [23].

His scientific work encompassed various physical disciplines, like the theory of light (optics), acoustics and the motion as well as the inner friction of fluids. In 1842/1843, he published some papers on the motion of incompressible fluids, [24]. In the paper titled "On the theories of the internal friction of fluids in motion" (1845) he presented his elaboration on the extension of the three-dimensional inviscid momentum equations of hydrodynamics (Euler equations) to flows with friction. After deducing the correct equations of motion, Stokes discovered that he was not the first to obtain them since Navier, Poisson and Saint-Venant, see Chap. 2, had already performed this extension. But the use of different physical assumptions in the frame of his derivation of the equations justified their publication. Since then, the fluid dynamic equations with frictions are named the Navier-Stokes equations. Further, he has contributed to the mathematical physics with outcomes now known as Stokes' theorem and Stokes' law [25].

18.17 William John Macquorn Rankine

William John Macquorn Rankine was a Scottish engineer and physicist. He was born on July 5, 1820, in Edinburgh and died on December 24, 1872, in Glasgow. Since he was not in good health as a child, most of his education took place at home with private tutors. For a short while he visited also the Glasgow High School in 1830. For two years from 1836–1838 he studied at the University of Edinburgh without obtaining a degree before leaving the university in 1838. Thereafter, he became an apprentice to the engineer John Benjamin MacNeill, where he worked for a long time on engineering projects. He was appointed to the regius chair of civil engineering and mechanics at Glasgow in 1855. William Rankine founded the Scottish Institution of Civil Engineering and became its first president in 1857. During his career as an engineer he developed a strong interest in mathematical and physical problems. He worked in the field of thermodynamics (heat), introduced the term energy, subdivided this term in kinetic and potential energies and formulated thermodynamic cycles similar to Rudolf Clausius, [26]. The two scientists, Rudolf Clausius and William Rankine, approached the second law of thermodynamics from the point of view of the transformation from one kind of energy to the other. But whereas Clausius considered the conversion between heat and work and the flow of heat from high temperature to low temperature, Rankine concentrated on the transition from kinetic to potential energies.

In the context of fluid dynamics the Rankine-Hugoniot equations are well-known, which are important relations describing discontinuous solutions in supersonic flow fields [27, 28]. These findings are the basis of modern shock wave theory. He became a fellow of the Royal Society of Edinburgh in 1849 and of the Royal Society of London in 1853. Furthermore, he was elected to the American Academy of Arts and Sciences in 1856 as well as to the Swedish Academy of Sciences in 1868. He was awarded an honorary degree from Trinity College in Dublin (1857).

18.18 Hermann Ludwig Ferdinand von Helmholtz

Hermann Ludwig Ferdinand v. Helmholtz

The German scientist and philosopher Hermann Ludwig Ferdinand von Helmholtz made extensive contributions to physiology, physics and mathematics. He was born on August 31, 1821, in Potsdam and died on September 8, 1894, in Berlin, Germany. After graduating from the high school (gymnasium) he received a medical education at the Friedrich Wilhelm Medical Institute in Berlin, where he prepared his doctoral thesis in 1843. Considering Helmholtz's later contributions to mathematics, it would be reasonable to expect that he took courses in mathematics at the university at that time. But he did not, rather he studies mathematics on his own, reading the works of modern mathematicians. When he has left the university, he spent five years as a military doctor at a regiment in Potsdam.

From 1848 on, he held various academic posts in the field of physiology in Berlin (1848), Kaliningrad (Königsberg) (1849), Bonn (1855) and Heidelberg (1858). During these years, his scientific interests progressed from physiology to physics. He was appointed professor of physics at the University of Berlin in 1871 and later became (1882) the first director of the Physico-Technical Institute in Berlin. Besides his outstanding publication *"Handbuch der physiologischen Optik"* (Handbook of Physiological Optics) (1867) [29], he formulated the law of energy conservation in the paper *"Über die Erhaltung der Kraft"* (On the Conservation of Force) (1847), [30]. In this publication, he showed that the assumption that work cannot be continually produced from nothing leads to the conservation of kinetic energy and he also demonstrated that in situations where energy appears to be lost, it is actually converted into heat. In the field of fluid dynamics, he discovered the laws about vortices in inviscid flows (1858) leading to the conclusion of their amazing stability, [31]. His engagement in the disciplines of thermodynamics and electrodynamics directed to the introduction of the term "free energy". The Gibbs-Helmholtz equation was a later outcome of these activities. In 1883, he was ennobled by the German Emperor (Deutscher Kaiser) Wilhelm I. .

18.19 James Clerk Maxwell

James Clerk Maxwell

James Clerk Maxwell was a Scottish physicist. Born on June 13, 1831, in Edinburgh, Scotland and death on November 5, 1879, in Cambridge, England. His education started in 1841, when he entered the Edinburgh Academy. After six years he continued his studies at the University of Edinburgh, where he read voraciously on a variety of subjects. During this time his exceptional abilities were recognized. In 1850, he went to the University of Cambridge, where William Hopkins was his mathematical teacher. After an early appointment in 1856 to the professorship of natural philosophy at Marischal College in Aberdeen, he joined King's College, London, in 1860 as a professor in the same scientific discipline. An election to the Royal Society followed in 1861. His most fruitful period was between 1860 and 1865, when he published the two classical papers on the electromagnetic field theory [32, 33], which were his most fundamental contribution to the physics. The four partial differential equations, now known as Maxwell's equations, are the substance of this theory. Furthermore, he made significant contributions to the kinetic theory of gases, by applying independently of Ludwig Boltzmann, the methods of probability and statistics for the description of the properties of molecules (Maxwell-Boltzmann distribution). This theory represented a change from the concept of certainty, heat is viewed as flowing from hot to cold, to one of statistics, molecules at high temperatures have only a high probability of moving towards those at low temperatures. The textbook about thermodynamics, *"Theory of Heat"*, was published by him in 1871 [34]. Maxwell resigned from his professorship at King's College in 1865 and returned to the family estate in Glenlair, but came back to Cambridge in 1871, after being elected as the first Cavendish professor of physics.

Many physicists regard Maxwell as the scientist, who had the strongest impact on the 20th century physics, in particular the special relativity theory and the quantum mechanics.

18.20 Ernst Mach

Ernst Mach

Ernst Mach was a Czech scientist, at that time part of the Austro-Hungarian Monarchy, with interests in mathematics, physics, philosophy and later the physiology of sensory perceptions. He was born in Chrlice (Czech Republic) on February 18, 1838, and died on February 19, 1916, in Haar, Germany. Ernst Mach was educated at home until the age of 14 and then, he briefly attended a gymnasium. He entered the University of Vienna, Austria, in 1855, where he received his doctorate in physics in 1860. For a number of years he taught mechanics and physics in Vienna, was appointed awhile as professor in Graz (1864) and accepted 1867 a post as professor of experimental physics at the Charles University of Prague, where he remained until 1895. After that, Ernst Mach returned to the University of Vienna as professor of inductive philosophy. He retired from this post in 1901, when he was appointed to the Austrian parliament. His scientific work has covered a lot of aspects of the physics of optical waves, like interference, diffraction, refraction, polarization, etc. It is worth mentioning that he proposed the physical principle, that inertia results from a relationship of an object with all the rest of the matter in the universe. Mach advanced the concept that all knowledge is derived from sensation. Thus, phenomena under scientific investigation can only be understood in terms of sensations, or experiences. This means that all statements in natural science have to be empirically verifiable. Such rigorous criteria of verifiability led him to reject the metaphysical idea of the absoluteness of space and time. This prepared the way for A. Einstein's relativity theory.

In the field of gasdynamics his work on the flow structure of projectiles moving with supersonic speeds (the first identification of shock waves) was of seminal importance until this day [35, 36]. The ratio of the flight velocity u of a flight object to the speed of sound a bears his name in honor of his fundamental contribution to the understanding of supersonic flows.

18.21 Osborne Reynolds

Osborne Reynolds

Osborne Reynolds was a British mathematician and physicist. On August 23, 1842, he was born in Belfast, Ireland and died on February 21, 1912, in Watchet, England. His education took place in a school in Dedham. After completing his schooling, he took an apprenticeship with an engineering firm in 1861. Later, he entered Queens' College in Cambridge, studying mathematics and graduating in 1867. In the following year (1868), he applied for and was elected to the newly installed chair of engineering at Owen College - later to become the Victoria University of Manchester. There, he remained as a professor of engineering until 1905, [37, 38]. His scientific interests were versatile. Besides his well-known contributions to fluid dynamic problems, he also published papers on the physics of gases, liquids and granular materials as well as on solar and cometary matters. After 1873, Osborne Reynolds primarily focused on fluid mechanics, and in this field his contributions received a status of seminal importance.

In 1883, he published his famous paper: *An experimental investigation of the circumstances which determine whether the motion of water in parallel channels shall be direct or sinuous and of the law of resistance in parallel channels.* This paper discusses the fundamental experimental observations for laminar and turbulent flows and their dependency on the dimensionless quantity ul/ν, now called the Reynolds number Re [39], see also [40]. In 1886, he published fundamental work on the theory of hydrodynamic lubrication, making him the dominant figure in the evolution of tribology. Three years later he developed a theoretical model for turbulent flow regarding the formulation of the Navier-Stokes equations, the so called Reynolds Averaged Navier-Stokes equations. Osborne Reynolds became a Fellow of the Royal Society in 1877 and won the Royal Medal eleven years later. He was elected President of the Manchester Literary and Philosophical Society in 1888 and received the Dalton Medal in 1903.

18.22 Pierre Henri Hugoniot

Pierre Henri Hugoniot

Pierre Henri Hugoniot was a French mathematician and artillerist, birth in Allenjoei, France on June 5, 1851, and death in Nantes, France in February 1887. He entered the Ëcole Polytechnique in Paris in 1868 and graduated in military engineering in 1872. He became professor of mechanics and ballistics at the Lorient Artillery School (1879–1882) and assistant director of the Central Laboratory of the Marine Artillery (1882–1884). On January 1884, he was appointed Captain of the Marine Artillery.

Since the beginning of the nineteenth century, numerous well-known scientists, including Simëon Denis Poisson (1808), George Gabriel Stokes (1848), John William Strutt (also known as Lord Rayleigh) (1877), William Thompson (also known as Lord Kelvin) (1880), Bernhard Riemann (1860), have worked on the theory of sound, in particular the propagation of sound waves of finite amplitude. They all found that the theory allowed for waves, which become more and more steep (as the running time of the wave increases) resulting in a discontinuous wave surface (shock). They treated the physical problem as an adiabatic one, which was the main deficiency of their thinking. Moreover, they denied that a discontinuous solution exists in nature. William Rankine (1870) was the first to recognize that shock transition alters the temperature and the heat conduction, thus behaves non-adiabatic. Then, Pierre Henri Hugoniot developed the jump conditions for a one-dimensional discontinuous flow by evaluating the general conservation equations for mass, momentum and energy for non-adiabatic processes in discontinuous surfaces. He found the famous correlation between the pressure, the density and the internal energy before and behind a discontinuity $e' - e = \frac{p'+p}{2}(\frac{1}{\rho} - \frac{1}{\rho'})$, today called the Hugoniot relation or the Hugoniot adiabatic curve, [41]. This relation is the basis of modern shock wave theory [42]. It seems that Pierre Henri Hugoniot was unaware of William Rankine's prior work on this topic, see Sect. 18.17.

18.23 Martin Hans Christian Knudsen

Martin Hans Christian Knudsen

Martin Hans Christian Knudsen was a Danish physicist. He was born on February 15, 1871, in Hasmark and died on May 27, 1949, in Copenhagen, Denmark. He studied physics at the University of Copenhagen, where he graduated in 1896. Knudsen responded to one of the prize questions posed by the University of Copenhagen on electrical sparks and won the university's gold medal in 1895. In 1912, he accepted a position as professor at this university, where he remained until his retirement in 1941. From 1927 to 1928, he served as rector of the university. In 1927, he was one of the participants of the fifth Solvay Conference on Physics, which was held at the International Solvay Institute for Physics in Belgium. His scientific work encompassed two main interests, namely the kinetic theory of gases at low pressures (rarefied gases, molecular flows) and the hydrography. He was the Danish delegate in the "Central Committee for Oceanic Research of the International Council for Exploration of the Sea" for the long period of 1902–1947, the last fourteen years serving as its vice-president [43]. Knudsen also acted as president of the International Association of Physical Oceanography (1930–1936). In particular, within the frame of the kinetic theory of gases he provided the first indirect experimental confirmation of Maxwell's law of velocity distribution. He measured the flow of gases through a small hole and found that the results were in good accordance with the formula calculated on the basis of Maxwell's law, see Sect. 18.19. Moreover, he performed an extensive series of basic investigations on the behavior of gases at very low pressure and designed for example an absolute manometer now called Knudsen manometer. Today, he is known for the Knudsen effect, the Knudsen gas, the Knudsen cell etc., and in particular, the Knudsen number, which characterizes the different flow regimes ranging from continuum to free molecular flow, [44].

Knudsen was a member of several Academies of Sciences, among them the Royal Danish, the German and the Royal Prussian ones.

18.24 Ludwig Prandtl

Ludwig Prandtl

Ludwig Prandtl was born on February 4, 1875, in Freising and died on August 15, 1953, in Göttingen. He attended an elementary school in the autumn of 1881 and began his secondary education at the gymnasium in Freising, Bavaria, in September 1885. Later, he went to the Ludwig Gymnasium in Munich, where he graduated in 1894. At the Technical University in Munich, he studied mechanical engineering which he completed in 1898. Ludwig Prandtl earned his doctorate in 1901, while working as an assistant of August Föppl a famous professor of theoretical mechanics at that time. In the same year, at the age of 26, he was appointed as a professor of mechanics at the University of Hannover. Afterwards, the famous mathematician Felix Klein persuaded Ludwig Prandtl to accept a post as professor in Göttingen (1904), where he remained until he died in 1953. In Göttingen he got the promise of setting up a wind tunnel for the execution of practical investigations. This tunnel became operational in 1908. In 1915, another project was initiated to build a more powerful wind tunnel that would give Prandtl more scope for research. It was completed in 1917 and had a performance of 300 horsepower. Already in 1904, Ludwig Prandtl had published the ground-breaking article *"Über Flüssigkeitsbewegung bei sehr kleiner Reibung"* (About fluid motion with very small friction.) [45], in which the concept of the boundary layer (Grenzschicht) was introduced. He provided further outstanding contributions in the fields of gasdynamics, turbulence modelling, wing theory, similarity theory, etc. [46]. When the *Gesellschaft für Angewandte Mathematik und Mechanik (GAMM)* was founded in 1922, he became the Society's first president. Ludwig Prandtl received many honors for his contributions to the physics of fluid dynamics. Among them are the Gold Medal of the Royal Society of London (1929) and the Daniel Guggenheim Medal (1930). More than twenty academic institutions elected him as a member.

Theodore von Kármán stated about him in 1954 [47]:

"The man who gave modern wing theory its practical mathematical form was one of the most prominent representatives of the science of mechanics, and especially fluid mechanics, of all time, Ludwig Prandtl".

18.25 Theodore von Kármán

Theodore von Kármán

Theodore von Kármán was a Hungarian-born aerospace engineer, mathematician and physicist. His birth was on May 11, 1881, in Budapest, Hungary and his death on May 7, 1963, in Aachen, Germany. He was tutored partly at home and later entered the Minta Gymnasium in Budapest to complete his education. At the University of Budapest he studied mechanical engineering and graduated in this discipline in 1902. After working as an assistant in hydraulics at this university for three years, he left Budapest in order to study at Göttingen. In 1908, he received his doctorate for the mathematical model explaining the buckling of large structures. The analysis of measurements in a newly constructed wind tunnel by Ludwig Prandtl in Göttingen led him to find the mathematical description of a vortex street. A vortex street consists of alternating double rows of vortices behind a blunt body. Today, such vortex streets bear his name: The von Kármán Vortex Street [48]. After his experience in Göttingen he became interested in aeronautics. In 1913, the Technical University of Aachen offered him the post as director of the Aerodynamic Institute, which he held until 1930. There, he began theoretical work on aircraft design, which was facilitated by a newly installed wind tunnel. In 1930, the political events in Germany persuaded him to leave Aachen and to accept a post as director of the Aeronautical Laboratory of the California Institute of Technology. Three years later, he founded the U.S. Institute for Aeronautical Sciences, which conducted research in fluid mechanics, turbulence theory and supersonic flight. Regarding the developments in Germany, efforts were exerted in the field of rocket theory, which has also led to the foundation of the Jet Propulsion Laboratory with von Kármán as its director. In 1949, he retired from these posts and further acted as professor emeritus. After World War II, he established within the frame of NATO the Advisory Group of Aerospace Research and Development (AGARD).

References

1. Drake, S.: Galileo's first telescopic observations. J History Astronomy **7**, 153–168 (1976)
2. Kepler, J.L.: Astronomia Nova: New Astronomy (translated by W. Donahue). Cambridge (1992)
3. Aiton, E.J.: Kepler and the Mysterium cosmographicum. Sudhoffs Arch. **61**(2), 173–194 (1977)
4. Röd, W.: Descartes: Die Genese des Cartesianischen Rationalismus. C.H. Beck, Munich (1995)
5. Clarke, D.M.: Descartes: A Biography. Cambridge University Press, Cambridge (2006)
6. Mahoney, M.S.: The Mathematical Career of Pierre de Fermat (1601–1665). Princeton (1994)
7. Bell, A.E.: Christian Huygens and the Development of Science in the Seventeenth Century. London (1947)
8. Newton, I.: Philosophiae Naturalis Principia Mathematica (Mathematical Principles of Natural Philosophy). Jussu Societatis Regiae ac typis Josephi Streater, London (1687)
9. Finster, R., van den Heuvel, G.: Gottfried Wilhelm Leibniz. Rowohlt Taschenbuch Verlag, Reinbeck bei Hamburg, Germany (1990)
10. Kempe, M.: Die beste aller möglichen Welten. S. Fischer Verlag, Frankfurt (2022)
11. Bernhard, H.: The Bernoulli Family. In: Wussing, H., Arnold, W. (eds.) Biographien bedeutender Mathematiker, Berlin (1983)
12. Bernoulli, D.: Hydrodynamica, sive de viribus et motibus fluidorum commentarii. Argentorati, sumptibus Johannis Reinholdi Dulseckeri (1738)
13. Bernhard, H.: Euler. In: Wussing, H., Arnold W. (eds.) Biographien bedeutender Mathematiker. Berlin (1983)
14. Grimsley, R.: Jean d'Alembert, 1717–83. Oxford (1963)
15. Krbek von, F.: Geschichte des Prinzips von d'Alembert. Wiss. Zeit. Univ. Greifswald. Math.-Nat. Reihe, vol. 2, pp. 15–22 (1953)
16. O'Connor, J.J., Robertson, E.F.: Siméon-Denis Poisson, Biography. https://mathshistory.st-andrews.ac.uk/Biographies/Poisson/ (2002)
17. O'Connor, J.J., Robertson, E.F.: Jean Claude Barré de Saint-Venant, Biography. https://mathshistory.st-andrews.ac.uk/Biographies/Saint-Venant/ (2000)
18. Anderson, J.D.: A history of Aerodynamics and Its Impact on Flying Machines. Cambridge University Press, New York (1997)
19. Hamilton, W.R.: Theory of systems of rays. Trans. Royal Irish Acad. **15**, 69–174 (1828)
20. Hamilton, W.R.: On a new species of imaginary quantities connected with a theory of quaternions. Proc. Royal Irish Acad. **2**, 424–434 (1844)
21. Grassmann, H.G.: Die lineale Ausdehnungslehre. Otto Wigand, Leipzig (1844)
22. Fearnley-Sander, D.: Hermann Grassmann and the creation of linear algebra. Amer. Math. Monthly **86**, 809–817 (1979)
23. Voit, C.: Nekrolog auf Sir George Gabriel Stokes. Sitzungsberichte der mathematisch-physikalischen Klasse der K.B. Akademie der Wissenschaften, München, Band XXXIII (1903)
24. Stokes, G.G.: On the steady motion of incompressible fluids. Trans. Cambridge Philosophical Soc. **7**, 439–453 (1842)
25. Mathematical and Physical Papers by George Gabriel Stokes. https://www.cambridge.org/core/books/mathematical-and-physical-papers/ 58B93740612EFCB1C365A1DCC3
26. Hutchison, K.W.J.M.: Rankine and the rise of thermodynamics. British J. Hist. Sci. **14**, 1–26 (1981)
27. Rankine, W.J.M.: Proc. Roy. Soc., vol. 21, London (1873)
28. Rankine, W.J.M.: Proc. Am. Acad. Arts Sci. **1**, 276–278 (1873)
29. Helmholtz, H.: Handbuch der physiologischen Optik. Leopold Voss, Leipzig (1867)
30. Helmholtz, H.: Über die Erhaltung der Kraft. Reimer, Berlin URN: b4-200905192283 (1847)
31. Helmholtz, H.: Über Integrale der hydrodynamischen Gleichungen, welche den Wirbelbewegungen entsprechen (On the Integrals of Hydrodynamic Equations to Which Vortex Motion Conform). J. für die reine und angewandte Mathematik **55**, 25–55 (1858)
32. Maxwell, J.C.: A dynamical theory of the electromagnetic field. Philos. Trans. Roy. Soc. **155**, 459–512 (1865)

33. Maxwell, J.C.: The scientific papers of James Clerk Maxwell. ISBN 978-1-108-01225-6 (2011)
34. Maxwell, J.C.: Theory of Heat. (9th Ed.) Courier Dover Publications. ISBN 978-0-486-41735-6 (2001)
35. Mach, E., Salcher, P.: Photographische Fixierung der durch Projectile in der Luft eingeleiteten Vorgänge. Sitzungsberichte der Kaiserlichen Akademie der Wissenschaften **95**(Part 2), 764–781 (1887)
36. Mach, E., Salcher, P.: Optische Untersuchung der Luftstrahlen. Sitzungsberichte der Akademie der Wissenschaften, XCVIII Band, Wien (1889)
37. Jackson, J.D.: Osborne Reynolds–scientist, engineer and pioneer. Proc. Roy. Soc. Lond. A **451**, 49–86 (1995)
38. Fairclough, C.: Happy birthday, Osborne Reynolds (2017) https://www.comsol.com/blogs/happy-birthday-osborne-reynolds/
39. Reynolds, O.: An experimental investigation of the circumstances which determine wether the motion of water shall be direct or sinuous, and of the law of resistance in parallel channels. Royal Society. Phil, Trans (1883)
40. Reynolds, O.: On the dynamical theory of incompressible viscous fluids and the determination of the criterion. Royal Society. Phil, Trans (1895)
41. Salas, M.D.: The curious events leading to the theory of shock waves. In: Invited Lecture at the 17th Shock Interaction Symposium, Rome (2006)
42. Chéret, R.: The Life and Work of Pierre-Henri Hugoniot. Shock waves, vol. 2, Springer Verlag, pp. 1–4 (1992)
43. Adam, H., Steckelmacher, W.: Vacuum science and technology. Pioneers of the 20th Century. (Paul A. Redhead, Ed.), AIP Press (1994)
44. Hirschel, E.H.: Basics of Aerothermodynamics, 2nd edn. Springer Verlag, Berlin/Heidelberg (2015)
45. Prandtl, L.: Über Flüssigkeitsbewegung bei sehr kleiner Reibung. Verh. d. III. Intern. Math. Kongress, Heidelberg, Germany (1904) (B.G. Teubner Verlag, Leipzig, pp. 485–491 (1905)
46. Meier, G.E.A., (Ed.). Ludwig Prandtl, ein Führer in der Strömungslehre. Vieweg Verlag (2000)
47. Rotta, J.C.: Die Aerodynamische Versuchsanstalt in Göttingen, ein Werk Ludwig Prandtls. Verlag Vandenhoeck & Ruprecht, Göttingen (1990)
48. Kármán v., Th., Edson, L. Die Wirbelstraße. Mein Leben für die Luftfahrt. Hamburg,: (Original Edition: The wind and beyond. Little Brown and Company, Boston/Toronto) (1968)

Appendix A
Symbols

Only the important symbols are listed. In general the page number is indicated, where a symbol appears first or is defined. Dimensions are given in terms of the SI basic units.

A.1 Latin Letters

		page	
$\mathbb{A}, \mathbb{B}, \mathbb{C}$	matrices: non-conservative Euler equations, Cartesian coordinates	169	[–]
$\mathsf{A}^{\xi}, \mathsf{A}^{\eta}, \mathsf{A}^{\zeta}$,	matrices: non-conservative Euler equations, arbitrary coordinates	172	[–]
$\mathbf{a}$	neutral vector field function	41	[–]
a_x, a_y, a_z	components of **a** in Cartesian coordinates	43	[–]
a_r, a_φ, a_z	components of **a** in cylindrical coordinates	50	[–]
a_θ, a_φ, a_r	components of **a** in spherical coordinates	56	[–]
C_K	$= \frac{\kappa_\infty T_\infty}{\mu_\infty v_\infty^2}$ constant	96	[–]
$C_{\xi^2}, C_{\xi \cdot v_\xi}, C_{\xi \cdot v}$	abbreviations: TLA Eq. (10.4)	130	[–]
$\hat{C}_{\xi \cdot v_\xi}, \hat{C}_{\xi \cdot v}, \hat{C}_{v \cdot v_\xi}$	abbreviations: TLA Eq. (10.5)	131	[–]
c	speed of sound	30	[m/s]
c_R	speed of sound for real gas	172	[m/s]
c	neutral scalar field function	41	[–]
c_p	(mass) specific heat at constant pressure	38	[m^2/s^2 K]
c_v	(mass) specific heat at constant volume	123	[m^2/s^2 K]
c_{in}	unsteady shock speed	200	[m/s]
c_1, c_2	constant velocities of advection equation	207	[m/s]
D	normal velocity of a discontinuous surface	211	[m/s]

C. Weiland, *The Hierarchy of Fluid Dynamic Equations*,
https://doi.org/10.1007/978-3-031-71198-5

D_i	multicomponent mass diffusion coefficient of species i	74	$[\mathrm{m}^2/\mathrm{s}]$
$\check{D}_i$	$= D_i + \frac{1}{Sc_t}\frac{\mu_t}{\rho}$ abbreviation	125	$[\mathrm{m}^2/s]$
D_{ij}	binary diffusion coefficient	118	[–]
$\mathbf{E}, \mathbf{F}, \mathbf{G}$	vector: inviscid part, Eq. (5.1)	69	[–]
$\mathbf{E}_{visc}, \mathbf{F}_{visc}, \mathbf{G}_{visc}$	vector: viscous part, Eq. (5.1)	69	[–]
$\mathbf{E}_S, \ldots$	vector: species conservation "inviscid part" , Eq. (9.3)	118	[–]
$\mathbf{E}_{S,visc}, \ldots$	vector: species conservation "viscous part" , Eq. (9.3)	118	[–]
$\hat{\mathbf{E}}, \hat{\mathbf{F}}, \hat{\mathbf{G}}$	vector: inviscid part of RANS equations, Eq. (5.17)	74	[–]
$\hat{\mathbf{E}}_{visc}, \hat{\mathbf{F}}_{visc}, \hat{\mathbf{G}}_{visc}$	vector: viscous part of RANS equations, Eq. (5.17)	74	[–]
$\hat{\mathbf{E}}_C, \hat{\mathbf{F}}_C, \hat{\mathbf{G}}_C$	vector: inviscid part of RANS equations; non-equilibrium thermodynamics, Eq. (5.19)	76	[–]
$\hat{\mathbf{E}}_{C,visc}, \ldots$	vector: viscous part of RANS equations; non-equilibrium thermodynamics, Eq. (5.19)	76	[–]
$\hat{\mathbf{E}}_T, \ldots$	vector: turbulent "inviscid part", Eq. (8.5)	109	[–]
$\hat{\mathbf{E}}_{T,visc}, \ldots$	vector: turbulent "viscous part", Eq. (8.5)	109	[–]
$\mathbf{E}_{vib}, \ldots$	vector: "inviscid part" of vibrational energy, Eq. (7.11)	100	[–]
$\mathbf{E}_{vib,visc}, \ldots$	vector: "viscous part" of vibrational energy, Eq. (7.11)	100	[–]
Ec	Eckert number	156	[–]
e	total energy per unit mass	30	$[\mathrm{m}^2/\mathrm{s}^2]$
e_i	internal energy per unit mass	30	$[\mathrm{m}^2/\mathrm{s}^2]$
$e_{i,i}$	internal energy of species i per unit mass	118	$[\mathrm{m}^2/\mathrm{s}^2]$
$e_{v,j}$	vibrational energy of molecular species j	120	$[\mathrm{m}^2/\mathrm{s}^2]$
e_e	electron energy per unit mass	122	$[\mathrm{m}^2/\mathrm{s}^2]$
$\bar{e}$	total energy per unit volume	33	$[\mathrm{kg/s}^2\ \mathrm{m}]$
F, G	boundaries of computational domain	136	[m]
$\mathbf{f}_e$	vector: external forces per unit mass	30	$[\mathrm{m/s}^2]$
$\mathbf{g}$	vector: gravitational acceleration	45	$[\mathrm{m/s}^2]$
$\mathbf{H}_{visc}$	vector: viscous part of TLA Eq. (10.2)	129	[–]
H	altitude of flight trajectory	182	[m]
h	enthalpy per unit mass	30	$[\mathrm{m}^2/\mathrm{s}^2]$
h_0	total enthalpy per unit mass	30	$[\mathrm{m}^2/\mathrm{s}^2]$
h_i	enthalpy of species i	33	$[\mathrm{m}^2/\mathrm{s}^2]$
$\mathbb{I}$	identity matrix	172	[–]
$I_{i''}$	ionization potential of species i''	122	[–]
$\mathbb{J}, \mathbb{K}, \mathbb{L}$	matrices: quasi-conservative Euler equations, Cartesian coordinates	169	[–]
J	Jacobian of coordinate transformation	82	[–]

J_{lm}	$= \frac{\partial E_l}{\partial Q_m}$ Jacobian in tensor notation	193	[–]
$\mathbf{j}_i$	vector: diffusion mass flux of species i	33	[kg/m^{2}s]
$\mathsf{K}^\xi, \mathsf{K}^\eta, \mathsf{K}^\zeta$	matrices: quasi-conservative Euler equations, arbitrary coordinates	170	[–]
k	turbulent kinetic energy	72	[m^2/s^2 K]
k	number of reaction products of reaction r	117	[–]
k_c	equilibrium constant	117	[–]
$k_{f,r}$	forward reaction rate of reaction r	117	[–]
$k_{b,r}$	backward reaction rate of reaction r	117	[–]
$\mathbb{M}$	$= \frac{\partial Q}{\partial X}$ matrix: transformation non-conservative conservative formulation	172	[–]
M	mass	32	[kg]
M_i	mol mass of species i	117	[kg]
M_∞	freestream Mach number	39	[–]
m	$= \rho v_x$ dependent momentum variable	193	[kg/m^2 s]
$\dot{n}_{e,i''}$	number density of species i''	122	[–]
n_1, n_2, n_3	eigenvectors Eq. (16.3)	208	[–]
$\mathbb{P}_k$	matrices: constructed by the right eigenvectors of A^k	174	[–]
Pr	Prandtl number	73	[–]
Pr_t	turbulent Prandtl number	73	[–]
p	static pressure	30	[N/m^2]
p_i	partial pressure of species i	30	[N/m^2]
p_e	pressure of electrons	122	[N/m^2]
$\mathbf{Q}$	vector: conservative variables, Eq. (5.1)	69	[–]
$\hat{\mathbf{Q}}$	vector: conservative variables of RANS equations, Eq. (5.17)	74	[kg/m^3]
$\mathbf{Q}_S$	vector: species density, Eq. (9.3)	118	[kg/m^3]
$\hat{\mathbf{Q}}_T$	vector: turbulent variables, Eq. (8.5)	109	[–]
$\hat{\mathbf{Q}}_C$	vector: conservative variables of RANS equations; nonequilibrium thermodynamics, Eq. (5.19)	76	[kg/m^3]
$\mathbf{Q}_{vib}$	vector: vibrational energies, Eq. (7.11)	100	[–]
$Q_{v,j}$	vibrational coupling term	78	[kg/s^3 m]
Q_e	electron coupling term	78	[kg/s^3 m]
Q_I	energy due to ionization by electron impact	122	[kg/s^3 m]
$Q_{T-v,j}$	energy exchange between translational and vibrational modes	120	[kg/s^{3}m]
Q_{T-e}	energy exchange due to elastic collision	122	[kg/s^3 m]
$Q_{e-v,j}$	energy exchange between electronic and vibrational modes of species j	120	[kg/s^3 m]
Q_{e-v}	energy exchange between electronic and vibrational modes	122	[kg/s^3 m]
Q_{D_j}	energy removed at dissociation		

	or gained at recombination of species j	120	[kg/s^3m]
Q_{v-v_j}	energy exchange between different vibrational modes	120	[kg/s^3 m]
$\mathbf{q}$	vector: molecular transport of energy and mass diffusion	30	[kg/s^3]
$\mathbf{q}_\xi$	characteristic vector	226	[–]
q_i	components of **q** in tensor notation	70	[kg/s^3]
$q_{e,k}$	conduction of electron energy	122	[kg/s^3]
$q_{rad,g}$	energy due to gas radiation	122	[kg/s^3 m]
q_{gw}	heat flux of the gas	222	[kg/s^3]
q_w	heat flux into the wall	222	[kg/s^3]
$q_{rad,w}$	heat flux due to surface radiation	222	[kg/s^3]
$\mathbf{q}_{v,j}$	vector: conduction of vibrational energy of molecular species j	120	[kg/s^3]
q_H	heat sources	30	[kg/s^3m]
q_1, q_2	variables of the advection equation	207	[–]
$\mathcal{R}$	universal gas constant	122	[m^2/s^2 K]
R	specific gas constant	123	[m^2/s^2 K]
Re	Reynolds number	94	[–]
r, φ, z	cylindrical coordinates	41	[m]
r, θ, φ	spherical coordinates	41	[m]
$\mathbf{r}_\xi$	characteristic vector	225	[–]
$\mathbf{S}_{NS}$	vector: external sources, Eq. (5.1)	69	[–]
$\mathbf{S}_S$	vector: source term of species conservation, Eq. (9.3)	118	[kg/m^3s]
$\mathbf{S}_C$	vector: source term of conservation of chemical and thermal species, Eq. (5.19)	76	[kg/m^3s]
$\hat{\mathbf{S}}_\mathbf{T}$	vector: turbulent sources, Eq. (8.5)	109	[–]
$\mathbf{S}_{vib}$	vector: source terms of vibrational energies, Eq. (7.11)	100	[–]
Sc	Schmidt number	94	[–]
Sc_t	turbulent Schmidt number	74	[–]
S_F	sources acting at the surface	30	[–]
S_V	volume sources due to field function ϕ	30	[–]
s	entropy per unit mass	30	[m^2/s^2 K]
$\mathbb{T}_k$	matrices: constructed by the right eigenvectors of K^k	174	[–]
T	temperature of the fluid	30	[K]
T_w	wall temperature	222	[K]
T_r	recovery temperature	222	[K]
T_{ra}	radiation adiabatic temperature	222	[K]
$T_{v,j}$	vibrational temperature of molecular species j	120	[K]
$\mathbf{t}$	eigenvector Eq. (17.28)	238	[–]
$\mathbf{U}_i^*$	vector: velocity of diffusion of species i	117	[m/s]

$\mathbf{U}_1, \mathbf{U}_2$	vector: velocity in discontinuous flow	209	[m/s]
U_k	$= k_x v_x + k_y v_y + k_z v_z + k_t$ contravariant velocity vector in k system	174	[m/s]
$U_{1,n}, U_{2,n}$	normal velocity components of $\mathbf{U}_1, \mathbf{U}_2$	209	[m/s]
$U_{1,t}, U_{2,t}$	tangential velocity components of $\mathbf{U}_1, \mathbf{U}_2$	209	[m/s]
$V_{1,n}, V_{2,n}$	normal velocity components for a system at rest	211	[m/s]
V_x, V_y, V_z	components of contravariant velocity vector	86	[m/s]
$\mathbf{v}$	vector: velocity	30	[m/s]
v_i, v_j	components of $\mathbf{v}$ in tensor notation	70	[m/s]
v_x, v_y, v_z	components of $\mathbf{v}$ in Cartesian coordinates	45	[m/s]
v_r, v_φ, v_z	components of $\mathbf{v}$ in cylindrical coordinates	53	[m/s]
v_θ, v_φ, v_r	components of $\mathbf{v}$ in spherical coordinates	61	[m/s]
$\mathbf{X}$	vector: non-conservatve variables Eq. (14.10)	169	[–]
$[X_i]$	molar concentration of species i	117	[1/m^3]
X_i	mol fraction of species i	119	[–]
x, y, z	Cartesian coordinates	41	[m]
x_i, x_j, x_k	coordinates in tensor notation	72	[–]
$\bar{y}$	stretching the y coordinate Eq. (14.45)	186	[m]
Z	$= Z(\rho, T)$ compressibility factor	123	[–]
$\bar{z}$	stretching the z coordinate Eq. (14.45)	186	[m]

A.2 Greek Letters

The Greek alphabet reads

lower case: $\alpha, \beta, \gamma, \delta, \epsilon, \zeta, \eta, \theta, \vartheta, \iota, \kappa, \lambda, \mu, \nu, \xi, o, \pi, \rho, \sigma, \tau, \phi, \varphi, \chi, \upsilon, \psi, \omega$.
upper case: $\Gamma, \Delta, \Theta, \Lambda, \Xi, \Pi, \Sigma, \Upsilon, \Phi, \Psi, \Omega$.

		page	
α	angle of attack	39	[^{0}]
α_i	source term of species conservation equation	94	[kg/m^3 s]
$\beta_x, \beta_y, \beta_z$	abbreviations (laminar flow) Eq. 6.5	86	[–]
$\hat{\beta}_x, \hat{\beta}_y, \hat{\beta}_z$	abbreviations (turbulent flow) Eq. 6.9	89	[–]
$\Gamma_{1,2,3}$	Riemann invariants	208	[m/s]
Γ	potential of external forces	30	[m^2/s^2]
γ	ratio of specific heats	73	[–]
δ	boundary layer thickness	156	[m]
$\bar{\bar{\delta}}$	neutral tensor field function	41	[–]

ε	emisson coefficient	99	[–]
κ	thermal conductivity	74	[kgm/s^3 K]
$\kappa_{v,j}$	frozen thermal conductivity of vibrational energy	120	[kgm/s^3 K]
κ_e	frozen thermal conductivity of electron energy	122	[kgm/s^3 K]
$\check{\kappa}$	$= \kappa(1 + \frac{Pr\mu_t}{Pr_t \mu})$ total thermal conductivity	74	[kgm/s^3 K]
Λ_ξ	$= \frac{d\xi}{d\tau}$ eigenvalues of $\mathsf{A}^\xi(\mathbf{X})$	225	[m/s]
λ_k^i	eigenvectors, $i = 1, ..., 5, k = \xi, \eta, \zeta$	50	[m/s]
λ	bulk viscosity	46	[kg/s m]
μ	molecular viscosity	30	[kg/s m]
μ_t	turbulent viscosity	73	[kg/s m]
μ^*	$= \mu + \frac{\mu_t}{\sigma_k}$ abbreviation	75	[kg/s m]
$\check{\mu}$	$= \mu + \mu_t$ total viscosity	74	[kg/s m]
ν_e	collision frequency	122	[–]
$\nu'_{i,r}$	stoichiometric coefficient of forward reaction	117	[–]
$\nu''_{i,r}$	stoichiometric coefficient of backward reaction	117	[–]
ξ, η, ζ	non-orthogonal curvilinear (arbitrary) coordinates	82	[–]
Φ	volume integraton of scalar field function	30	[–]
ϕ	potential of the velocity field	184	[m^2/s]
ϕ	scalar field function	30	[–]
φ	perturbation potential of the velocity field	186	[m^2/s]
φ	dependent variable	70	[–]
$\bar{\varphi}$	mean part of Reynolds averaging	70	[–]
φ'	fluctuating part of Reynolds averaging	70	[–]
$\hat{\varphi}$	mean part of Favre averaging	70	[–]
φ''	fluctuating part of Favre averaging	70	[–]
ρ	density	93	[kg/m^3]
ρ_i	density of species i	93	[kg/m^3]
ρ_e	density of electrons	122	[kg/m^3]
$\Sigma(\xi, \eta, \zeta, t)$	function of discontinuous surface	211	[–]
Σ_R	abbreviation (real gas), $\mathbb{M}$, $\mathbb{M}^{-1}$ matrices	174	[–]
Σ	abbreviation (perfect gas), $\mathbb{M}$, $\mathbb{M}^{-1}$ matrices	174	[–]
σ_k	constant of turbulence model	74	[–]
$\bar{\bar{\sigma}}$	total stress tensor	32	[kg/s^2 m]
$\bar{\bar{\tau}}$	shear stress tensor	30	[kg/s^2 m]
τ_{ij}	components of the shear stress tensor in tensor notation	46	[kg/s^2 m]
$\tau_{xx}, \tau_{yy}, \tau_{zz}, \tau_{xy}, \tau_{xz}, \tau_{yz}$	elements of viscous stress tensor $\bar{\bar{\tau}}$ in Cartesian coordinates	69	[kg/s^2 m]
$\hat{\tau}_{ij}^{turb}$	components of Reynolds stress tensor in tensor notation	108	[kg/s^2 m]
ψ_R	abbreviation (real gas), $\mathbb{J}, \mathbb{K}, \mathbb{L}$ matrices	169	[–]
ψ	abbreviation (perfect gas), $\mathbb{J}, \mathbb{K}, \mathbb{L}$ matrices	169	[–]
ψ	flux of the field function	30	[–]

Ω_1	abbreviation (real gas), $\mathbb{A}$, $\mathbb{B}$, $\mathbb{C}$ matrices	170	[–]
Ω_2	abbreviation (real gas), $\mathbb{A}$, $\mathbb{B}$, $\mathbb{C}$ matrices	170	[–]
ω	specific dissipation rate	94	[1/s]
ω_i	mass fraction of species i	74	[–]

Appendix B
Acronyms

ALE	Arbitrary Lagrange-Euler formulation
ARIANE 5	European rocket managed by ESA
CPU	Central Processor Unit
CRV	Crew Return Vehicle
DAVIS-VOL	Daimler-Benz finite-volume Navier-Stokes code
DES	Detached Eddy Simulation
DLR	German Aerospace Center
DNS	Direct Numerical Simulation
DES	Detached Eddy Simulation
EADS	European Aeronautic Defence and Space Company
ESA	European Space Agency
ESRANGE	European Space and Sounding Rocket Range (Sweden)
HERMES	European winged re-entry space plane
HM60	Vulcain Rocket Motor of ARIANE 5
HOPPER	SSTO European system concept
ISS	International Space Station
LES	Large Eddy Simulation
MSTP	Manned Space Tranportation Programme (Europe)
DES	Detached Eddy Simulation
NASA	National Aeronautics and Space Administration
N-S	Navier-Stokes (equations)
ONERA	National Aerospace Research Center (France)
PHOENIX	Flight demonstrator for SSTO system HOPPER (Germany)
PNS	Parabolized Navier-Stokes (equations)
RANS	Reynolds Averaged Navier-Stokes (equations)
RSS	Restricted Separation Nozzle
SPACE-SHUTTLE	US winged re-entry space plane
TNS	Thin-Layer Navier-Stokes (equations)
TWG	Transonic Windtunnel Göttingen (Germany)

C. Weiland, *The Hierarchy of Fluid Dynamic Equations*,
https://doi.org/10.1007/978-3-031-71198-5

URANS	Unsteady Reynolds Averaged Navier-Stokes (equations)
X-24A	NASA: experimental vehicle
X-33, X-34, X-37	NASA: experimental vehicles
X-38	ESA-NASA: joint project

Index

A
Abbett, M., 18, 38, 192
Abe, T., 121
Abgrall, R., 39
Adam, H., 269
Aiton, E.J., 248
Allmaras, S.R., 108
Altenbach, H., 12
Amtsberg, J., 161
Anderson, D.A., 170, 224
Anderson, J.D., 2, 3, 10, 15, 116, 159, 259
Archimedes, 1, 9
Aris, R., 41, 70
Aristotle, 1, 9
Averaging process
  Favre, 71
  Reynolds, 71

B
Babenko, K.I., 18
Baldwin, B.S., 107
Barber, R.W., 242
Becker, E., 30
Becker, R., 35
Behr, R., 24, 100, 110
Bell, A.E., 251
Bernhard, H., 254, 255
Bernoulli
  Daniel, 13, 254
  Jacob, 12
  Johann, 13
Berry, S.A., 110
Best, J.T., 145
Bird, R.B., 12, 30, 41
Blunt body flow solution, 39
Blunt body problem, 191
Blunt delta wing flow, 132
Boundaries
  impermeable surface, 227, 237
  shock fitting, 227, 232
  subsonic inflow, 227, 228
  subsonic outflow, 227, 229
  supersonic inflow, 227, 232
  supersonic outflow, 227, 232
Boundary condition
  characteristic variable, 224, 237, 239
  inviscid flow, 223
  no slip, 222, 241
  physical, 224
  post correction technique, 238
  radiation adiabatic wall temperature, 100
  thermal, 222
  viscous flow, 240
Boundary layer approach for
  continuity equation, 157
  energy equation, 158
  momentum equations, 157
Boundary layer equations
  derivation of, 155
  general, 15
Boundary value problem, 221
Boussinesq hypothesis, 72, 105
Breitsamter, Ch., 3, 134, 212
Bronstein, I.N., 41
Buelow, P.E., 135
Buoyancy, 9

C
Candler, G.V., 119
Caporicci, M., 21, 240
Cauchy, A.L., 14
Cavalieri, B.F., 10

C. Weiland, *The Hierarchy of Fluid Dynamic Equations*,
https://doi.org/10.1007/978-3-031-71198-5

Cebeci, T., 107
Chakravarthy, S.R., 170, 224
Characteristic, 208, 215
  direction, 224
  theory, 19, 224
Chaussee, D.S., 135, 145
Chemistry
  oxygen-hydrogen, 118
  oxygen-nitrogen, 118
Chéret, R., 268
Chorin, A.J., 206
Clarke, D.M., 249
Combustion chamber flow, 93
Conservation
  of energy, 33
  of mass, 31
  of momentum, 32
Continuity equation
  Cartesian coordinates, 45
  cylindrical coordinates, 53
  differential form, 34
  incompressible, 149
  spherical coordinates, 61
Coordinate transformation
  arbitrary, 82
  cylindrical, 49
  spherical, 56
Couette flow, 3
Courant, R., 4, 19, 168
Cousteix, J., 15, 134, 160
Crocco's momentum equation, 35

D
D'Alembert, J.-B., 13, 256
D'Ambrosio, D., 102
Dash, S.M., 144
Da Vinci, L., 10
Davis, R.W., 151
De Saint-Venant, J.C., 2, 14
Désidéri, J.-A., 39
Descartes, R., 10, 249
Diagonalization of Euler equations, 170, 224
Differential form, 194
Discontinuity, 206
  contact, 19, 83, 213, 236
  strong, 208
  weak, 211
Dissipation function, 38
Drake, S., 247

E
Eckert, M., 2, 10, 15, 105
Edson, L., 271
Edwards, C.H., 2, 11
Edwards, J.A., 145
Eigenvalue, 170, 208, 224, 237
Eigenvector, 170, 207, 224, 238
Emerson, D.R., 242
Energy equation
  Cartesian coordinates, 47
  cylindrical coordinates, 54
  enthalpy form, 38
  entropy form, 38
  excited electron modes, 121
  internal energy form, 37
  spherical coordinates, 63
  total energy form, 36
  total enthalpy form, 37
  vibrational, 96, 100, 119
Eriksson, L.-E., 21, 240
Erlebacher, G., 109
Euler equations, 192, 207
  chemical non-equilibrium real gas, 181
  conservative, 192
  conservative form in vector notation
    arbitrary coordinates, 167
    Cartesian coordinates, 166
  continuity, 165
  differential version, 192
  energy, 165
  energy with entropy, 166
  energy with total enthalpy, 166
  equilibrium real gas, 180
  first quasi-linear form, 170
  incompressibility condition, 165
  integral version, 194
  momentum, 165
  non-conservative, 38, 193, 224
  non-conservative form in vector-matrix notation, 169
  quasi-conservative, 193, 225
  quasi-conservative form in arbitrary coordinates, 170
  quasi-conservative form in vector notation, 167
  quasi-linear forms, 170
  second quasi-linear form, 172
Euler's homogeneity relation, 168
Euler, L., 2, 13, 165, 255
Euler solution
  chemical non-equilibrium, 103, 182
  space vehicle Hermes equilibrium real gas, 181
  thermal and chemical non-equilibrium, 104

F
Fairclough, C., 267
Favre averaged
 continuity equation, 71
 energy equation, 72
 momentum equations, 72
Fearnley-Sander, D., 261
Fermat de, P., 10, 250
Fey, M., 39
Fick's law, 116
Finite difference scheme, 38, 167
Finster, R., 253
Fischer, J., 124, 180
Fitzgerald, S.M., 110
Flow reattachment, 100
Fluid dynamic equations
 Cartesian coordinates, 42
 component form, 41
 cylindrical coordinates, 49
 differential form, 34
 integral form, 31
 spherical coordinates, 56
 symbolic form, 30
Flux vector splitting method, 170
Forces
 external, 32
 internal, 32
 volume, 32
Formulation
 conservative, 192
 Euler, 87
 Lagrange, 87
 Lagrange-Euler, 87
 non-conservative, 193
 quasi-conservative, 193
Frohn, A., 241
Fujita, K., 121

G
Galilei, G., 10, 247
Gersten, K., 15, 159
Glowinski, R., 39
Gnoffo, P.A., 119
Gordon, S., 118, 180, 235
Grassmann, H.G., 261
Griffith, B.J., 145
Grimsley, R., 256
Groh, A., 241
Gross, A., 23, 109, 116, 213
Grossmann, B., 18
Gupta, R.N., 119

H
Haeberle, J., 100
Hagen-Poiseuille problem, 3
Haidinger, F.A., 83, 110
Hamilton, W.R., 260
HÄnel, D., 106, 127, 241
Hartmann, G., 21, 83, 100, 181, 212, 240
Heat fluxes
 due to wall radiation, 222
 into the wall, 222
 of the gas, 222
Helmholtz v., H.L.F., 15, 152, 264
Helmholtz vorticity equation, 151
Hilbert, D., 4, 19, 168
Hirsch, C., 30, 74, 135
Hirschel, E.H., 2–4, 10, 15, 18, 20, 24, 33, 46, 74, 83, 87, 100, 106, 116, 119, 134, 159, 160, 180, 212, 222, 241, 242, 269
Homentropic, 182
Hooke's law, 12
Horvath, T.J., 110
Hugoniot, P.H., 268
Hussaini, M.Y., 109
Hutchison, K., 263
Huygens, Ch., 1, 10, 251

I
Ievalts, J.O., 135
Infinitesimal calculus, 10
Initial value problem, 221
Inouye, M., 18
Integral form, 194
Integral method, 160
International Space Station ISS, 180
Irrotational flow field, 182
Isentropic, 182, 184

J
Jackson, J.D., 267
Jacobians, 168, 170, 178
Janus, J.M., 170, 224
Jeltsch, R., 39

K
Karmann, P., 39
Kármán v., Th., 271
Kempe, M., 253
Kepler, J., 10, 248
Keraus, R., 124, 180
Kirtley, K.R., 107

Klingbeil, E., 41, 56, 64, 70
Knudsen, M.H.Ch., 269
Koppenwallner, G., 242
Kordulla, W., 15, 134, 160
Krause, E., 4, 18, 20, 30, 159
Krbek von, F., 256
Kreiss, H.O., 109
Kruger Jr., C.H., 116
Kuczera, H., 180
Kühn, A., 161
Kutler, P., 18

L
Lagrange, J.L., 13
Laplace operator, 150
Laplace, P.S., 14
Laval nozzle, 16
Lawrence, S.L., 135, 145
Lax, P.D., 192, 199
Leibniz, G.W., 11, 253
LeVeque, R.J., 206
Lightfoot, E.N., 12, 30, 41
Lockerby, D.A., 242
Lomax, H., 18, 107
Low density, 93
Low separation, 100
Lueroth, J., 11
Lyubimov, A.N., 17, 18, 38, 49, 192

M
MacCormack, R.W., 119
Mach, E., 266
Mahoney, M.S., 250
Mallet, M., 39
Marconi, F., 18
Mariotte, E., 1, 10
Marsden, J.E., 206
Matrices
  $\mathbb{A}, \mathbb{B}, \mathbb{C}$, non-conservative Euler equations perfect gas, 169
  $\mathbb{A}, \mathbb{B}, \mathbb{C}$, non-conservative Euler equations real gas, 170
  A^k, non-conservative Euler equations arbitrary coordinates perfect gas, 172
  $\mathbb{P}$ of non-conservative eigenvectors perfect gas, 175
  $\mathbb{P}$ of non-conservative eigenvectors real gas, 176
  $\mathbb{T}$ of quasi-conservative eigenvectors perfect gas, 177
  $\mathbb{T}$ of quasi-conservative eigenvectors real gas, 178
  Jacobians $\mathbb{J},\mathbb{K},\mathbb{L}$ real gas, 167
  Jacobians $\mathbb{M}$ perfect gas, 173
  Jacobians $\mathbb{M}$ real gas, 173
  Jacobians K^k perfect gas arbitrary coordinates, 171
  Jacobians K^k real gas arbitrary coordinates, 171
Maus, J.R., 145
Maxwell, J.C., 242, 265
McBride, B.J., 118, 180, 235
Meier, G.E.A., 270
Menne, S., 21, 39, 83, 102, 181, 182, 212, 240
Merkle, C.L., 127
Method of characteristics, 170
Molvik, G.A., 127
Momentum equations
  Cartesian coordinates, 45
  cylindrical coordinates, 53
  differential form, 34
  incompressible, 150
  incompressible, vector notation, 150
  spherical coordinates, 61
Moore, E.F., 151
Moretti, G., 18, 38, 192
Morrison, J.H., 137
Mundt, Ch., 124, 180

N
Navier, C.L.M.H., 2, 258
Navier-Stokes equations
  incompressible, 149
  laminar, 69
  laminar, arbitrary coordinates, 83
  laminar, arbitrary coordinates; thermal non-equilibrium, 100
  parabolized, 16
  RANS chemical non-equilibrium, 76, 93, 96
  RANS perfect gas, 74
  RANS thermal and chemical non-equilibrium, 76
  Reynolds-averaged (RANS), 74
  thin layer (TLA), 16, 127
  turbulent (RANS), 15
  turbulent (RANS), arbitrary coordinates, 87
Navier-Stokes solution
  chemical non-equilibrium, 103
Newton method, 110
Newton, I., 2, 11, 252
Newton's law, 12

No slip, 155
Nozzle extension flow, 93, 110

O
O'Connor, J.J., 257, 259
Oertel, H., 116, 214
Onofri, M., 21, 240
Orszag, S.A., 107
Otsu, H., 121

P
Pandolfi, M., 102, 238
Panel method, 160, 163
Park, C., 116
Pekkari, L.O., 21, 240
Périaux, J., 39
Pfitzner, M., 18, 39, 83, 167, 182, 211, 236, 238
PNS equations
  laminar, perfect gas, arbitrary coordinates, 137
  laminar, perfect gas, Cartesian and cylindrical coordinates, 136
  turbulent, perfect gas, arbitrary coordinates, 142
Poisson equation, 150
Poisson, S.D., 14, 257
Polz, G., 161
Popp, M., 21, 240
Potential equation, 16, 110, 182
  first order transonic small perturbation, conservative, 187
  first-order transonic small perturbation, non-conservative, 188
  full, conservative, 184
  full, non-conservative, 185
  perturbations forms, 186
  perturbed, non-conservative, 186
  perturbed, Prandtl-Glauert, 187
  second order transonic small perturbation, conservative, 188
  transonic Guderley-v. Kármán, 188
  transonic small perturbation, 187
Potential flow, 212
Potential solution
  three-dimensional, Butler wing, 185
Prabhu, D., 135
Prandtl, L., 15, 106, 155, 270
Prandtl-Glauert potential equation, 187
Pressure relation in incompressible flow, 150
Prince, S.A., 145
Pulliam, T.H., 18, 127
Purtell, L.P., 151

Q
Quentin, J., 161

R
Rakich, J.V., 135
Rankine-Hugoniot condition, 185
Rankine-Hugoniot equations, 16, 138, 210
Rankine, W.J.M., 263
Real gas flow, 115
Recirculation region, 241
Reese, J.M., 242
Reinhardt, W.A., 18
Reynolds, O., 14, 267
Reynolds stress tensor, 72
Riedelbauch, S., 132, 236
Riemann
  invariant, 208
  problem, 206
  solver, 19, 170
Riemann, B., 14, 206
Rizzi, A., 2, 3, 10, 18, 134, 212
Robertson, E.F., 257, 259
Röd, W., 249
Roe, P.L., 19, 170
Rotta, J.C., 15, 105, 270
Rusanov, V.V., 17, 18, 38, 49, 192

S
Sacher, P., 180
Saint-Venant de, J.C., 259
Salas, M.D., 170, 205, 224, 268
Salcher, P., 266
Sarkar, S., 109
Schiff, L.B., 145
Schlichting, H., 15, 159
Schröder, W., 39, 83, 100, 181, 212, 240
Schwane, R., 127
Seiner, J.M., 144
Semendjajew, K.A., 41
Separation bubble, 100
Shear layer, 15, 19, 83, 211, 213, 236
Shear stress tensor
  Cartesian coordinates, 48
  cylindrical coordinates, 56
  elements, 140
  molecular transport of momentum, 106
  spherical coordinates, 64
  tensorial notation, 46
Shinn, J.L., 119

Shock
bow, 38, 137
capturing, 19, 192, 211
embedded, 38, 240
fitting, 18, 192, 211, 232
layer, 241
normal, 213
oblique, 213
surface, 211, 233
tube problem, 214
velocity, 216
wave, 19, 83, 213, 236
Similarity number of
Eckert, 156
Prandtl, 156
Reynolds, 156
Schmdt, 94
Sine-squared law, 12
Sinha, N., 144
Slooff, J., 186
Smith, A.M.O., 107
Smoluchowski, M., 242
Sod, G.A., 214
Sommerfeld, A., 13, 35, 41, 165
Space flight trajectory
flow regimes, 182
Space shuttle orbiter, 182
velocity-altitude diagram, 182
X-38 space vehicle, 182
Space marching method, 135
Space vehicles
Hermes, 180
Hope X, 180
PHOENIX, 180
Space shuttle orbiter, 180
winged, 182
X-33, X-34, X-37, 180
X-38, 180
Spalart, P.R., 108
Species conservation
binary diffusion coefficient, 118
enthalpy, 118
Euler equations chemical non-equilibrium real gas, 181
internal energy, 118
symbolic form, 116
turbulent flow, 118
vector notation, 117
Species continuity equation
chemical non-equilibrium, 93, 94
chemical non-equilibrium, arbitrary coordinates, 124
Split-coefficient matrix method, 170
Srinivasan, S., 124, 180
Staudacher, W., 3, 134, 212
Steckelmacher, W., 269
Steger, J.L., 18, 127, 145, 170
Stokes, G.G., 3, 262
Stokes problem, 3
Stokes relation, 46
Stress tensor
Reynolds, 106
total, 32
viscous shear, 32
Stuart, W.E., 12, 30, 41
Surzhikov, S.T., 121
Suzuko, K., 121
Szema, K.Y., 145

T

Tannehill, J.C., 124, 135, 145, 180
Teichmann, H., 41, 70
Thermodynamic equation of state, 166
Thermodynamic state
chemical non-equilibrium, 116
chemical reactions, 115
dissoziations, 115
electron coupling, 79
electronic energy of conduction, 122
electron modes, 76, 115
energy amount due to gas radiation, 122
energy amount due to ionization, 122
energy changed: dissociation or recombination, 120
energy exchange: collisions electrons and heavy particles, 122
energy exchange: different vibrational modes, 120
energy exchange: electronic-vibrational, 120, 122
energy exchange: translational-vibrational, 120
equilibrium, 168, 211, 236, 240
excitation of molecular vibrations , 115
excited electron modes, 121
ionization, 115
non-equilibrium, 76, 93, 213, 240
perfect, 69, 123
recombinations, 115
thermal non-equilibrium, 96, 119
thermodynamic equilibrium, 122
vector of heat transfer, 79
vibrational coupling, 79
vibrational energy of conduction, 120
TLA equations

laminar, perfect gas, arbitrary coordinates, 130
laminar, perfect gas, Cartesian coordinates, 128
turbulent, chemical non-equilibrium, arbitrary coordinates, 132
turbulent, perfect gas, arbitrary coordinates, 131
turbulent, perfect gas, Cartesian coordinates, 129
Torricelli, E., 10
Transport of heat
chemical non-equilibrium, laminar, 119
chemical non-equilibrium, turbulent, 119
Turbulence model
algebraic, 144
algebraic mixing length idea, 105
dimensionless, 93
mixing length of Baldwin-Lomax, 107
mixing length of Cebeci-Smith, 107
mixing length of Prandtl, 106
one equation, 105
one equation of Baldwin-Barth, 108
one equation of Spalart-Allmaras, 108
two equation, 98, 105
two equation, arbitrary coordinates, for PNS equations, 144
two equation compressibility corrections, 110
two equation of Wilcox, 108
two equation vector notation
arbitrary coordinates, 111
Cartesian coordinates, 109
Turbulence simulation
Detached Eddy Simulation (DES), 105
Direct Numerical Simulation (DNS), 105
Large Eddy Simulation (LES), 105
Turbulent
heat flux, 73
kinetic energy, 108
mass diffusion coefficient, 74
Prandtl number, 73, 94
pressure, 73
Schmidt number, 74, 94
specific dissipation rate, 108
total energy, 73
Turbulent flow
general, 71, 240
rocket motor nozzle, 110
X-38 space vehicle, 110
Typical boundary layer solution
helicopter fuselage model, 160
Typical flow solution
incompressible square cylinder flow, 151
X-38 space vehicle, two equation turbulence model, 110
Typical PNS solution, 145
Typical RANS flow solution
experimental nozzle, 21
HOPPER/PHOENIX re-entry vehicle, 100
Vulcain2 nozzle extension, 20
X-38 re-entry vehicle, 23
Typical TLA solution, 132

U
Upwind method, 170

V
Van den Heuvel, G., 253
Van Leer, B., 19, 170
Variables
conservative, 167
primitive, 167
Vector-matrix notation, 67, 193
Vector notation, 67, 192
Venkatapathy, E., 135
Vincenti, W.G., 116
Vinokur, M.J., 19, 83
Viviand, H., 19, 83
Voit, C., 262
Vortex sheet, 15, 211
Vorticity diffusion , 153
Vorticity transport equation
general, 152
incompressible inviscid, 152
incompressible viscous, 152
Voskresenskii, G.P., 17

W
Wallwitz v., G., 4
Warming, R.F., 18, 170
Warnatz, J., 118, 119
Warsi, Z.U.A., 19
Weatherill, N.P., 20
Weber, C., 100, 110
Weiland, C., 13, 18, 21, 23, 24, 30, 38, 39, 49, 56, 74, 83, 87, 91, 100, 102, 110, 119, 159, 167, 168, 180, 182, 183, 186, 187, 192, 211, 213, 222, 236, 240, 241
Weilmuenster, K.J., 124, 180
White, J.A., 137

Whitfield, D.L., 170, 224
Wilcox, D.C., 74, 105, 109
Williams, M.J., 145
Wolf, D.E., 144

Y
Yakhot, V., 107
Yee, H.C., 19, 170

Z
Zeiss, W., 24
Zeman, O., 109
Zierep, J., 30, 159, 187
Zurmühl, R., 174